T0329811

Intelligent IoT for the Digital World

Intelligent IoT for the Digital World

Incorporating 5G Communications and Fog/
Edge Computing Technologies

Yang Yang
ShanghaiTech University and Peng Cheng Lab, China

Xu Chen
Sun Yat-sen University, China

Rui Tan
Nanyang Technological University, Singapore

Yong Xiao
Huazhong University of Science and Technology, China

This edition first published 2021

The right of Yang Yang, Xu Chen, Rui Tan and Yong Xiao to be identified as the authors of this work has been asserted in accordance with law.

Registered Offices
John Wiley & Sons, Inc., 111 River Street, Hoboken, NJ 07030, USA
John Wiley & Sons Ltd, The Atrium, Southern Gate, Chichester, West Sussex, PO19 8SQ, UK

Editorial Office
The Atrium, Southern Gate, Chichester, West Sussex, PO19 8SQ, UK

For details of our global editorial offices, customer services, and more information about Wiley products visit us at www.wiley.com.

Wiley also publishes its books in a variety of electronic formats and by print-on-demand. Some content that appears in standard print versions of this book may not be available in other formats.

Library of Congress Cataloging-in-Publication Data

Names: Yang, Yang, 1974- author.
Title: Intelligent IoT for the digital world / Yang Yang, ShanghaiTech
   University and Peng Cheng Lab, China, Xu Chen, Sun Yat-sen University, China, Rui Tan,
   Nanyang Technological University, Singapore, Yong Xiao, Huazhong University
   of Science and Technology, China.
Description: First edition. | Hoboken, NJ : John Wiley & Sons, Inc., 2021.
   | Includes bibliographical references and index.
Identifiers: LCCN 2020039767 (print) | LCCN 2020039768 (ebook) | ISBN
   9781119593546 (cloth) | ISBN 9781119593553 (adobe pdf) | ISBN
   9781119593560 (epub)
Subjects: LCSH: Internet of things.
Classification: LCC TK5105.8857 .Y355 2021 (print) | LCC TK5105.8857
   (ebook) | DDC 005.67/8–dc23
LC record available at https://lccn.loc.gov/2020039767
LC ebook record available at https://lccn.loc.gov/2020039768

Cover Design: Wiley
Cover Image: produced by Julie Liu

Set in 9.5/12.5pt STIXTwoText by SPi Global, Chennai, India

C9781119593546_120321
Printed and bound by CPI Group (UK) Ltd, Croydon CR0 4YY

Contents

Preface

Since the concept of "pairing technology" or "digital twin" was first introduced by NASA for the space exploration and aviation industry[1,2], we have been on our way towards the digital world. Over the last decade, this digitization progress has been greatly accelerated, thanks to a broad spectrum of research and development (R&D) activities on internet of things (IoT) technologies and applications. According to the latest Cisco Annual Internet Report published in February 2020, the number of networked devices in the world will reach 29.3 billion by 2023, which includes 14.7 billion Machine-to-machine (M2M) or IoT devices[3]. While the joint report by Ovum and Heavy Reading, announced in May 2019, gave a more optimistic number, i.e. 21.5 billion IoT devices by 2023[4]. This explosive growth of IoT devices and connections clearly shows the technical trends and business opportunities of Machine-type Communications (MTC), such as the Ultra Reliable Low Latency Communications (URLLC) scenario specified in 3GPP Release 16 standard[5], and the massive MTC (mMTC) scenario based-on Narrowband IoT (NB-IoT) technology and endorsed by the latest International Mobile Telecommunications-2020 (IMT-2020) 5G standard. Both IoT and MTC-focused scenarios have been widely recognized and welcomed by different industry verticals. For examples, sensor devices for medical and health monitoring applications are usually light, reliable, energy efficient, wearable, or even implantable. Camera devices for security and public safety applications are usually hidden, durable, massively deployed, synchronized, and always connected. Robot devices for intelligent manufacturing applications are accurately positioned, constantly tracked in three dimensions, busy scheduled, fully utilized, and always ready for instant action. Besides the basic functions of sensing and transmitting data, some "smart" devices possess additional computing and storage resources, thus are capable of data screening and preprocessing tasks with localized tiny artificial intelligence (AI) algorithms[6].

These massive connected devices or things deployed in our neighborhood are continuously generating more and more data, which is collected at different time scales, from different sources and owners, and has different formats and characteristics for different

---

1 Bernard Marr, What is Digital Twin Technology and Why is it so Important?, Forbes, March 6, 2017.
2 Cisco, Cisco Annual Internet Report (2018–2023), February 2020.
3 Scott Buchholz and Bill Briggs, Tech Trends 2020, Deloitte, January 7, 2020.
4 Keynote by Alexandra Rehak and Steve Bell, IoT World, May 14, 2019.
5 Yigang Cai, 3GPP Release 16, IEEE Communications Society Technology Blog, 10 July 2020.
6 10 Breakthrough Technologies 2020, MIT Technology Review, February 26, 2020.

purposes and applications. As the UK mathematician Clive Humby pointed out in 2006, "Data is the new oil. It is valuable, but if unrefined it cannot really be used." According to the IDC's "Data Age 2025" white paper[7], sponsored by SEAGATE, the total data in our world is increasing at a Compounded Annual Growth Rate (CAGR) of 61% and will reach an astonishing 175 zettabytes (ZB) by 2025, among which 90 ZB data will be contributed by a variety of IoT devices. In order to realize its full potential, we must tackle the big challenge of integrating, processing, analyzing, and understanding such a huge volume of heterogeneous data from a wide range of applications and industrial domains. From the perspective of technology developers and service providers, more high-quality data, in terms of variety, accuracy and timeliness, will definitely contribute to deeper knowledge and better models for characterizing and representing our complex physical world, thus enabling us to develop very precise digital twins for everything and everyone in the digital world. This digitization process will effectively reshape our human societies, accelerate overall economic growth, and generate huge commercial values, since disruptive data-driven business models and cross-domain collaboration opportunities will be created and fully utilized for exploring all sorts of potential markets. According to a study of IDC, the worldwide revenue for big data and business analytics was about USD189.1 billion in 2019, and will quickly increase to USD274.3 billion by 2022[8]. The importance of data for future economic growth and social development has been widely recognized. In April 2020, the Chinese government officially defined "data" as a new factor of production, in addition to the traditional ones such as land, labor, capital, and entrepreneurship (or enterprise). It is clear that the digital world will rely on all kinds of trusted data to make timely, correct and fair decisions, and eventually, to provide comprehensive, objective and efficient solutions for every industrial sector and government department. Besides the volume, variety and value of data, IDC has predicted that, by 2025, about 30% of the world data, i.e. 52.5 ZB, will be consumed in real-time[9]. For example, intelligent manufacturing and interactive entertainment applications both have high requirements on data velocity, hence local data should be quickly processed and analyzed by the devices/things themselves, or at nearby edge/fog computing nodes. Last but not least, data privacy and veracity are protected by not only advanced technologies such as a trusted execution environment (TEE), secure multi-party computation (MPC), differential privacy, blockchain, confidential computing, and federated learning, but also specific laws such as the General Data Protection Regulation (GDPR) in the EU since May 2018 and the California Consumer Privacy Act (CCPA) in the US since January 2020. Both laws aim at protecting every consumer's personal data, anywhere at any time, and set strict rules on information disclosure and transparency, thus to enforce that collecting people's data will act in the best interest of those people.

The traditional three-layer IoT architecture consists of (i) the perception layer with devices or things, i.e. sensors and actuators, for collecting data and taking actions; (ii) the network layer with gateways, access points, switches and routers for transmitting data and control signals; and (iii) the application layer with clouds and data centers for analyzing

---

7 David Reinsel, John Gantz, and John Rydning, Data Age 2025, The Digitization of the World: from Edge to Core, IDC and SEAGATE, November 2018.
8 Worldwide Big Data and Analytics Spending Guide, IDC, April 2019.
9 David Reinsel, John Gantz, and John Rydning, Data Age 2025, The Digitization of the World: from Edge to Core, IDC and SEAGATE, November 2018.

and exploiting cross-domain data, and developing and managing intelligent IoT services. As IoT devices and their data are growing explosively, this architecture cannot satisfy a series of crucial service requirements on massive simultaneous connections, high bandwidth, low latency, ultra-reliable under bursty traffic, end-to-end security and privacy protections. In order to tackle these challenges and support various IoT application scenarios, a user-centric flexible service architecture should be implemented, so that feasible micro-services in the neighborhood can be identified and assembled to meet very sophisticated user requirements[10]. This desirable data-driven approach requires a multi-tier computing network architecture that not only connects centralized computing resources and AI algorithms in the cloud but, more importantly, utilizes distributed computing resources and algorithms in the network, at the edge, and on IoT devices[11]. Therefore, most data and IoT services can be efficiently processed and executed by intelligent algorithms using local or regional computing resources at nearby sites, such as empowered edge and distributed cloud[12]. In doing so, a large amount of IoT data need not to be transmitted over long distances to the clouds, which means lower communication bandwidth, lower service delay, reliable network connectivity, lower vulnerability to different attacks, and better user satisfaction in all kinds of industrial sectors and application scenarios.

Based on the above analyses, we believe a pyramid model could best describe the fundamental relationships between these three elements, i.e. data (as raw material), computing (as hardware resource) and algorithms (as software resource) jointly constitute the triangular base to support a variety of user-centric intelligent IoT services at the spire by using different kinds of smart terminals or devices. This book aims at giving a state-of-the-art review of intelligent IoT technologies and applications, as well as the key challenges and opportunities facing the digital world. In particular, from the perspectives of network operators, service providers and typical users, this book tries to answer the following five critical questions.

1. What is the most feasible network architecture to effectively provide sufficient resources anywhere at any time for intelligent IoT application scenarios?
2. How can we efficiently discover, allocate and manage computing, communication and caching resources in heterogeneous networks across multiple domains and operators?
3. How do we agilely achieve adaptive service orchestration and reliable service provisioning to meet dynamic user requirements in real-time?
4. How do we effectively protect data privacy in IoT applications, where IoT devices and edge/fog computing nodes only have limited resources and capabilities?
5. How do we continuously guarantee and maintain the synchronization and reliability of wide-area IoT systems and applications?

Specifically, Chapter 1 reviews the traditional IoT system architecture, some well-known IoT technologies and standards, which are leveraged to improve the perception of the physical world, as well as the efficiency of data collection, transmission and analysis. Further,

---

10 N. Chen, Y. Yang, T. Zhang, M. T. Zhou, X. L. Luo, and J. Zao, "Fog as a Service Technology," IEEE Communications Magazine, Vol. 56, No. 11, pp. 95–101, November 2018.
11 Y. Yang, "Multi-tier Computing Networks for Intelligent IoT," Nature Electronics, vol. 2, pp. 4–5, January 2019.
12 David W. Cearley, Top 10 Strategic Technology Trends for 2020, Gartner, October 21, 2019.

a pyramid model concentrated on user data, distributed algorithms and computing resources is proposed and discussed. This model is based on the multi-tier computing network architecture and applies a data-driven approach to coordinate and allocate most feasible resources and algorithms inside the network for effective processing of user-centric data in real-time, thus supporting various intelligent IoT applications and services, such as information extraction, pattern recognition, decision making, behavior analysis and prediction. As 5G communication networks and edge/fog/cloud computing technologies are getting more and more popular in different industrial sectors and business domains, a series of new requirements and key challenges should be carefully addressed for providing more sophisticated, data-driven and intelligent IoT services with usable resources and AI algorithms in different application scenarios. For instance, in a smart factory, 4G/5G mobile communication networks and wireless terminals are ubiquitous and always connected. A large variety of industrial IoT devices are continuously monitoring the working environment and machines, and generating massive data on temperature, humidity, pressure, state, position, movement, etc. This huge amount of data needs to be quickly analyzed and accurately comprehended with domain-specific knowledge and experiences. To satisfy the stringent requirements on end-to-end service delay, data security, user privacy, as well as accuracy and timeliness in decision making and operation control, the proposed new model and architecture are able to fully utilize dispersive computing resources and intelligent algorithms in the neighborhood for effectively processing massive cross-domain data, which is collected and shared through intensive but reliable local communications between devices, machines and distributed edge/fog nodes.

Chapter 2 presents the multi-tier computing network architecture for intelligent IoT applications, which comprises not only computing, communication and caching (3C) resources but also a variety of embedded AI algorithms along the cloud-to-things continuum. This architecture advocates active collaborations between cloud, fog and edge computing technologies for intelligent and efficient data processing at different levels and locations. It is strongly underpinned by two important frameworks, i.e. Cost Aware Task Scheduling (CATS) and Fog as a Service Technology (FA^{2}ST). Specifically, CATS is an effective resource sharing framework that utilizes a practical incentive mechanism to motivate efficient collaboration and task scheduling across heterogeneous resources at multiple devices, edge/fog nodes and the cloud, which are probably owned by different individuals and operators. While FA^{2}ST is a flexible service provisioning framework that is able to discover, orchestrate, and manage micro-services and cross-layer 3C resources at any time, anywhere close to end users, thus guaranteeing high-quality services under dynamic network conditions. Further, two intelligent application scenarios and the corresponding technical solutions are described in detail. Firstly, based on edge computing, an on-site cooperative Deep Neural Network (DNN) inference framework is proposed to execute DNN inference tasks with low latency and high accuracy for industrial IoT applications, thus meeting the strict requirements on service delay and reliability. Secondly, based on fog computing, a three-tier collaborative computing and service framework is proposed to support dynamic task offloading and service composition in Simultaneous Localization and Mapping (SLAM) for a robot swarm system, which requires timely data sharing and joint processing among multiple moving robots. Both cases are implemented and evaluated in real experiments, and a set of performance metrics demonstrates the effectiveness of the proposed multi-tier computing network and service architecture in supporting intelligence IoT applications in stationary and mobile scenarios.

Under this architecture, Chapter 3 investigates cross-domain resources management and adaptive allocation methods for dynamic task scheduling to meet different application requirements and performance metrics. Specifically, considering a general system model with Multiple Tasks and Multiple Helpers (MTMH), the game theory based analytical frameworks for non-splittable and splittable tasks are derived to study the overall delay performance under dynamic computing and communication (2C) resources. The existence of a Nash equilibrium for both cases is proven. Two distributed task scheduling algorithms are developed for maximizing the utilization of nearby 2C resources, thus minimizing the overall service delay and maximizing the number of beneficial nodes through device/node collaborations. Further, by taking storage or caching into consideration, a fog-enabled 3C resource sharing framework is proposed for energy-critical IoT data processing applications. An energy cost minimization problem under 3C constraints is formulated and an efficient 3C resources management algorithm is then developed by using an iterative task team formation mechanism. This algorithm can greatly reduce energy consumption and converge to a stable system point via utility improving iterations. In addition, based on the fundamental trade-off relationship between service delay and energy consumption in IoT devices/nodes, an offload forwarding mechanism is developed to promote collaborations of distributed fog/edge nodes with different computing and energy resources. The optimal trade-off is achieved through a distributed optimization framework, without disclosing any node's private information, nor lengthy back-and-forth negotiations among collaborative nodes. The proposed mechanism and framework are evaluated via an extensive simulation of a fog-enabled self-driving bus system in Dublin, Ireland, and demonstrate very good performance in balancing energy consumption among multiple nodes and reducing service delay in urban mobile scenarios.

After 3C and energy resources are properly managed, Chapter 4 concentrates on dynamic service provisioning in multi-tier computing networks. Firstly, at the network edge, an online orchestration framework is proposed for cross-edge service function chaining to maximize the holistic cost-efficiency through joint optimization of resource utilization and traffic routing. By carefully combining an online optimization technique with an approximate optimization method, this framework runs on top of geographically dispersed edge/fog nodes to tackle the long-term cost minimization problem with future uncertain information. In this way, the benefits of service function chaining are fully unleashed for configuring and providing various intelligent services in an agile, flexible, and cost-efficient manner. Secondly, inside a computing network using renewable energy, a network slicing framework for dynamic resource allocation and service provisioning is proposed, where a regional orchestrator timely coordinates workload distribution among multiple edge/fog nodes, and provides necessary slices of energy and computing resources to support specific IoT applications with Quality of Service (QoS) guarantees. Based on game theory and the Markov decision process, an effective algorithm is developed to optimally satisfy dynamic service requirements with available energy and network resources under randomly fluctuating energy harvesting and workload arrival processes. Thirdly, across multiple networks, a multi-operator network sharing framework is proposed to enable efficient collaborations between resource-limited network operators in supporting a variety of IoT applications and high-speed cellular services simultaneously. This framework is based on the Third Generation Partnership Project (3GPP) Radio Access Network (RAN) sharing architecture, and can significantly improve the utilization of network resources, thus effectively reducing the overall operational costs of multiple networks. Both the network slicing and multi-network sharing frameworks are evaluated by using more than 200 base station (BS) location data from two mobile operators in the city of Dublin, Ireland. Numerical results show they can greatly improve the workload processing capability and almost double the total number of connected IoT devices and applications.

Chapter 5 addresses the privacy concerns in public IoT applications and services, where IoT devices are usually embedded in a user's private time and space, and the corresponding data is in general privacy sensitive. Unlike resourceful clouds that can apply powerful security and privacy mechanisms in processing massive user data for training and executing deep neural network models to solve complex problems, IoT devices and their nearby edge/fog nodes are resource-limited, and therefore, have to adopt light-weight algorithms for privacy protection and data analysis locally. With low computing overhead, three approaches with different privacy-preserving features are proposed to tackle this challenge. Specifically, random and independent multiplicative projections are applied to IoT data, and the projected data is used in a stochastic gradient descent method for training a deep neural network, thus to protect the confidentiality of the original IoT data. In addition, random additive perturbations are applied to the IoT data, which can realize differential privacy for all the IoT devices while training the deep neural network. A secret shallow neural network is also applied to the IoT data, which can protect the confidentiality of the original IoT data while executing the deep neural network for inference. Extensive performance evaluations based on various standard datasets and real testbed experiments show these proposed approaches can effectively achieve high learning and inference accuracy while preserving the privacy of IoT data.

Chapter 6 considers clock synchronization and service reliability problems in wide-area IoT networks, such as long-distance powerlines in a state power grid. Typically, the IoT systems for such outdoor application scenarios obtain the standard global time from a Global Positioning System (GPS) or the periodic timekeeping signals from Frequency Modulation (FM) and Amplitude Modulation (AM) radios. While for indoor IoT systems and applications, the current clock synchronization protocols need reliable network connectivity for timely transmissions of synchronization packets, which cannot be guaranteed as IoT devices are often resource-limited and their unpredictable failures cause intermittent network connections and synchronization packet losses or delays. To solve this problem, a natural timestamping approach is proposed to retrieve the global time information by analyzing the minute frequency fluctuations of powerline electromagnetic radiation. This approach can achieve sub-second synchronization accuracy in real experiments. Further, by exploiting a pervasive periodic signal that can be sensed in most indoor electromagnetic radiation environments with service powerlines, the trade-off relationship between synchronization accuracy and IoT hardware heterogeneity is identified, hence a new clock synchronization approach is developed for indoor IoT applications. It is then applied to body-area IoT devices by taking into account the coupling effect between a human body and the surrounding electric field generated by the powerlines.

Extensive experiments show that this proposed approach can achieve milliseconds clock synchronization accuracy.

Finally, Chapter 7 concludes this book and identifies some additional challenging problems for further investigation.

We believe all the challenges and technical solutions discussed in this book will not only encourage and enable many novel intelligent IoT applications in our daily lives but, more importantly, will deliver a series of long-term benefits to businesses, consumers, governments, and human societies in the digital world.

Yang Yang
ShanghaiTech University and
Peng Cheng Laboratory, China
August 19, 2020

Acknowledgments

We would like to thank all the people who have made contributions to this book. In particular, we want to acknowledge the enormous amount of help we received from Yang Liu, Zening Liu, Lei Zhu, Zhouyang Lin, and Guoliang Gao at School of Information Science and Technology, ShanghaiTech University, China; Zhi Zhou, Liekang Zeng, and Siqi Luo at Sun Yat-sen University, China; Anqi Huang and Rong Xia at Huazhong University of Science and Technology, China; Linshan Jiang, Guosheng Lin, Jun Zhao, Mengyao Zheng, Dixing Xu, Chaojie Gu, and Zhenyu Yan at Nanyang Technological University, Singapore; Xin Lou and Yang Li at Advanced Digital Sciences Center, Singapore; David Yau at Singapore University of Technology and Design; Peng Cheng at Zhejiang University, China; and Jun Huang at Massachusetts Institute of Technology, USA.

In addition, we want to express our gratitude to our colleagues for their kind support and encouragement. Last but not least, this book has been partially supported by the National Key Research and Development Program of China under grant 2020YFB2104300, the Program for Guangdong Introducing Innovative and Entrepreneurial Teams under grant 2017ZT07X355, the Singapore Ministry of Education AcRF Tier 1 grant 2019-T1-001-044, and the National Natural Science Foundation of China (NSFC) under grant 62071193.

Acronyms

2C	Computing and Communication
3C	Computing, Communication and Caching
3GPP	the Third Generation Partnership Project
5G	Fifth Generation
ADC	analog-to-digital converter
AFE	analog front-end
AI	Artificial Intelligence
AM	Amplitude Modulation
ASTA	Arrivals See Time Averages
BHCA	Busy Hour Call Attempts
BN	Busy Node
BPF	band-pass filter
BR	Bandwidth Reservation
BS	Base Station
CAGR	Compounded Annual Growth Rate
CATS	Cost Aware Task Scheduling
CCPA	California Consumer Privacy Act
CDC	Cloud data center
CDF	Cumulative Distribution Function
CNoT	common notion of time
CSP	Clock Synchronization Protocol
C-V2X	cellular vehicle-to-everything
DNN	deep neural network
DRL	deep reinforcement learning
DSRC	Dedicated Short Range Communications
EDC	Edge Data Center
eMBB	enhanced mobile broadband
FFT	fast Fourier transform
FM	Frequency Modulation
FN	Fog Node
FTSP	Flooding Time Synchronization Protocol
GAP	generalized assignment problem
GDPR	General Data Protection Regulation
GPS	Global Positioning System

GRP	Gaussian random projection
H2M	human-to-machine
HDP	Health Device Profile
HetVNET	Heterogeneous Vehicle NETwork
HN	Helper Node
ICN	Information-Centric Networking
IIR	infinite impulse response
IMT-2020	International Mobile Telecommunications-2020
IMU	Inertial Measurement Unit
IoIT	Internet of Intelligent Things
IoT	Internet of Things
LPCC	linear prediction coefficients
LPWA	Low power wide area
LTE-V	Long Term Evolution for Vehicle
M2M	Machine-to-machine
MANET	Mobile ad hoc network
MCC	Mobile Cloud Computing
MEC	Mobile/Multi-access Edge Computing
mMTC	massive machine type communications
MPC	Multi-party Computation
MRF	mean removal filter
MTC	Machine-type Communications
MTMH	Multiple Tasks and Multiple Helpers
NB-IoT	Narrowband IoT
NGMN	Next-Generation Mobile Network
NOMA	Non-Orthogonal Multiple Access
PDF	probability density function
PLL	Phase-locked Loop
POMT	Paired Offloading of Multiple Tasks
POST	Parallel Offloading of Splittable Tasks
PRACH	Physical Random Access Channel
PTP	Precision Time Protocol
QoS	Quality of Service
R&D	Research and Development
RACH	Random Access Channel
RAN	Radio Access Network
RBS	Reference-Broadcast Synchronization
RMSE	root mean square error
RTT	Round-trip-time
SD-WAN	software-defined wide area network
SLAM	Simultaneous Localization and Mapping
SoA	Service-oriented Architecture
TDMA	Time Division Multiple Access
TEE	Trusted Execution Environment
TN	Task Node

TPSN	Timing-sync Protocol for Sensor Networks
UE	User Equipment
URLLC	Ultra Reliable Low Latency Communications
UWB	Ultra WideBand
VANET	Vehicular Ad-hoc NETwork
VR/AR	virtual/augment reality
ZB	zettabytes
ZCD	zero crossing detection

1

IoT Technologies and Applications

1.1 Introduction

As a new dimension to the world of information and communication technologies (ICTs), the concept of the internet of things (IoT) aims at providing "connectivity for anything", "by embedding short-range mobile transceivers into a wide array of additional gadgets and everyday items, enabling new forms of communication between people and things, and between things themselves", according to the seventh International Telecommunication Union (ITU) Internet Reports 2005 (International Telecommunication Union (ITU), 2005). In recent years, different IoT technologies and standards have been actively developed for different industrial sectors and application scenarios, such as smart city, intelligent transportation system, safety, and security system, intelligent agriculture, environment monitoring, smart factory, intelligent manufacturing, smart home, and healthcare. Various IoT-centered business models and value chains have become, consolidated, and popular; these IoT applications are effectively accelerating the digitalization and transformation progresses of traditional industries (Union, 2016). As a result, they have generated tremendous economic and social benefits to our society, as well as huge impacts on people's daily life.

Sensors, machines, and user devices are the "things", which are usually equipped with limited energy resources and simple sensing, computing, communication, motion, and control capabilities. By using a large number of sensors in the field, a typical IoT system can effectively expand its service coverage and capability in sensing, collecting, and processing different types of raw data obtained from the physical world. Most redundant data with low value will be aggregated or filtered for saving scarce communication resources, i.e. bandwidth, storage,and energy. Selected data with potential values, e.g., characteristics of unexpected events, will be transmitted from different sites through multi-hop communication networks to a centralized computing facility, such as a data center, for further in-depth investigation. New information will be extracted, or new events will be discovered, from this more comprehensive analysis of massive data from a much larger area across multiple sites and locations.

In the early days, IoT systems were usually developed according to rigid rules or requirements. The main purpose is to improve the perception of the physical world, as well as the efficiency of data collection and analysis, in different IoT applications such as environment monitoring and emergency management. As a well-known application-driven IoT architecture, the ISO/IEC 30141-IoT Reference Architecture is often adopted in system designs

Intelligent IoT for the Digital World: Incorporating 5G Communications and Fog/Edge Computing Technologies,
First Edition. Yang Yang, Xu Chen, Rui Tan, and Yong Xiao.
© 2021 John Wiley & Sons Ltd. Published 2021 by John Wiley & Sons Ltd.

and service deployments (Union, 2016). Data acquisition involves all kinds of sensors, such as RFID, MEMS, bar code, and video camera. However, due to dynamic application scenarios and environments, the key function and challenge for IoT systems are high-quality data collection (transmission) through wireless ad hoc networks. Many wireless access and networking technologies have been developed for ensuring timely and reliable connectivity and transmission for data collection at low cost and low energy consumption (Yang et al., 2006b,a, Zhao et al., 2015). In addition to the existing standards for mobile communications, the internet, and broadcasting networks, a series of wireless communication technologies have been developed for supporting IoT data transmissions in various application scenarios, e.g. RFID, Wi-Fi, NFC, ZigBee, LoRa, and Sigfox (Jia et al., 2012, Li et al., 2011, Vedat et al., 2012, Alliance, 2012, Augustin et al., 2016, Sigfox, 2018a). By collaboratively analyzing data from multiple sensors in different areas, a more comprehensive perception of the actual environment and a timely understanding of the exact situation will be achieved, thus enabling better decision making and performance optimization for particular industrial operations.

Nowadays, a series of advanced technologies on smart sensing, pervasive computing, wireless communication, pattern recognition, and behavior analysis have been widely applied and integrated for supporting more and more sophisticated IoT applications, such as intelligent transportation system and intelligent manufacturing. Such complex applications can significantly improve system automation and efficiency in massive data analysis and task execution. To achieve this goal, the key function and challenge for IoT systems is accurate information extraction, which heavily depends on domain-specific knowledge, valuable experience, and technical know-how contributed by real experts and field workers. In order to make IoT systems more accessible and deployable, the fourth-generation (4G) and fifth-generation (5G) mobile communication standards have specified several important IoT application scenarios, i.e. Narrowband IoT (NB-IoT) in 4G massive Machine Type Communications (mMTC) and ultra-reliable and low latency communications (URLLC) in 5G (3GP, 2017, 3GPP, 2016a, Yang et al., 2018). Furthermore, the latest developments in cloud computing and big data technologies enable fast and accurate analysis of huge volumes of structured and non-structured data, creating lots of business opportunities for the development of more sophisticated and intelligent IoT systems. The continuous progression and widespread deployment of IoT technologies have been transforming many industrial sectors and commercial ecosystems. Now, IoT applications and services are becoming indispensable to our daily lives and business models, e.g., remote control of door locks, lights, and appliances at homes and offices, real-time modeling of resource consumption patterns and streamline business processes in factories, constant surveillance for property security, public safety and emergency management in cities.

To meet the fast-growing demands of various IoT applications and services for different businesses and customers, leading ICT companies, such as Amazon, Google, Microsoft, Cisco, Huawei, Alibaba, and JD, have launched their own cloud-based IoT platforms for data-centric services. However, these enterprise-level platforms are not designed for data sharing, nor service collaboration. General concerns of data security and customer privacy strictly prevent the attempts of connecting and integrating them for much bigger commercial benefits and global influences. Besides, it is even harder to overcome the existing barriers of vertical industries and realize cross-domain information exchanges for minimizing the redundancies and fragments at different but related IoT applications.

In the future, when artificial intelligence (AI) technologies are widely adopted in most industrial sectors and application domains, new links will be established between those domain-specific island-like solutions. In most cases, they are not used to share original data, but only to exchange necessary knowledge that is purposely learned from separated/protected datasets for customized applications. To realize this ambitious vision, the key function and challenge for future IoT systems is innovative knowledge creation, which requires high-quality data, super-intelligent algorithms, and more computing resources everywhere. Centralized cloud computing alone cannot support this fundamental change, while dispersive fog computing technologies will fill the computational gap along the continuum from cloud to things. Therefore, future intelligent IoT systems will fully utilize the best available computing resources in their neighborhood and in the cloud to calculate novel effective mechanisms and solutions, which are acceptable and executable across different enterprise platforms, industrial sectors, and application domains. In this way, those domain-specific IoT systems are not closed or isolated any more, they will become much more powerful and influential by working collaboratively, thus significantly saving global resources, improving overall performance, and maximizing potential benefits in all systems. There is no doubt that future IoT applications and services will be shifting from data-centric to knowledge-centric. Very soon, they will become better than human-designed ones, since they are taught by us and powered by accumulated data, sophisticated AI algorithms, endless computing resources, and fast evolution. Eventually, they will help us not only search and identify useful information from massive raw data, but more importantly, discover, and create new knowledge to expand our horizons and capabilities.

The rest of this chapter is organized as follows. Section 1.2 reviews some well-known and emerging IoT standards and technologies. Section 1.3 introduces intelligent IoT technologies, including an intelligent user-centered IoT network, in which data, computing power, and intelligent algorithms are distributed around its users. Typical IoT applications are summarized in Section 1.4. New requirements and challenges of IoT systems are analyzed in Section 1.5. Finally, Section 1.6 concludes this chapter.

1.2 Traditional IoT Technologies

1.2.1 Traditional IoT System Architecture

The IoT is a platform where every day devices become smarter, every day processing becomes more intelligent, and every day communication becomes more informative. While the IoT is still seeking its own shape, its effects have already started in making incredible strides as a universal solution media for the connected scenario. Architecture specific study does always pave the conformation of the related field. IoT is a dynamic global network infrastructure with the capability of self-configuring based on standards and interoperable protocols. IoT enables physical and virtual "things" to use intelligent interfaces and seamlessly integrate into an information network (van Kranenburg and Dodson, 2008).

A multi-layer technology is used to manage the connected "things" is known as the IoT system. It brings the physical or virtual devices to the IoT network, and provides the various services to the devices by using machine to machine communication. An traditional IoT system architecture is comprised of various functional layers to enable IoT applications, including asensing layer, network layer, and application layer. This section will introduce these functional layers.

1.2.1.1 Sensing Layer

The sensing layer plays the role of interface between the physical world and information world, and is the foundation of IoT. This layer consists of a physical layer system such as smart sensors and devices, and communicates with the network layer. The main function is to identify and track objects. To achieve this function, several technologies including RFID, bar code technology, sensor technology, positioning technology, or other information sampling technology can be implemented. With the development of science and technology, sensors are becoming more and more intelligent. The smart sensors have numerous advantages over conventional sensors, such as low cost and power, flexible connection, high-reliability band efficient, and less cable communication. IoT systems are based on the data that provide actuation, control, and monitoring activities. IoT devices can exchange data with other connected devices and applications, or collect data from other devices. They can also process the data locally, or send the data to a centralized server or use cloud-based servers to process the data. They can perform some tasks locally and other tasks within IoT infrastructure based on temporal and spatial constraints (i.e. memory, processing capabilities, communication latencies, and speeds, and deadlines). IoT devices may contain multiple interfaces for communicating with other devices (wired and wireless). IoT devices can also be of many types, such as wearable sensors, smart watches, LED lights, cars, and industrial machines. IoT devices and systems should have the ability to dynamically adapt to changing environments and take actions based on their operating conditions, user environment or perceived environment. For example, consider a pollution surveillance system consisting of multiple sensor nodes. Existing IoT supports different hardware platforms, such as Arduino, Intel Galileo Gen, Intel Edison, Beagle Bone Black, and Raspberry Pi. The platforms are classified according to key parameters, including processor, GPU, operating voltage, clock speed, bus width, system memory, flash memory, EEPROM, supported communications, development environment, programming language, and I/O connection. For example, the Arduino Uno platform supports a lightweight system with 2 KB of memory, which can be used to build sensor networks economically. Intel Edison has provided better performance, which can support 1 GB system memory and could support more applications flexibly.

1.2.1.2 Network Layer

The network layer performs communication between the device and a remote server. The IoT environment consists of an enormous number of smart devices, but with many constraints. Limited processing capability, storage volume, battery life, and communication range are among of these constraints. Therefore, the IoT implementation requires a communication protocol that can efficiently manage these conditions (Al-Sarawi et al., 2017, Farhan et al., 2017).

The protocols form the backbone of IoT systems, and enable network connection and coupling with application programs. Communication protocols allow devices to exchange data over a network. The protocols define data exchange format, data encoding, device addressing scheme, and routing of packets from source to destination. Other functions of the protocols include sequence control, flow control, and retransmission of lost packets. There are many wireless communication technologies with various parameters (Ahmed et al., 2016, Madakam et al., 2015). The communication technology of IoT can operate in sensor and backhaul network scenarios. Sensor network standards (such as ZigBee, RFID or Bluetooth) work over relatively short distances (i.e. tens of meters), with low data rates and low energy consumption. On the other hand, standards such as GPRS, LTE, WiMAX can work over long distances and provide high throughput. However, they consume more energy, require expensive and fixed base station infrastructure with proper communication connectivity (Lee and Kim, 2016, Ahmed et al., 2016).

IEEE 802.11: 802.11 is a collection of wireless local area network (WLAN) communication standards for wi-fi. For example, 802.11a operates in the 5 GHz band, 802.11b and 802.11g operate in the 2.4 GHz band, 802.11n operates in the 2.4/5 GHz bands, 802.11ac operates in the 5 GHz band, and 802.11ad operates in the 60 GHz band. These standards provide data rates from 1 Mb s^{-1} to 6.75 Gb s^{-1}. Wi-fi provides a 20 m indoor communication range and 100 m in the outdoors (Ray, 2018). The emerging IEEE 802.11ah is a promising communication standard that supports a massive number of heterogeneous devices in the IoT. It provides attractive features like improved scalability, low energy consumption, and large coverage area. In this chapter, the authors analyze the performance of IEEE 802.11ah, and compare it with a prominent alternative, the IEEE 802.15.4. The simulation results show that the new 802.11ah standard performs better than the 802.15.4 in terms of association time, throughput, delay, and coverage range (Hazmi et al., 2012).

IEEE 802.15.4: 802.15.4 is a collection of low-rate wireless personal area network (LR-WPAN) standards. It can be contrasted with other approaches, such as Wi-Fi, which offers more bandwidth and require more power. The emphasis is on very low-cost communication of nearby devices with little to no underlying infrastructure, intending to exploit this to lower power consumption even more. LR-WPAN standards provide data rates from 40 Kb s^{-1} to 250 Kb s^{-1}. These standards provide low-cost and low-speed communication to power-constrained devices. They operate at 868/915 MHz and 2.4 GHz frequencies at low and high data rates, respectively. Important features include real-time suitability by reservation of guaranteed time slots (GTS), collision avoidance through CSMA/CA, and integrated support for secure communications. Devices also include power management functions such as link quality and energy detection. The standard does have provisions for supporting time and rate sensitive applications because of its ability to operate in pure CSMA/CA or TDMA access modes (Pasolini et al., 2018).

Mobile communication: There are different generations of mobile communication standards. 2G is short for second-generation wireless telephone technology. One of the benefits of 2G is that 2G signals consume less battery power, so they help mobile batteries to last longer. Digital coding improves the voice clarity and reduces noise in the line. 2G digital signals are considered environment friendly. The use of digital data service assists mobile network operators to introduce short message service over the cellular phones. Digital encryption has provided secrecy and safety to the data and voice calls. The use of

2G technology requires strong digital signals to help mobile phones work. 3G is the third generation of mobile phone standards and technology, superseding 2G, and preceding 4G. It is based on the International Telecommunication Union (ITU) family of standards under the International Mobile Telecommunications programme, IMT-2000 (Bhalla and Bhalla, 2010). 3G was launched in Japan on October 2001 by NTT DoCoMo. 3G provided a good experience for the mobile user, and supports higher speed connection than the previous generations (and, 2014). The essential factor of this technology is to merge the wireless principles like time division multiple access (TDMA), a global system for mobile communication (GSM) and code division multiple access (CDMA). 4G refers to the fourth generation of cellular wireless standards. It is a successor to 3G families of standards. The nomenclature of the generations generally refers to a change in the fundamental nature of the service, non-backwards compatible transmission technology, and new frequency bands. IoT devices based on these standards can communicate over cellular networks. Data rates for these standards range from 9.6 Kb s^{-1} (2G) to 100 Mb s^{-1} (4G), which are available from the 3GPP websites. 5G is the next generation cellular network that aspires to achieve substantial improvement on quality of service, such as higher throughput and lower latency. 5G wireless system can bring as much as 1000 and above times the capability offered by today's mobile world.

1.2.1.3 Application Layer

The application layer supports various IoT applications, which can transitionally be deployed on cloud platforms. IoT cloud platforms are designed to be meant for particular application specific domains, such as, application development, device management, system management, heterogeneity management, data management, analytics, deployment, monitoring, visualization, and finally research purposes. An IoT system serves various types of functions such as services for device modeling, device control, data publishing, data analytics, and device discovery. Management block provides different functions to govern an IoT system to seek the underlying governance of an IoT system. A security functional block secures the IoT system by providing functions such as authentication, authorization, and privacy. The application layer is the most important one in terms of users as it acts as an interface that provides various functions to the IoT system. The application layer allows users to visualize and analyze the system status at the current stage of action, and sometimes predict futuristic prospects. It is obvious that there are many more platforms currently present in the market, such as the Google Cloud Platform, Microsoft Azure IoT Suite, IRI Voracity, Particle, ThingWorx, IBM Watson IoT, and Amazon AWS IoT Core. For example, Google Cloud provides a multi-layered secure infrastructure. It helps in improving operational efficiency. It provides predictive maintenance for equipment, solutions for smart cities and buildings, and real-time asset tracking. The Microsoft Azure IoT solution is designed for different industry needs. It can be used from manufacturing to transportation to retail. It provides solutions for remote monitoring, predictive maintenance, smart spaces, and connected products.

Available architectures explore multiple opportunities to seek the advantageous part of IoT while encouraging the developer and user groups to get application specific solutions. However, the central issue of these architectures is the lack of full interoperability of interconnected things at the abstraction level. This leads to many proclaimed problems, such as degraded smartness of high degree, less adaptability, limited anonymity, poor behavior of the system, reduced trust, privacy, and security. IoT architectures do pose several network oriented problems due to their limitations in a homogeneity approach.

1.2.2 IoT Connectivity Technologies and Protocols

The IoT refers to the inter connection and exchange of data among devices/sensors. Currently, with the explosive growth of IoT technologies, an increasing number of practical applications can be found in many fields, including security, asset tracking, agriculture, smart metering, smart cities, and smart homes. Short-range radio technologies e.g., radio-frequency identification (RFID), near field communication (NFC), and ZigBee are widely used for building automation, automotive, and monitoring devices. For example, wi-fi based on the IEEE 802.11 standards are used in most office environments. However, short-range radio technologies are not adapted for scenarios that require long range transmission.

Cellular networks are widespread and ubiquitous, covering 90% of the world's population, and other technologies like wi-fi don't have the same scale, requiring users to search for and connect to a local network. RF providers, and wireless infrastructure companies and carriers have made massive investments in cellular networks to provide a secure and reliable service to as many customers as possible. By leveraging existing infrastructure and mature technology, cellular IoT can connect millions of IoT devices with little additional investment. Solutions based on cellular communications (e.g., 2G, 3G, and 4G) can provide larger coverage, but they consume excessive energy.

IoT application requirements have driven the emergence of a new wireless communication technology: low power wide area network (LPWAN) (Mekki et al., 2019). LPWAN is increasingly gaining popularity in industrial and research communities because of its low-power, long-range, and low-cost communication characteristics. It provides long-range communication up to 10–40 km in rural zones and 1–5 km in urban zones. In addition, it is highly energy efficient (i.e. 10+ years of battery lifetime) and inexpensive, with the low cost of a radio chip-set (Mekki et al., 2019). In summary, LPWAN is highly suitable for IoT applications that only need to transmit tiny amounts of data in long range. Many LPWAN technologies have arisen in the licensed as well as unlicensed frequency bandwidth. Among them, Sigfox, LoRa, and NB-IoT are today's leading emergent technologies that involve many technical differences(Mekki et al., 2019).

A myriad of IoT connectivity solutions are available to support a wide range of IoT applications with miscellaneous requirements. Therefore, in order to select an optimal technology for each application, various factors, such as power consumption, security issues, deployment cost, communication range, data rate, and latency, are required to be considered. A comparison of some typical IoT connecting solutions (i.e. RFID, NFC, Zigbee, LoRa, Sigfox, and NB-IoT) using a pre-specified factor is given in Table 1.1.

The three main technical requirements for any enterprise looking into IoT connectivity technology are coverage, energy efficiency, and data rate. No single technology can excel in all these aspects, as they are conflicting objectives and every radio technology has to make trade-offs. All IoT applications need good coverage to connect the devices but some need to cover only certain indoor areas while others require extensive coverage in rural or remote regions. A technology with long range is better suited to connecting devices scattered in a

Table 1.1 Comparison of IoT connecting technologies.

Technology	RFID	NFC	BLE	Zigbee	6LoWPAN	LoRa	Sigfox	NB-IoT	MIOTY
Range	1–5 m	1–10 cm	1–10 m	75–100 m	100 m	2–15 km	3–50 km	10–15 km	15–20 km
Bandwidth	2–26 MHz	14 kHz	1–2 MHz	2 MHz	3 MHz	<500 kHz	100 Hz	180 kHz	200 kHz
Frequency band	125–134.2 kHz, 13.56–433 MHz, 860–960 MHz	13.56 MHz	2.4 GHz	868 MHz, 915 MHz, 2.4 GHz	2.4 GHz, Sub GHz	868 MHz, 915 MHz, Sub 1 GHz	915–928 MHz, Sub 1 GHz	700 MHz, 800 MHz, 900 MHz	133–966 MHz
Data rate	4–640 kbps	100–424 kbps	1 Mbps	250 kbps	250 kbps	50 kbps with FSK	Less than 100 bps	200 kbps	Sub 1 kbps
Latency	1–10 ms	100 ms	6 ms	~15 ms	2–6 ms	1–10 ms	10–30 ms	40 ms–10 s	10 ms–10 s
Modulation scheme	OOK, FSK, PSK	ASK	DQPSK, DPSK	O-QPSK	QPSK, BPSK	FSK	GFSK, DBPSK	BPSK, QPSK	BPSK, MSK
Battery lifetime	Battery free	Battery free	1–5 days	1–5 years	1–2 years	<10 years	<10 years	>10 years	up to 20 years
Application	Materials management, attendee tracking	Payment, ticketing	IoT device authentication, localization	Smart home, healthcare	Smart city infrastructures	Air pollution monitoring, fire detection	Smart meter, pet tracking	Street lightning, agriculture	Dense IoT network scenarios, smart city
Advantages	High speed and convenience, and very low cost	Convenient to use	Supported by most smartphones	Highly reliable and scalable	Massive connection without complex routing	Highly immune to interference, adaptive data rate	High reliability, low device complexity	Better coverage range and coverage	Low packet error rate, high energy efficiency
Limitation	Brings security and privacy concerns	Limited data rates	Limited range and battery life	Short range, communication security issues	Limited range and data rages	Longer latency, not acknowledges all packets	Multiple transmissions, suffer from interference	No hand-off support, low interference immunity, lacks in ACK	Low data rate
Standard body	ISO/IEC	ISO/IEC	Bluetooth SIG	ZigBee Alliance	IETF	LoRa Alliance	Sigfox	3GPP	MIOTY Alliance

wide area. The energy efficiency of a connectivity technology has a significant impact on the lifetime or the maintenance cycle of IoT devices relying on battery or energy harvesting and is dependent on range, topology and complexity of the connectivity technology. The overall energy consumption of the device also depends on the usage of the application, such as the frequency and duration of message transmission. Data rate requirements for IoT applications vary from hundreds of bit per second (bps) for metering to several megabits per second (Mbps) for video surveillance. In this section, we will introduce traditional IoT connectivity technologies from two categories, low-power and short-range connectivity technologies, and low data rate and wide-area connectivity technologies. Low-power and short-range connectivity technologies include RFID, NFC, Zigbee, and BLE. Low data rate and wide-area connectivity technologies include 6LoWPAN, Sigfox, LoRa, and NB-IoT. Then, we will introduce several emerging IoT technologies, which improve the performance of IoT connectivity in various aspects, such as latency and accessibility. Finally, we will introduce requirements and challenges for intelligent IoT services from the perspective of network provider and users.

1.2.2.1 Low-power and Short-range Connectivity Technologies
Radio Frequency IDentification

The roots of radio frequency identification (RFID) date back to World War II. Germans, for instance, used an interesting maneuver in which their pilots rolled their planes as they return to base, so it would change the reflecting radio signal. This simple procedure alerted the ground radar crew of German planes returning and not allied aircraft. It can be considered one of the first passive ways to identify an object by means of a radio frequency signal, which was known as "identify friend or foe (IFF)". In the 1960s and 1970s, RFID systems were still embedded within the context of "secret technology". As an example, Los Alamos National Laboratory was asked by the Energy Department of United States of America to develop a system for tracking nuclear materials and control sensitive materials. In the 1990s, MIT's Auto-ID Center developed the global standard for RFID and other sensors, which described the IoT as a system where the internet is connected to the physical world through ubiquitous sensors. In the 2010s, the decreased cost of equipment and tags, increased performance to a reliability of 99.9% and a stable international standard brought a major boost in the use of RFID systems.

There are two major initiatives regarding RFID standardization: the International Standard Organization (ISO) and EPCglobal. The ISO uses a cross-industry perspective with a generic approach. Meanwhile EPCglobal adopts a more application-specific approach. The widely recognized outcome of EPCglobal is the establishment of the electronic product code (EPC), a unique code for item numbering, for identification of objects by using a similar approach to barcode numbering. ISO works closely with International Electro-technical Commission (IEC) which is responsible for a general type of RFID standards covering issues of air interface, data content, conformance, and performance. ISO/IEC standards cover certain areas of technology. For instance, ISO 18000 is a series of standards that define air interface standards consisting of channel bandwidth, EIRP, modulation, data coding, and bit rate, and ISO 15418 is a standard that defines data content. There are also many separate standards that had already been developed for livestock tracking (ISO 11785, ISO 11784 and ISO 14223) (Jia et al., 2012, Adhiarna and Rho, 2009).

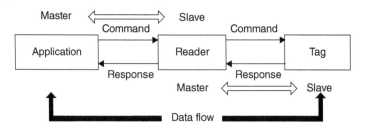

Figure 1.1   The architecture of an RFID system.

A typical RFID system consists of three main components: RFID tags, reader, and an application system (Finkenzeller, 2010, Jia et al., 2012), as shown in Figure 1.1. RFID uses electromagnetic fields to automatically identify and track tags attached to objects. The tags contain electronically stored information. The RFID tags are known as transponders (transmitter/responder), which are attached to the objects to count or identify. Tags could be either active or passive. Active tags are those that have partly or fully battery powered, have the capability to communicate with other tags, and can initiate a dialogue of their own with the tag reader. Passive tags collect energy from a nearby RFID reader's interrogating radio waves. Active tags have a local power source (such as a battery) and may operate hundreds of meters from the RFID reader. Tags consist mainly of a coiled antenna and a microchip, with the main purpose of storing data. The reader is called as transceiver (transmitter/receiver) made up of a radio frequency interface (RFI) module and control unit. Its main functions are to activate the tags, structure the communication sequence with the tag, and transfer data between the application software and tags. The application system is known as data processing system, which can be an application or database, depending on the application. The application software initiates all readers and tags activities. RFID provides a quick, flexible, and reliable way for electronically detecting, tracking, and controlling a variety of items. RFID systems use radio transmissions to send energy to a RFID tag while the tag emits a unique identification code back to a data collection reader linked to an information management system. The data collected from the tag can then be sent either directly to a host computer, or stored in a portable reader and up-loaded later to the host computer.

RFID technology has many advantages. The RFID tag and reader should not have LOS to make the system work, and a RFID reader is capable of scanning multiple tags simultaneously. Unlike barcodes, RFID tags can store more information. Moreover, it follows the instructions/commands of the reader, and provides the location to the reader along with its ID. RFID technology is versatile in nature and hence smaller and larger RFID devices are available as per application. Tags can be read only as well as read/write, unlike barcodes. However, RFID technology also has disadvantages. Active RFID is costly due to the use of batteries. Privacy and security are concerns with the use of RFID on products as it can be easily tapped or intercepted. RFID devices need to be programmed which requires some time. The external electromagnetic interference can limit the RFID remote reading. The coverage range of passive tags is limited to around 3 m.

RFID takes the market in many different areas including inventory management, personnel, asset tracking, controlling access to restricted areas, supply chain management, and counterfeit prevention. The addition of other sensors around AIDC (automated identification and data capture) technologies such as infrared detectors, radiation, humidity, and others in RFID applications contributed to the development of IoT by extending it to reach intelligent services and providing local capabilities for actuation. For example, in manufacturing, RFID technology offers many applications in the automotive industry. The RFID-based anti-theft device is a protection device installed in many vehicles. RFID also offers great promise for the assembly and manufacturing process of automobiles, especially for flexible and flexible production planning, spare parts, and inventory management. RFID technology not only helps automate the overall assembly process, but also significantly reduces costs and shrinkage, and provides better service for automotive users, including more efficient replacement parts ordering and automatic generation of maintenance reminders. The benefits that RFID brings to the automotive industry, including production processes and end users, are visibility, traceability, flexibility, and enhanced security. In the supply chain, managers will be able to monitor the status of shipments like a crate filled with fruit. With sensors, RFID tags, and RFID readers, the manager sees the exact location of the crate inside the warehouse, the fruit's point of origin, days until expiration, and temperature in real-time. A visible, and transparent process improves efficiency, reduces waste, and allows traceability. If a shipment is determined to be unsuitable for consumption due to disease or other circumstances, the source or cause of the defection will quickly be discovered because of the great wealth of information available.

In summary, the adoption of RFID is spurring innovation and the development of the IoT, which are commonplace throughout households, offices, warehouses, parks, and many other places. Industry and government mandates are regulating RFID technologies leading to accepted standards across industries allowing for interoperability among devices. Additionally, the cost and size of devices continue to decrease which allows companies to embed smaller, common items with RFID chips and sensors. Although promising, RFID is not without its challenges, which arise from electromagnetic interference, security, and privacy issues. Communication between tags and readers are inherently susceptible to electromagnetic interference. Simultaneous transmissions in RFID lead to collisions as readers and tags typically operate on the same wireless channel. Therefore, efficient anti-collision protocols for identifying multi-tags simultaneously are of great importance for the development of large-scale RFID applications Due to its cost and resource constraint limitations, RFID systems do not have a sufficient security and privacy support. Many researchers and scientists work to implement low cost security and privacy protocol to increase the applicability. Lots of lightweight solutions have been proposed for RFID, but they are still expensive and vulnerable to the security and do not fully resolve the security issues.

Near Field Communication

Near field communication (NFC) is a short-range wireless communication technology. NFC technology uses magnetic coupling to send and receive signals. When two NFC enabled devices are close enough (from touch to 10 cm), they create an electromagnetic field between them. That electromagnetic field allows active NFC devices to power up and communicate with the passive NFC device. The active NFC device then picks up on variations in signal levels specific to the passive device and reads those variations as a signal. A detector and decoder circuit in the active NFC device is then used to comprehend

Figure 1.2 NFC system architecture. Source: Vedat Coskun, Busra Ozdenizci, and Kerem Ok. The survey on near field communication. Sensors, 15(6):13348–13405, jun 2015. doi: 10.3390/s150613348. Licensed under CC BY 4.0

the passive NFC signal and extract the relevant information. NFC technology builds on RFID, which uses an ISO/IEC standard. NFC was approved as an ISO/IEC standard in 2003, and is standardized in ECMA-340 and ISO/IEC 18092. These standards specify the modulation schemes, coding, transfer speeds and frame format of the RF interface of NFC devices, as well as initialization schemes and conditions required for data collision-control during initialization for both passive and active NFC modes. They also define the transport protocol, including protocol activation and data-exchange methods. NFC incorporates a variety of existing standards including ISO/IEC 14443 Type A and Type B.

The possible interaction styles among NFC devices provide three different operating modes, as shown in Figure 1.2 (Coskun et al., 2015). Three types of NFC devices are involved in NFC communication: smartphones, NFC tags, and NFC readers. In reader/writer mode, an active NFC device can read, write or change the stored data in a passive NFC tag. This mode is just like the traditional RFID technology, where a terminal reads the data from the embedded chip on the smart card. The maximum possible data rate in this mode is 106 kbps. In peer-to-peer node, two active devices can exchange small amount of data between them. As both devices are battery-powered, they can establish radio link in between them. They set up a bi-directional, half duplex channel having a maximum data rate of 424 kbps. In card emulation node, the active NFC devices work as a smart card based on ISO/IEC 14443 Type A and Type B. This model is compatible with the pre-existing smart-card industry.

NFC has advantages in IoT application. One of the notable benefits or advantages of NFC revolves around its simplicity and expansive applications. It is easier to set up and deploy than Bluetooth because it does not require pairing or manual configuration. The connection is automatic and takes a fraction of a second. It also uses less power than other types of wireless communication technologies. Another remarkable advantage of NFC is that it supports the widespread application of contactless payment systems. Several companies have implemented payment transactions based on this technology. However, NFC also has disadvantages. It is too expensive for companies to purchase and maintain related machines and other equipment in IoT applications, so small companies could find it difficult to sustain their existing turnover and enhance their profits. Installing the hardware and software and hiring technicians to maintain the same could result in spiraling expenses for the concerned company. A critical limitation or disadvantage of NFC is that it is not as effective and efficient as Bluetooth or Wi-Fi Direct when it comes to data transfer rates. NFC can only send and receive very small packets of data. Its maximum transfer rate is 424 kbps while Bluetooth 2.1 has a transfer rate of 2.1 Mbps. While NFC transactions are undoubtedly more secure than regular credit card payments, this technology is not completely free from risk. Rapid evolution in technology always comes with an equally powerful negative consequence. Mobile phone hacking is now rampant and attackers are coming out with newer methods to gain unauthorized access into users' personal, social security and financial data stored therein. This makes the entire system vulnerable and insecure. The obvious lack of security could discourage both users and companies from warming to this technology in the near future.

NFC offers great and varied promise in services such as payment, ticketing, gaming, crowd sourcing, voting, navigation, and many others. NFC technology enables the integration of services from a wide range of applications into one single smartphone. NFC technology is typically used for payments and marketing applications today. Payment using NFC integrated smart cards offers easier payment compared to conventional multiple step payment process. Top payment services like Visa and MasterCard are offering NFC embedded smart cards to customers. NFC with smart cards can be used for fast payments at grocery shops, parking tickets, adding shopping points, and redeeming coupons with just a single tap of the card. All the major banks around the globe offer smart cards with NFC chips integrated. NFC integrated system can be used in medicine and healthcare activities. NFC offers greater accuracy and convenience in prescribing medicine, easier check-in, payments, checking status of patients, tracking records by embedding NFC tags to patient's charts. NFC integrated devices can be easily paired and configured. Medical professionals can easily check schedules and access medical devices and equipment.

NFC is an emerging technology of the last decade. Even though it remains a comparatively newborn technology, NFC has become an attractive research area for many researchers and practitioners due to its exponential growth and its promising applications and related services. From the technical point of view, some security issues in NFC technology have already been solved and standardization is mostly provided as well. However, there are still unsolved security issues. For example, new protocols/mechanisms on off-line and on-line authentication of NFC tags should be studied. NFC specific alternative key exchange protocols should be proposed to prevent various attacks on RF communication.

BLE

Bluetooth low energy (BLE, Bluetooth 4, Bluetooth Smart) is an innovative technology, developed by the Bluetooth Special Interest Group (SIG), which aims to become the best alternative to the huge number of standard wireless technologies already existing and widespread on the market. Bluetooth wireless technology is an inexpensive short-range radio technology that does not require the use of proprietary cables between devices such as laptops, smartphones, cameras, and printers, with an effective range of 10–100 m. In addition Bluetooth usually communicates at a speed below 1 Mbps, and it uses the specifications of the IEEE 802.15.1 standard. A group of Bluetooth devices sharing a common communication channel is called Piconet. The Piconet can support two to eight devices for data sharing at a time, and the data can be text, pictures, video, and sound. The Bluetooth SIG is composed of more than 1000 companies including Intel, Cisco, HP, Aruba, Intel, Ericsson, IBM, Motorola, and Toshiba.

BLE was first introduced in 2010, and its goal is to extend Bluetooth applications to the applications of power-constrained devices such as wireless sensors, in which the amount of data transmission is small and communication rarely occurs. This differs from conventional Bluetooth applications, such as audio and data streaming, which require a large amount of data transmission and frequent interaction between two communication devices. In addition, device cost is more important for wireless sensor controls than for audio streaming. To address these application requirements, BLE introduces a new radio, which is a derivative of the conventional Bluetooth, and new interfaces. In Bluetooth Classic, there are 79 channels, each with a channel bandwidth of 1 MHz and a raw symbol rate of 1 Msymbol/s. The modulation scheme could be Gaussian frequency shift keying (GFSK), quadrature phase shift keying (4PSK), or 8PSK. For BLE, the modulation scheme is GFSK with raw data rate of 1 Msymbols/s with 2 MHz channel bandwidth, which is double that of Bluetooth Classic.

Zigbee

ZigBee is a low-cost, low data rate, and short distance wireless ad hoc networks standard which contains a set of communication protocols. ZigBee is developed by ZigBee Alliance based on IEEE 802.15.4 reference stack model and mainly operates in two different frequency bands: 2.4 GHz and 868/915 MHz. The original idea of ZigBee-style can be tracked to the end of 1990s, when proposed by the IEEE 802.15 group. After that, IEEE 802.15.4 (TG) group was devoted to the bottom standards. In 2001, ZigBee Alliance was founded by Honeywell and some other companies which aims at creating, evolving, and promoting standards for ZigBee. In 2004, the ZigBee Alliance published the ZigBee 1.0 (a.k.a. ZigBee 2004) standards. Then, the ZigBee 2006, which revised a former version, was published in 2006. In 2007, the alliance published ZigBee PRO standard which contains two sets of advanced commands. In 2009, the standard was more flexible and had remote control capability named ZigBee RF4CE. From 2009, ZigBee adopted IETF's IPv6 standard as the standard of Smart Energy and was committed to forming a globally unified network.

In general, the system architecture of ZigBee is four layers, as shown in Figure 1.3 (Farahani, 2011). As mentioned before, ZigBee is developed at the top of the IEEE 802.15.4 standard. The ZigBee standard is built on the two lower layers: the physical (PHY) layer and the medium access control (MAC) sub-layer which are defined by IEEE 802.15.4, then

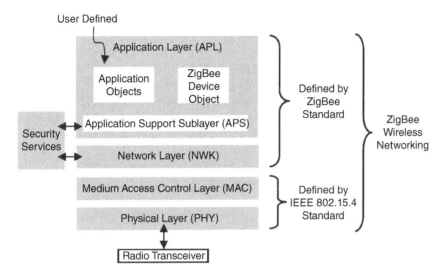

Figure 1.3 Zigbee system architecture. Source: From Shahin Farahani. ZigBee wireless networks and transceivers. Newnes, 2011. © 2011, Newnes.

ZigBee Alliance provides the network (NWK) layer and the framework for the application layer (Alliance, 2012).

PHY: The IEEE 802.15.4 standard defines the PHY layer of ZigBee. The PHY layer is the lowest layer, and defines the interface between the PHY channel and the MAC sub-layer. It also provides a data service and a PHY layer management service. The PHY layer data service directly controls and communicates with the radio transceiver (transmitter and receiver). The PHY layer management service maintains the database which is related to the PHY layer parameters. The protocol stipulates that the PHY layer operate in two separate frequency bands: 2.4 GHz and 868/915 MHz.

MAC: The IEEE 802.15.4 standard defines the MAC layer of ZigBee. The MAC layer provides the interfaces between the PHY layer and the NWK layer and controls access to the radio channel using a CSMA-CA mechanism. The MAC layer is responsible for establishment, maintenance, and termination of wireless links between devices, and it also leads a super-frame structure and transmitting bacon frame into the protocol system.

NWK: The network layer provides interfaces between the MAC layer and the APL layer and is responsible for managing the network and routing. The NWK layer supports three topologies: star, tree, and mesh topologies. Managing the network includes network establishment, end device discovery, join, and departure, and these operations are controlled by the ZigBee coordinator. The ZigBee coordinator is not only responsible for assigning a network address for devices in a network but is also responsible for discovering and maintaining the path in the network due to the terminal devices having no ability for routing. Routing is to choose the path to forward information to the target devices.

APL: The APL layer is the highest protocol layer in the ZigBee standard. The APL layer can be divided into three parts: application support sub-layer (APS), application framework, and ZigBee device object (ZDO). The APL layer is responsible for mapping the variety of applications to the ZigBee network, and it mainly includes: service discovery, convergence of multiple data stream, and security.

ZigBee has many advantages, and the main characteristics of ZigBee are low data rate, low power consumption, low complexity, high security, and support for a variety of network topologies. However, there still exit some disadvantages. The cost for the end devices is difficult to reduce at present, which makes it is not cheap when deploying a large number of end devices. The communication distance is about 75–100 m, because the communication frequency mainly operates in 2.4 GHz, which is difficult to penetrate through blocks. So obstacles seriously affect the communication distance.

ZigBee is used to provide services such as small area monitors, security, discovery, profiling, and so on for industrial control, household automatic control, and other places where sensor network-based applications are deployed. For example, ZigBee can be used in the building energy consumption monitoring system due to low cost, device sparsity, low energy consumption and self-organized characteristics. The end devices equipped with different sensors are used to monitor the temperature, humidity, voltage, and so on. The end devices can also collect the data from water meters, gas meters and electricity meters. These data, which are gathered from a variety of end devices, will be sent to the upper computer, then the policy will be made by the special system to achieve goals such a energy consumption monitoring, temperature control, and energy-saving operation management.

In summary, ZigBee provides short distance, low complexity, low energy consumption, low data rate, and low cost technology for wireless networks, and it effectively compensates for the vacancies in the low-cost, low-power, and low-rate wireless communication market. ZigBee also has many facets that can be improved, and we believe that if improvements in ZigBee technology don't stop there will be more and more ZigBee-based applications in our lives.

1.2.2.2 Low Data Rate and Wide-area Connectivity Technologies
6LoWPAN
The IoT is an emerging paradigm in which smart objects are seamlessly connected to the internet and can potentially collaborate to achieve common goals such as supporting innovative home automation services. IPv6 over a low-power personal area network (6LoWPAN) is an interesting protocol that supports the implementation of IoT in resource-constrained environments. 6LoWPAN devices are vulnerable to attacks from wireless sensor networks and internet protocols. 6LoWPAN is a protocol definition to enable IPv6 packets to be carried on top of low power wireless networks, specifically IEEE 802.15.4. The concept was born from the idea that internet protocols could and should be applied to even the smallest of devices. The initial goal was to define an adaptation layer – "IP over Foo" to deal with the requirements imposed by IPv6, such as the increased address sizes and the 1280 byte MTU. Its emergence provoked the expansion of LR-WPAN. The bottom layer of 6LowPAN technology espouses PHY and MAC layer standards of IEEE802.15.4, and 6LowPAN desires IPv6 as the networking technology. Its goal market primarily is wireless sensor networks. The 6LowPAN is used because it is based on IPv6. Lo in 6LowPAN stands for low power. IP communications and low power consumption is usually contradictory. WPAN stands for wireless personal area networks. A WPAN is a personal area network for connecting devices around a person. A popular WPAN is Bluetooth. Bluetooth is used to

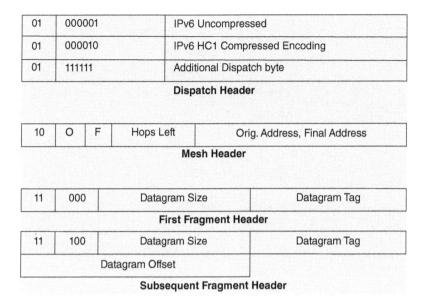

01	000001	IPv6 Uncompressed
01	000010	IPv6 HC1 Compressed Encoding
01	111111	Additional Dispatch byte

Dispatch Header

10	O	F	Hops Left	Orig. Address, Final Address

Mesh Header

11	000	Datagram Size	Datagram Tag

First Fragment Header

11	100	Datagram Size	Datagram Tag
	Datagram Offset		

Subsequent Fragment Header

Figure 1.4 6LoWPAN header layout.

interconnect our computer accessories or our audio equipment like Bluetooth headset or hands free kit. 6LoWPAN provides larger scale networks than Bluetooth. 6LoWPAN can create meshed networks with higher distance. By using 868/915 MHz instead of 2400 MHz, the coverage in buildings is much better.

In view of the addresses and security that are claimed in wireless sensor networks, plus the additional expansion of IPv6, initiating IPv6 protocol into embedded equipment has turn out to be an unavoidable propensity. However, the payload length sustained by MAC in IPv6 is greatly larger than one afforded by the 6LowPAN bottom layer. In order to put into practice the seamless connection of the MAC layer and network layer, the 6LowPAN working group recommend adding or toting up an adaptation layer between the MAC layer and the network layer to accomplish the header compression, fragmentation, reassembly, and mesh route forwarding. The reference model of the 6LowPAN Protocol Stack is shown in Figure 1.4.

The fragmentation header includes 4 bytes for the first fragment and 5 bytes for subsequent fragments, as shown in Figure 1.4. It supports a larger fragmentation and payloads than the size of the 802.15.4 frame and also includes fields to specify the size of the original datagram as well as a sequence number for ordering the received packets. The datagram tag field is used to identify all of the fragments of a single original packet. Each 6LowPAN header includes a type identifier and the most common headers are identified with a predefined prefix for each. As shown in Figure 1.4, the dispatch header (1 byte) is used to define the type of header to follow. The dispatch header is identified by the first two bits set to either 00 or 01. In order to provide a means for coexistence with non-6LoWPAN networks, the bit pattern 00 is reserved to identify these non-6LoWPAN frames. The remaining 6 bits indicate if the following field is an uncompressed IPv6 header or an HC1 header defining the IPv6 compressed header. Only 5 of the 64 dispatch header types have thus far been

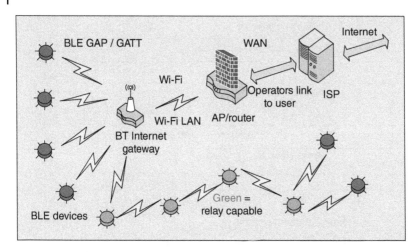

Figure 1.5 A BLE network.

defined. The special value of all ones indicates that the header contains an additional byte to allow for 256 more header types (Mulligan, 2007).

Although 6LoWPAN is not widely known like some other standards, such as RFID and NFC, 6LoWPAN uses IPv6, based on which 6LoWPAN is different from other standards and has obvious advantages. As the world has more and more IoT devices, 6LoWPAN provides advantages for low-power wireless sensor networks and other forms of low-power wireless networks. Due to its flexibility and convenience, 6LowPAN has broad market prospects. 6LowPAN technology is able to support IoT devices, especially in the field of smart city and industrial wireless. 6LowPAN can be used to achieve reorganization of fragments and route optimization. Therefore, when 6LowPAN technology is fully feasible and persistent, it must bring great convenience to people's work and life.

Figure 1.5 presents a network architecture that shows how BLE can be used in IoT applications. Multiples BLE devices are connected by a BT internet gateway or relay nodes. A router could forward the gateway through wireless or wired networks, such as wi-fi and fiber. The upper layers of the network indicates a traditional star network, and the lower layers show how the start network can work in conjunction with a mesh network.

Since IoT applications are broad and diverse, it is certain that the market requires various wireless technologies. However, in the end there will be market winners and losers for a specific application domain. Since Bluetooth is pervasive in smartphones and personal computers, it is gaining ground in home automation applications even though the effort to be a truly low-power technology is still in progress. However, whether Bluetooth Low Energy or Smart, or whatever you call it, will be a dominant technology for applications that only require small amounts of data communications is yet to be seen over time.

LoRa

LoRa (short for long range) is an emerging technology, which operates in a non-licensed band below 1 GHz for long-range communication. LoRaWAN defines the communication protocol and the system architecture, while LoRa defines the physical layer. LoRa Technology offers compelling features for IoT applications including long range, energy efficient and secure data transmission. The technology can be utilized by public, private or hybrid networks and provides longer communication range than cellular networks. The bandwidth has been established to ensure data rates from 0.3 kbps up to 50 kbps, which is not much compared with IEEE 802.11 but enough for the majority of applications in automation and data collection field, and also ensures maximization of the battery life in the case of mobile or autonomous terminals. The concept is really affordable for IoT applications, especially because of the reduce cost of implementation in long range conditions.

LoRa was invented by a startup in France called Cycleo whose employees are veterans of big semiconductor companies who wanted to build a long range low power communication device. They filed a patent in 2008 titled "Fractional-N Synthesized Chirp Generator" and another in 2013 titled "Low Power Long Range Transmitter". Later this company was acquired by another French company named Semtech that's into manufacturing of analogue and mixed-signal semiconductors. Semtech's LoRa Technology has amassed over 600 known uses cases for smart cities, smart homes and buildings, smart agriculture, smart metering, smart supply chain and logistics, and more, with 97 million devices connected to networks in 100 countries and growing. While Semtech provides the radio chips featuring LoRa technology, the LoRa Alliance, a non-profit association and the fastest growing technology alliance, drives the standardization and global harmonization of LoRaWAN, which is a MAC protocol for LoRa(Augustin et al., 2016). To fully define the LoRaWAN protocol, and to ensure interoperability among devices and networks, the LoRa Alliance develops and maintains documents to define the technical implementation, including MAC layer commands, frame content, classes, data rates, security, and flexible network frequency management, and so on.

A LoRa server architecture consists of end nodes, a LoRa gateway, a LoRa network server, and a LoRa application server, as shown in Figure 1.6. LoRa end nodes are the sensors or application where sensing and control takes place. These nodes are often placed remotely. The nodes are the devices sending data to the LoRa network server. These devices could be

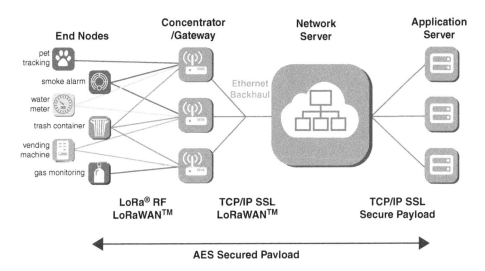

Figure 1.6 LoRa network architecture.

for example sensors measuring air quality, temperature, humidity, location, and so on. The LoRa gateways are different from cellular communication where mobile devices are associated with the serving base stations. The gateways receive data from the devices and typically run an implementation of the packet-forwarder software. This software is responsible for the interface with the LoRa hardware on the gateway. The LoRa server component provides the LoRaWAN network server component, responsible for managing the state of the network. It has knowledge of devices active on the network and is able to handle join-requests when devices want to join the network. When data is received by multiple gateways, LoRa network server will de-duplicate this data and forward it once to the LoRaWAN application server. When an application server needs to send data back to a device, the LoRa network server will keep these items in queue until it is able to send to one of the gateways. LoRa application server provides an API which can be used for integration or when implementing your own application-server. The LoRa application server component implements a LoRaWAN application server compatible with the LoRa server component. It provides a web-interface and APIs for management of users, organizations, applications, gateways, and devices.

Everyday municipal operations are made more efficient with LoRa Technology's long range, low power, secure, and GPS-free geolocation features. By connecting city services such as lighting, parking, waste removal, and more, cities can optimize the use of utilities and personnel to save time and money. LoRa Technology and smart city IoT networking can offer street light solutions that increase energy efficiency and reduce city operating costs. LoRa solutions are easy to implement into existing infrastructure and allow smart monitoring of the grid over a LoRaWAN network. LoRaWAN street light controller is LoRaWAN-alliance network compatible street light control system for street lights. The system provides a unique identity for every light, allows independent control of street lights on a calendar and timer basis, and allows instant manual control of lights from a software control system. LoRa Technology's low-power qualities and ability to penetrate dense building materials make it an ideal platform for IoT-connected smart home and building devices. In addition, the long range capabilities make it possible for LoRa-enabled sensors to track assets that stray from home. Sensors in smart home and building applications can detect danger, optimize utility usage, and improve the safety and convenience of everyday living. LoRa-enabled products can include thermostats, sprinkler controllers, door locks, leakage monitors, and smoke alarms. These devices connect to a building's network and allow consistent, remote monitoring to better conserve energy and predict when maintenance is necessary, saving property managers' money. LoRa Technology has the capacity to function in high density environments, such as in large enterprise buildings or campuses, and can handle thousands of unique messages per day.

LoRa has advantages in IoT applications. LoRa uses industrial, scientific, and medical (ISM) bands 868/915 MHz which is globally available. It has very wide coverage range about 5 km in urban areas and 15 km in suburban areas. It consumes less power and hence batteries will last for a longer duration. Single the LoRa Gateway device is designed to take care of thousands of end devices or nodes, it is easy to deploy due to its simple architecture. LoRa uses the adaptive data rate technique to vary output data rate output of end devices. This helps in maximizing battery life as well as overall capacity of the LoRaWAN network. The physical layer uses a spread spectrum modulation technique derived from chirp spread spectrum (CSS) technology. This delivers orthogonal transmissions at different data rates. Moreover, it provides processing gain and hence transmitter output power can be reduced with same RF link budget and hence will increase battery life. However, LoRa also has disadvantages. The LoRaWAN network size is limited based on a parameter called the duty cycle. It is defined as the percentage of time during which the channel can be occupied. This parameter arises from the regulation as key limiting factor for traffic served in the LoRaWAN network. LoRaWAN supports limited size data packets, and has longer latency. So it is not an ideal candidate to be used for real-time applications requiring lower latency and bounded jitter requirements.

In summary, the use of LoRa is strongly influenced by the characteristics of the environment in which the technology is implemented. The type of IoT application to be used needs to pay attention to these things. The use of LoRa in open areas such as rural areas has advantages in energy consumption, wide range, and flexibility in network development. However, for applications that require specific requirements such as the amount of data exchanged per specific time period then the network configuration needs to be considered primarily for indoor implementation. The LoRa network cannot transmit large amounts of data for a wider range of territories. LoRa technology is influenced by obstacles such as tall buildings and trees, which leads to an increase in packet loss levels in the zone. The use of GPS in the LoRa module has not been reliable, especially for real-time position tracking software applications. The limitations identified in LoRa technology become an opportunity for future research.

Sigfox

Sigfox was introduced by a French global network operator founded in 2009 and builds wireless networks to connect low-power objects such as IoT sensors and smartwatches, which need to be continuously powered on and emitting small amounts of data (Sigfox, 2018a,b). Sigfox wireless technology is based on LTN (low throughput network). It is a wide area network based technology which supports low data rate communication over larger distances. It is used for M2M and IoT applications which transmit only a few bytes per day. By employing ultra-narrow band in the sub-GHz spectrum, Sigfox efficiently uses the frequency band and has very low noise levels, leading to very low power consumption, high receiver sensitivity, and low cost antenna design. Sigfox developed a simple communications protocol, running in the license-free ISM bands at 868 and 915 MHz. It has very low cost, standard chips, and has a usable range of 5–10 km and a battery life that can support years of low data-rate transmission. Unlike the roll-out of cellular connectivity, where you could start with the areas of greatest use, typically capital cities, the IoT customer base is much more diverse. There are plenty of applications in agriculture and transport that need much wider coverage. Building their own network would have taken too long and cost too much, so they persuaded mobile operators to partner with them and install Sigfox's gateways on their existing towers, providing a fairly rapid coverage. Sigfox has doubled its connected devices from 2019 to 2020, going from around 6.9 million during the first half of 2019, to over 15.4 million at the start of 2020. According to Sigfox, the traffic on Sigfox's networks has increased to 26.5 million messages per day in late March 2020, from 24.6 million at the end of 2019 (https://enterpriseiotinsights. com/20200326/channels/news/sigfox-talks-1b23-and-maturing-of-iot).

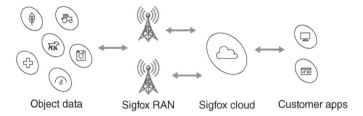

Figure 1.7 Sigfox network architecture.

Figure 1.7 depicts a simple Sigfox network architecture (Sigfox, 2018a,b). The Sigfox network consists of objects (end user devices), a Sigfox gateway or base stations, a Sigfox cloud, and application servers. Sigfox objects are connected with a gateway using star topology. There is a direct secure point to point link between Sigfox gateways and the Sigfox cloud. The cloud interfaces with servers using different protocols such as SNMP, MQTT, HTTP, IPv6, etc. as per end applications. Sigfox offers a software based communication solution, where all the network and computing complexity is managed in the cloud, rather than on the devices. All that together, devices connected through the Sigfox network only use the network when they are actually required to transmit data; in this procedure, much of the power consumption is reduced.

Sigfox has advantages in IoT applications. Sigfox has designed its technology and network to meet the requirements of mass IoT applications, a long device battery life-cycle, low device cost, a low connectivity fee, high network capacity, and long range. A device is not attached to a specific base station. Its broadcast messages are received by any base station in the range, and there is no need for message acknowledgement. UNB intrinsic ruggedness coupled with spatial diversity of the base stations offer great anti-jamming capabilities. UNB is extremely robust in an environment with spread spectrum signals. Low bit rate and simple radio modulation enable a 163.3 dB budget link for long range communications. Sigfox has tailored a lightweight protocol to handle small messages. Less data to send means less energy consumption, hence longer battery life. With its simple approach to connectivity, Sigfox provides extremely price-competitive connectivity subscriptions, and even more importantly, enables extremely simple and cost-efficient silicon modules. Sigfox is compatible with Bluetooth, GPS 2G/3G/4G and wi-fi. By combining other connectivity solutions with Sigfox, business cases and user experience can be drastically improved. However, Sigfox also has disadvantages. The narrow band spectrum emitted by a single Sigfox end device causes strong interference and collision to nearby existing wideband systems. More such sigfox devices will further enhance the interference. Sigfox supports one-way communication without acknowledgment. This necessitates multiple transmissions if the server does not receive data without errors. Due to the multiple transmissions, power consumption will increase which depends on number of re-transmissions. Due to low data rate support, it cannot be used for high data rate applications.

Nowadays, pallet tracking to determine the location of and goods condition are highly desirable in logistics. In this application, the most requirements are low cost sensors and long battery lifetime for asset tracking and status monitoring. In this case, Sigfox is a good solution. Logistics companies can have their own network so they have a guaranteed coverage in their facilities. Low cost Sigfox devices could be easily deployed on vehicles. Sigfox public base stations can be then used when vehicles are outside of the facilities or when goods arrive at customer locations. In the retail and hospitality industries, keeping guests satisfied and customers engaged is your number one priority. Now, the IoT is here to help. Sigfox-enabled IoT solutions for retail and hospitality change the game by keeping you connected to all aspects of your retail location, hotel, or restaurant powered by Sigfox's network dedicated exclusively to the IoT; the latest IoT solutions improve upon earlier versions of connectivity technology to provide a cost efficient, user-friendly experience.

In summary, Sigfox is rolling out the first global IoT network to listen to billions of objects broadcasting data, without the need to establish and maintain network connections. This unique approach in the world of wireless connectivity, where there is no signaling overhead, is a compact and optimized protocol. However, in order to support a myriad of devices in IoT, the interference between Sigfox devices and nearby existing wideband systems should be mitigated in further research. Further, new technologies and mechanisms should be investigated to reduce re-transmissions. The Sigfox system works well in a fixed location. There are issues such as interference and frequency inaccuracies in mobility environments, which also present challenges to further research and application.

NB-IoT

NB-IoT is a LPWAN technology based on narrowband radio technology. The technology provides improved indoor coverage, support for a massive number of low throughput devices, low delay sensitivity, ultra-low device cost, low device power consumption, and optimized network architecture. The technology can be deployed by utilizing resource blocks within a normal LTE carrier, or in the unused resource blocks within a LTE carrier's guard-band, or standalone for deployments in a dedicated spectrum. NB-IoT is standardized by the 3rd Generation Partnership Project (3GPP). Its specification was published at Release 13 of 3GPP on June 2016 (3GPP, 2016b). In December 2016, Vodafone, and Huawei deployed NB-IoT into the Spanish Vodafone network and sent the first message conforming to the NB-IoT standard to a device installed in a water meter. NB-IoT now has received strong support from Huawei, Ericsson, and Qualcomm. The objectives of NB-IoT are to ensure a device cost below US\$5, uplink latency below 10 s, up to 40 connected devices per household, a device with 164 dB coupling loss, and a ten-year battery life can be reached if the user equipment transmits 200 bytes of data a day on average. NB-IoT has entirely an extensive ecosystem that is available globally. This is primarily due to its support from more than 30 of the world's largest and top class operators. These operators have global communication coverage that serves above 3.4 billion customers and geographically serve over 90% of the IoT market (Huawei, 2016b).

NB-IoT network architecture is shown in Figure 1.8. In order to send data to an application, two optimizations for the cellular IoT (CIoT) in the evolved packet system (EPS) are defined, the user plane CIoT EPS optimization and the control plane CIoT EPS optimization (Schlienz and Raddino, 2016). Both optimizations may be used but are not limited to NB-IoT devices. In the control plane CIoT EPS optimization, data are transferred from the eNB (CIoT RAN) to the MME. From there, they may either be transferred via the serving gateway (SGW) to the packet data network gateway (PGW), or to the service capability exposure function (SCEF), which is only possible for non-IP data packets. From these nodes, they are finally forwarded to the application server (CIoT Services). DL data is transmitted over the same paths in the reverse direction. In this solution, there is no data radio bearer set up, and data packets are sent on the signaling radio bearer instead. Consequently, this solution is most appropriate for the transmission of infrequent and small data packets. The SCEF is a new node designed especially for machine-type data. It is used for delivery of non-IP data over a control plane and provides an abstract interface for the network services (authentication and authorization, discovery, and access network capabilities). With the user plane CIoT EPS optimization, data is transferred in the same way as the conventional data traffic, i.e. over radio bearers via the SGW and the PGW to the application server. Thus it creates some overhead on building up the connection, however it facilitates a sequence of data packets to be sent. This path supports both IP and non-IP data delivery.

NB-IoT has advantages in IoT applications. As it uses a mobile network it offers better scalability, quality of service, and security compared to unlicensed LPWAN such as LoRa/Sigfox. It offers long battery life due to low power consumption or current consumption. NB-IoT also offers extended coverage compare to GSM/GPRS systems, and co-exists with other legacy cellular systems such as GSM/GPRS/LTE. The NB-IoT compliant devices can be deployed/scheduled within any legacy LTE network. This helps them share capacity as well as other cell resources with the other wireless connected devices. NB-IoT modules are expected to be available at moderate cost. It offers better penetration of structures and better data rates compared to unlicensed band based standards (e.g. LoRaWAN and Sigfox). However, NB-IoT also has disadvantages. It offers lower data rates (about 250 Kbps download and 20 Kbps upload) compared to LTE. The bandwidth is about 200 KHz. Hence it is ideal to use NB-IoT for stationary devices. NB-IoT devices need to connect to an operator network via a licensed spectrum. Network and tower handoffs will be a problem, so NB-IoT is best suited for primarily static assets, like meters and sensors in a fixed location, rather than roaming assets.

The strong growth in the NB-IoT market has motivated many analyst firms to create forecasts showing the expected numbers of connections as well as the revenue potential. The global NB-IoT chipset market is expected to grow from $461million in 2020 to $2484million by 2025 at a compound annual growth rate (CAGR) of 40.0%, according to MarketsandMarkets Research Private Ltd (https://www.marketsandmarkets.com/Market-Reports/narrowband-iot-market-59565925.html). The NB-IoT market is a sub-set of this,

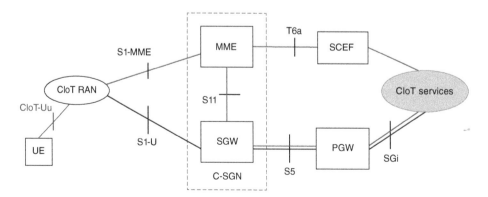

Figure 1.8 NB-IoT network architecture.

and it is important for operators to understand the revenue potential in the countries they operate in. The pet tracking use case is one application that helps the user to keep track of their pet's activities and, most importantly, location at all times. A small lightweight device placed around the neck of the pet embedded with an NB-IOT chip-set helps to send tracking information to the user's device. The NB-IOT device collects and sends location information leveraging GPS and location based services, and this can be done either periodically or in real-time based on the user's preferences. The user can then receive the information with a tracking route that is already integrated with the map. Furthermore, this device is embedded with several forms of alarms that can alert the user when the device battery is running low(Huawei, 2016a). Security has always been a very important aspect of human living; people at all times want to be guaranteed of home safety(Huawei, 2016a). Alarms and event detection will help to rapidly inform that user about a detected home intrusion. NB-IoT system will not only offer intelligent protection from intrusion but will also offer intelligence for detected events that can lead to a fire outbreak like a sudden increase in home temperature or smoke. Alarm and event detectors will make use of sensors placed in devices in ideal locations in the home that constantly communicates with the NB-IoT. This use case will make use of a very low data throughput and battery life of the devices will be ultra-critical.

In summary, NB-IoT is a promising technology for IoT applications. It can connect low-power IoT devices that are placed in weak coverage environments such as apartment basements. This is done by allowing devices to repeat signal transmissions while operating at very low power. Compared to LoRa and Sigfox, NB-IoT relies on the existing cellular infrastructure instead of new ones, thus the investments on a utility-dedicated communication infrastructure and the time required for deployment of applications is reduced. However, it is difficult to implement firmware-over-the-air (FOTA) or file transfers. Some of the design specifications for NB-IoT make it difficult to send larger amounts of data to a device.

1.2.2.3 Emerging IoT Connectivity Technologies and Protocols
MIOTY
Recently, BehrTech, an industrial IoT connectivity provider, has proposed a new IoT technology, called MIOTY, in the burgeoning IoT networking market. MIOTY is another LPWAN solution dedicated to private IoT networks (Behrtech, 2020). MIOTY was originally developed by the Fraunhofer Institute for Integrate Circuits (IIS), and subsequently licensed to BehrTech – is presented as a low-throughput tech for "last mile" industrial communications. As such it goes up against the likes of LoRaWAN, Sigfox, NB-IoT, and LTE-M, variously pushed by their backers as springboards for industrial IoT.

MIOTY is empowered with novel telegram-splitting ultra-narrowband (TS-UNB) to comply with specification from the European Telecommunications Standards Institute (ETSI). MIOTY is based on UNB technology with very narrow signal bandwidth to achieve long distance data communication between thousands of IoT devices and a base station. MIOTY works with standard transceivers. The basic concept of this technology is telegram splitting, which divides a compact telegram transmission into many equally sized radio bursts. TS-UNB, as defined by ETSI, splits the data packets to be transported in the data stream into small sub-packets at the sensor level. These sub-packets are then transmitted over fluctuating frequency and time. An algorithm in the base station permanently scans the

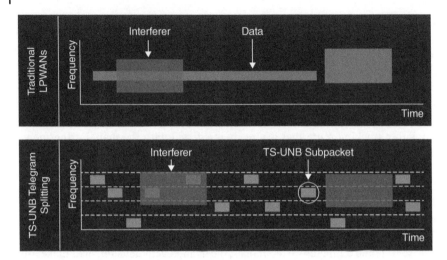

Figure 1.9 TS-UWB telegram splitting.

spectrum for MIOTY sub-packets and reassembles them into a complete message, as shown in Figure 1.9. For multiple access, the radio bursts are distributed over time and frequency. For correct decoding, only 50% of the radio bursts need to be collision free. This reduces the collision probability of telegrams and increases the tolerance against interference.

IoTivity

IoTivity is an open source framework that implements the Open Connectivity Foundation (OCF) standards providing easy and secure communications for IoT devices (Iotivity, 2020). The main goal of IoTivity is to provide seamless device-to-device connectivity regardless of the kind of operating system or communication protocol to satisfy various requirements of IoT. IoTivity is distributed with Apache license 2.0, thus anyone can use it, but revealing source codes based on it is not mandatory. IoTivity is available on multiple platforms and supports a variety of connectivity technologies, including wi-fi, ethernet, BLE, NFC and so on. It works on various OS, such as Linux, Android, Arduino, Tizen, and so on.

The IoTivity framework operates as middleware across all operating systems and connectivity platforms and has some essential building blocks. As shown in Figure 1.10, the discovery block supports multiple mechanisms for discovering devices and resources in proximity and remotely. Data transmission block supports information exchange and control based on a messaging and streaming model. Data management block supports the collection, storage and analysis of data from various resources. The CRUDN (create, read, update, delete, notify) block supports a simple request/response mechanism with create, retrieve, update, delete and notify commands. The common resource model block defines real world entities as data models and resources. Device management block supports configuration, provisioning, and diagnostics of devices. The messaging block of IoTivity is based on resource-based RESTful architecture model. Thus it presents everything (sensors or devices) as resources and uses the CRUDN model to manipulate resources by using IETF CoAP. The ID & addressing block supports OCF IDs and addressing for OCF entities, such as devices, clients, servers, and resources.

Figure 1.10 OCF core framework.

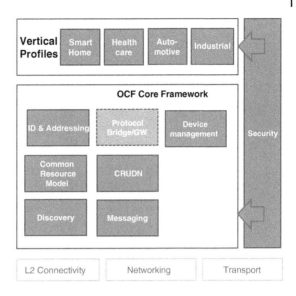

1.3   Intelligent IoT Technologies

Since IoT was proposed in 1999, its connotation has been in continuous development and expansion (Zhong et al., 2015). At the same time, the scale of devices has also increased at an unprecedented rate. According to the data released by "Statista", the total installed base of IoT connected devices is projected to reach 75.44 billion worldwide by 2025, a five-fold increase in ten years. Details are shown in Figure 1.11. Such a large number of IoT devices undoubtedly put forward higher requirements for the existing IoT (Datta and Sharma, 2017). In order to emphasize the level of intelligent IoT application, this

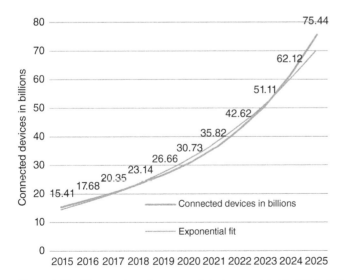

Figure 1.11 The number of connected devices worldwide 2015–2025.

Figure 1.12 A traditional cloud-based IoT architecture.

chapter introduces an intelligent system architecture that can better interpret the meaning and features of the current IoT.

With the rapid adaptation of modern smart technologies, IoT has gained much attention from the industry and the IT community in terms of networking and communication aspects. The number of connected devices to the internet will be huge in number. Traditional architectures and network protocols for IoT devices are not designed to support a high level of scalability, high amount of traffic and mobility together with typical cloud-based applications, such as environmental monitoring and public safety surveillance. In traditional cloud-based solutions, as shown in Figure 1.12, cloud servers are in charge of providing computation, algorithm, and data storage services for IoT users. Based on the services, IoT can provide various productions, services, and platforms for terminal users. However, due to a centralized server, traditional architectures have many issues such as high latency and low efficiency. Furthermore, it's difficult to manage these devices, generating an impressive amount of data as a whole, without having elasticity and flexibility inherently defined in the network. Thus, they are inefficient and have limitations to satisfy new requirements, such as medical monitoring devices, smart grids, transportation systems, and industrial and automation sectors.

Due to such a voluminous number of devices and new requirements, existing network architectures are no longer able to accommodate IoT devices. As users' requirements become more abstract, complex and uncertain, the cloud-based solutions can no longer meet the requirements of IoT users in various aspects, such as latency and computing efficiency. Thus, a generic and flexible multi-tier intelligent IoT architecture distributed computation becomes a potential solution, which is shown in Figure 1.13. In the IoT architecture, IoT applications are increasingly intelligent due to more meaningful data, distributed powerful processors and sophisticated algorithms. Typical IoT applications are also shifting from simple data sensing, collection and representation tasks towards complex information extraction and analysis. However, the applications usually follow rules and principles set by a specific industrial domain. Computing resources integrated

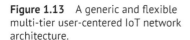

Figure 1.13 A generic and flexible multi-tier user-centered IoT network architecture.

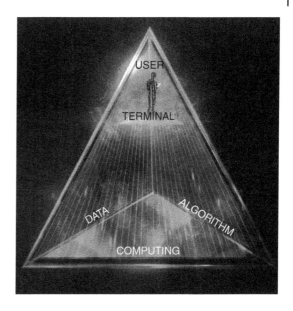

with environment cognition, big data and AI technologies, could be used to develop a user-centric approach in which different IoT services are autonomously customized according to specific applications and user preferences. Furthermore, the intelligent IoT architecture has not been well studied. How to find a scalable, flexible, interoperable, lightweight, energy-aware, and secure network architecture will be a challenge for researchers.

1.3.1 Data Collection Technologies

Section 1.2 mainly reviews the traditional IoT connectivity technologies and related protocols in detail. Due to the massive data exchanged between a large number of connected devices forming the IoT, the need to provide extremely high capacity, data rate and connectivity increase. Thus, 5G wireless networks are considered as a key driver for IoT (Alsulami and Akkari, 2018). In 2015, the International Telecommunication Union (ITU) released the 5G architecture and overall goals, defining three application scenarios for enhanced Mobile Broadband (eMBB), ultra-high reliability and low latency (uRLLC), and mass machine type communication (mMTC), as well as eight key performance indicators such as peak rate, traffic density, and so on, as is shown in Figure 1.14 (3GPP, 2014). Compared to 4G, 5G will offer at least 10 times the peak rate of 4G, transmission delays in milliseconds and connectivity of millions of meters per square kilometer.The initial commercial deployments of NR arc already under way during 2019, focusing on eMBB using the Release 15 version of the 3GPP specifications ("Rel-15"). The basis for uRLLC is inherent in the Rel-15 version of the 5G system, especially in respect of support for low latency. For the mMTC component, NR is complemented by the machine-type communications technologies known as LTE-M and NB-IoT already developed by 3GPP in Rel-13, which provide unrivaled low-power wide-area performance covering a wide range of data rates and deployment scenarios(Ghosh et al., 2019). In addition to that formal process, work has progressed on around 25 Release 16 studies, on a variety of topics:

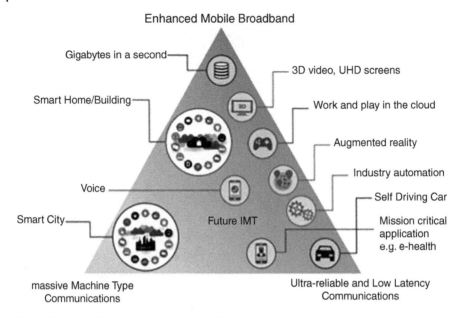

Figure 1.14 Application scenarios of intelligent IoT.

multimedia priority services, vehicle-to-everything (V2X) application layer services, 5G satellite access, local area network support in 5G, wireless and wireline convergence for 5G, terminal positioning and location, communications in vertical domains and network automation, and novel radio techniques. Further items being studied include security, codecs and streaming services, local area network interworking, network slicing and the IoT. 3GPP Release-16 Stage 3 was frozen in June 2020. The R16 standard is deployed for two other scenarios, namely URLLC and mMTC.

The technology of 5G has promoted the development of intelligent IoT. The following is a brief introduction of wireless basic technology and network technology about 5G.

1.3.1.1 mmWave

According to the protocol of 3GPP, 5G networks mainly use two frequencies: FR1 and FR2 bands. The frequency range of band FR1 is 450MHz to 6GHz, also known as the sub-6 GHz band. The frequency range of band FR2 is 24.25–52.6 GHz, often referred to as millimetre wave (mmWave) (Hong et al., 2017). The industry is very familiar with the band below 6 GHz, which is where the 4G LTE network runs. mmWave is relatively unknown, but in the process of 5G network construction, the advancement of mmWave technology will be the key. According to Shannon's formula, we can improve the spectral efficiency or increase the bandwidth to increase the rate (Agrawal and Sharma, 2016).

mmWave is known to have large bandwidths and high rates.The 4G LTE cellular system based on the sub 6GHz band has a maximum bandwidth of 100 MHz and a data rate of less than 1 Gbps. In the mmWave band, the maximum bandwidth available is 400 MHz, with data rates of 10 Gbps or more. In the 5G intelligent IoT era, such bandwidth performance can meet the needs of users for specific scenarios (Khurpade et al., 2018).

1.3.1.2 Massive MIMO

As the frequency used in mobile communication increases, the path loss also increases.However, if the size of the antenna is fixed, such as half wavelength or quarter wavelength, then the increase in carrier frequency means that the antenna is smaller. That means we can cram more and more high-frequency antennas into the same space. Based on this fact, we can compensate for the high frequency path loss by increasing the number of antennas without increasing the size of the antenna array. On the other hand, in the ideal propagation model, when the transmitting power at the transmitting end is fixed, the receiving power at the receiving end is proportional to the square of the wavelength, the transmitting antenna gain and the receiving antenna gain, and inversely proportional to the square of the distance between the transmitting antenna and the receiving antenna. At the millimeter band, the wavelength of radio waves is on the order of millimeters.The radio waves used in 2G/3G/4G are decimeter or centimeter waves. Because the receiving power is proportional to the square of the wavelength, the signal coverage of millimeter wave is small.

Based on the path loss and coverage issues described above, massive MIMO has been introduced. Massive MIMO is a key technology to improve system capacity and spectrum utilization in 5G. It was first proposed by researchers at Bell Laboratories in the United States, and it was found that when the number of base station antennas in a cell reaches infinity, the negative effects such as additive white Gaussian noise and Rayleigh fading can all be ignored, and the data transmission rate can be greatly improved (Gampala and Reddy, 2018). The beam formation of massive MIMO is different from conventional beam formation. Instead of the beam pointing straight at the terminal, it can point at the terminal from many different directions. The signal pre-processing algorithm can arrange the best route for the beam, and it can also send the data through the reflected path from the obstacle to the specified user under precise coordination.

1.3.1.3 Software Defined Networks

As more enterprises evolve their IoT proof-of-concept projects into live architectures, IoT won't be the only technology migration many of them are tackling, as enterprise networks today are also in the midst of a broader, more multi-faceted transformation.

IoT is coming into play just as enterprises are migrating beyond the hub-and-spoke architectures that have defined their networks for decades. In the traditional hub-and-spoke model, all services are processed in a centralized location, and all connectivity goes through that hub. All enterprise traffic from that hub might get backhauled through one or more MPLS links, but that model reflects a previous hardware-centric enterprise IT era, and doesn't allow flexibility to prioritize particular applications or traffic, to access applications from different locations and device types, or to host and process applications in one or more external clouds.

Multiple converging trends in recent years have begun to require a new network approach: the growth and variety of different devices – not just enterprise desktops, but smartphones, IoT sensors and other devices – connected to the network, the proliferation of more distributed networks and remote telecommuting, the ever-present need to reduce enterprise connectivity and hardware costs, the rise of new network connectivity technology options, like broadband internet access and 4G LTE, and an explosion in applications hosted in a variety of places, not just in an enterprise workstation or a corporate data center, but in a variety of potential cloud locations.

To address the challenges mentioned above, a new concept, software-defined wide area networking (SD-WAN), has emerged in recent years with the aim of simplifying all of this complexity and to help evolve enterprise networks into more flexible, programmable architectures that can meet the changing expectations of users.

SD-WAN accomplishes this by adding software overlay to the enterprise network that separates network control and management functions from the physical network, similar to what software-defined networking can do in a data center or public carrier network.

By separating the control from management in a variety of devices, network elements, and connectivity circuits that make up the network, an enterprise can create pool of total network capacity from these circuits to use as needed, while enabling visibility throughout the network. The visibility allows network managers, in turn, to dynamically identify the best possible paths for high-priority traffic, to allocate the necessary bandwidth and administer required security policies to ensure the quality and integrity of the most mission-critical services.

SD-WAN may play an important role to play in enterprise and industrial networks where IoT is starting to have a larger presence. While many IoT applications do not yet require large amounts of bandwidth on short notice, SD-WAN-based visibility into multiple enterprise connections and control of entire enterprise capacity pools will be able to dynamically allocate bandwidth for mission-critical IoT applications as they emerge, while also segmenting the most latency-sensitive and security-sensitive applications of the industrial IoT.

In addition, SD-WAN's ability to identify new devices coming onto the networks and allocate bandwidth to remote network users will serve enterprises well as they start to expand their IoT network presence throughout their WANs to branch offices and other distributed locations.

Also, SD-WAN can help enterprises manage network architectures in which the edge is becoming the center of data processing and analytics. SD-WAN will play a critical role in connecting, securing and pushing processing power closer to edge devices. This will increase the performance of IoT platforms by reducing latency for processing at the edge, and moving security processes – intrusion detection/prevention, DNS-layer security, and advanced malware protection – near the IoT devices. One of the original arguments for deploying SD-WAN was that it could help lower network expenses for enterprise by employing capacity pools that can help enterprises reduce their reliance on expensive MPLS links by maximizing use of available capacity from other circuits. As more devices and applications emerge in enterprises amid trends like IoT, however, the case for deploying SD-WAN has evolved to become just as much or more about application performance and security in complex network environments as it is about cost savings, according to Oswal. The technology can reduce the need to backhaul traffic from IoT devices all the way to the enterprise data center, instead, transporting that traffic on dedicated secure segments to edge processors that can filter and analyze much of the IoT device data, while transmitting only refined results to clouds for further analysis. This leads to less transport expense, but also faster, more secure application processing.

With its ability to segment traffic, visualize the best possible route paths, and apply appropriate security and network usage policies to different devices on the network could make sense for enterprises to have SD-WAN overlays in place before they go much further down their IoT road maps. That might not be something some enterprise managers have thought about as they pursued their IoT strategies, but putting the new software layer in place to simplify control of an existing network architecture could make it much easier to introduce a plethora of new IoT-connected devices to the network. Ultimately, IoT is just one emerging enterprise network architecture of many. It doesn't have a unique relationship with SD-WAN, but like other technologies in the enterprise, it can leverage SD-WAN for cost, performance, efficiency and security benefits.

1.3.1.4 Network Slicing

Network slicing (Trivisonno et al., 2017) is to cut the operator's physical network into a number of virtual networks, making each network adapted to different service needs, such as delay, bandwidth, security, reliability, and so on. Through network slicing, multiple logical networks are segmented on a separate physical network, thus avoiding the construction of a dedicated physical network for each service, which can greatly reduce the cost of deployment. Each network slice is logically isolated from the wireless access network to the core network for a wide variety of applications. Network slicing can be divided into at least three parts: a wireless network slice, a carrier network slice, and a core network slice. The core of network slicing (Khan et al., 2020) is network function virtualization (NFV), which separates the hardware and software parts from the traditional network. The hardware is deployed by a unified server, and the software is assumed by different network functions (NFs), to realize the requirements of a flexible assembly business. The basic technology of NFV is mainly cloud computing and virtualization. Virtualization technology can decompose common computing, storage, network, and other hardware devices into a variety of virtual resources in order to achieve hardware decoupling and realize network functions and dynamic flexible deployment as required. Cloud computing technology can achieve application flexibility and even the matching of resources and business load, which can not only improve the utilization of resources, but also ensure the system response speed. The three application scenarios in the existing 5G network architecture are the network slices, which are shown in Figure 1.15.

Network slicing is a logical concept that is based on reorganization of resources. According to the description in (NGMN, 2016), the network slicing consists of three layers: the service instance layer, the network slice instance layer, and the resource layer. The service instance layer represents the end user services or business services that can be supported. Each service is represented by a service instance. The network slice instance layer includes the network slice instances that can be provided. A network slice instance provides the network features that are required by the service instance. The resource layer provides all virtual or physical resources and network functions that are necessary to create a network slice instance (Zhang, 2019). In the context of a large number of network slices, there is a certain complexity in managing and orchestrating network slices. So far, there is no uniform standard for the management of network slices. At present, we can use the existing NFV MANO framework to manage 5G slices (Foukas et al., 2017). At the same time, the emergence of network slicing has also brought great security challenges, and 3GPP has identified many security risks related to 5G network slicing (3GPP, 2017). For example, because services are deployed on the same physical resource based on virtualization

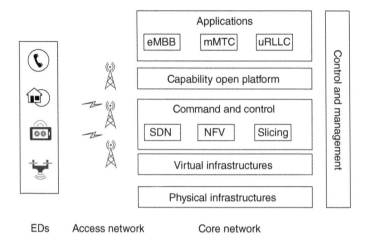

Figure 1.15 Network slices existing in 5G architecture.

technology, if an attacker hoards resources too much on one slice, the performance of other slices may not be guaranteed.

1.3.1.5 Time Sensitive Network

As awareness surrounding IoT standardization continues to grow, more eyes are being drawn to interoperability and network infrastructure solutions, including time-sensitive networking (TSN). TSN is a standard from the IEEE 802 committee and is designed to solve the need to process raw data in a time-critical fashion in addition to reducing latency and increasing robustness. To support new capabilities of IoT-enabled infrastructure, designers, engineers, and end users need to rely on time-synchronized and reliable networking.

TSN provides not only access to a tollway, or an express lane, but along with providing access, the signals along the way are all very tightly coordinated with time. Not only is there the benefit of a priority through the network, but it can actually guarantee end-to-end scheduling, and every light turns green at the right time.

The Avnu Alliance, an industry consortium driving open, standards-based deterministic networking, in addition to advancements made to TSN, is working with member companies to drive this next-generation standard and create an interoperable ecosystem through certification. Members are working within the Alliance to develop the foundational elements needed for industrial applications based on the common elements of AVB/TSN.

In TSN, there are some very interesting and compelling use cases for industry in many applications. However, there is also the possibility of getting onto a new track and using standardized technology, which is the target for most of the innovation in the networking space and leveraging these faster speeds.

TSN promises through standard silicon to converge the previously disparate technologies needed for standard ethernet communication, for deterministic high-speed data transfer, and for high accuracy time synchronization. These developments will create a common foundation that will impact numerous applications and markets ranging from machine control and asset monitoring to test cells and vehicle control. As IIoT adoption continues, increased amounts of data and widely distributed networks will require new standards for sharing and transferring critical information. Just as an ambulance or fire engine receives priority among other traffic during an emergency, the TSN standard ensures that critical, time-sensitive data is delivered on time over standard network infrastructure.

1.3.1.6 Multi-user Access Control

Effective access control for information collection is one of the major challenges in the development of IoT.

In most applications of the IoT, the data collected by the perception layer of the IoT nodes is usually protected by the secret keys of the nodes. When users want to access the resources, they need to apply to central authority(CA) for the corresponding keys and then decrypt it. Users need to interact with the CA for each access. With the increase of access nodes, users not only need to keep a large number of keys, but also need to communicate with the CA frequently (Feng and Zhu, 2016). This will not only burden users' storage and network communication overhead, but also bring security threats such as DDos attacks and man-in-the-middle attacks. One of the most arduous challenges in the evolution of the IoT is how to reduce communication between users and CA while accessing as many protected node resources as possible.When multiple users access resources protected by nodes in the same perception layer (Liu et al., 2018) they need to obtain the key to access this node. Once a user is attacked by the enemy and the key is invalid, other users cannot access it normally. How to ensure safe and effective access for multiple users is also a problem to consider.

Hierarchical access control schemes are usually used in practice. Through an interaction with the CA, the user can access the information resources of the corresponding level by using the obtained key, and obtain the keys of all levels below the level by using the partial order relationship between the levels to access as many levels of information resources as possible. Since Akl, when Taylor first proposed a hierarchical control scheme based on cryptography, the industry began to put forward different layered control schemes based on different application backgrounds and security requirements, using discrete logarithms to solve the problem and large prime decomposition problem to construct hierarchical access control scheme(Al-Dahhan et al., 2018). This has good expansibility. In addition, the remainder theorem is introduced into the hierarchical access control scheme, which has better storage capacity and a simpler key derivation process.

A single hash function is used to construct a layered key derivation algorithm and a lightweight layered access control scheme is designed, which can not only reduce the requirement of computing power and storage capacity of nodes, but also have dynamic extensibility. However, considering the special environmental requirements of the IoT perception layer, these control schemes cannot be directly applied to the IoT perception layer environment, and the storage and computing costs are too high, so they are not suitable for the resource-limited IoT perception layer environment. Therefore, the Merkle hash tree based hierarchical key acquisition scheme is widely used. This algorithm does not generate the protection keys of intermediate level nodes in the derivation process, which reduces the security risk of the derivation algorithm and has scalability. At this time, the algorithms of multi-user access control are still evolving.

1.3.1.7 Muti-hop Routing Protocol

Multi-hop routing (or multihop routing) is a type of communication in radio networks in which network coverage area is larger than the radio range of single nodes. Therefore, to reach some destination, a node can use other nodes as relays (Pešović et al., 2010).

Since the transceiver is the major source of power consumption in a radio node and long distance transmission requires high power, in some cases, multi-hop routing can be more energy efficient than single-hop routing (Fedor and Collier, 2007).

There are four typical applications of multi-hop routing – wireless sensor networks, wireless mesh networks, mobile ad hoc networks, and smart phone ad hoc networks.

A wireless sensor network (WSN) refers to a group of spatially dispersed and dedicated sensors for monitoring and recording the physical conditions of the environment and organizing the collected data at a central location. WSNs measure environmental conditions like temperature, sound, pollution levels, humidity, wind, and so on.

A wireless mesh network (WMN) is a communications network made up of radio nodes organized in a mesh topology. It is also a form of wireless ad hoc network (Toh, 2001). A mesh refers to rich interconnection among devices or nodes. Wireless mesh networks often consist of mesh clients, mesh routers and gateways. Mobility of nodes is less frequent. If nodes constantly or frequently move, the mesh spends more time updating routes than delivering data. In a wireless mesh network, topology tends to be more static, so that routes computation can converge and delivery of data to their destinations can occur. Hence, this is a low-mobility centralized form of wireless ad hoc network. Also, because it sometimes relies on static nodes to act as gateways, it is not a truly all-wireless ad hoc network.

A wireless ad hoc network (Toh, 1997) (WANET) or mobile ad hoc network (MANET) is a decentralized type of wireless network (Toh, 2001, Murthy and Manoj, 2004, Toh, 1997, Zanjireh and Larijani, 2015). The network is ad hoc because it does not rely on a pre-existing infrastructure, such as routers in wired networks or access points in managed (infrastructure) wireless networks(Zanjireh and Larijani, 2015). Instead, each node participates in routing by forwarding data for other nodes, so the determination of which nodes forward data is made dynamically on the basis of network connectivity and the routing algorithm in use (Zanjireh et al., 2013).

Smartphone ad hoc networks (SPANs, also smart phone ad hoc networks) are wireless ad hoc networks that use smartphones. Once embedded with ad hoc networking technology, a group of smartphones in close proximity can together create an ad hoc network. Smart phone ad hoc networks use the existing hardware (primarily Bluetooth and wi-fi) in commercially available smartphones to create peer-to-peer networks without relying on cellular carrier networks, wireless access points, or traditional network infrastructure. Wi-fi SPANs use the mechanism behind wi-fi ad-hoc mode, which allows phones to talk directly among each other through a transparent neighbor and route discovery mechanism. SPANs differ from traditional hub and spoke networks, such as wi-fi direct, in that they support multi-hop routing (ad hoc routing) and relays and there is no notion of a group leader, so peers can join and leave at will without destroying the network.

1.3.2 Computing Power Network

1.3.2.1 Intelligent IoT Computing Architecture

The data collected by IoT systems is growing exponentially. This will put forward a lot of requirements for the system, including data storage, computing power, and so on. Although the emergence of cloud computing (Armbrust et al., 2010) eases the pressure to some extent, if data is transmitted to the cloud center of remote geographic localization, it will cause a lot of problems. On the one hand, it could cause a lot of congestion for the network, especially when the network includes a great number of devices and only has limited communication resources; on the other hand, it can lead to a lot of delay. These problems will lead to a bad influence on the applications which need to make timely decisions. There will be a great quantity of computing resource pools distributed in future networks, providing various computing functions for numerous innovative applications in the 5G/AI era. Considering the significant trend of network and computing convergence evolution (NCC) in NET-2030, and the challenges arising from edge computing, it is necessary to research the "computing power network" (CPN) which supports high collaboration between computing and network resources, with optimal user experience. Based on the network-centric idea, collecting network resources, computing resources, storage resources, and algorithm resources information to the network control plane are needed to realize collaborative scheduling. So, offloading computing power to the edge has been proposed (Yu, 2016). A big problem is that the computing resources of the edge are very limited. It cannot meet the computational force requirements of the applications, which need plenty of computing resources. From the above content, we can find that a lot of early work was too one-sided from the initial consideration of only cloud center resources to the latter which focuses too much on edge resources (Guo et al., 2018). They did not consider how to better integrate multi-level collaboration. The authors (Wang et al., 2019b) (Yao et al., 2019) proposed the EdgeFlow system for IoT applications, which is a multi-tier data flow processing system. This system makes full use of the computing and transmission resources of the whole network. However, there are still some application tasks between terminal devices, mobile edge computing servers, and cloud centric servers that cannot achieve maximum resource utilization efficiency. Therefore, it is necessary to build more layers of servers to facilitate the data offloading layer by layer in order to make full use of the resource. In the process of intelligence, the new computing architecture is derived in the end, which is depicted in Figure 1.16. Of course, this involves a lot of problems and challenges. First of all, the total discharge of the tasks in each layer are interlinked. Secondly, the remaining

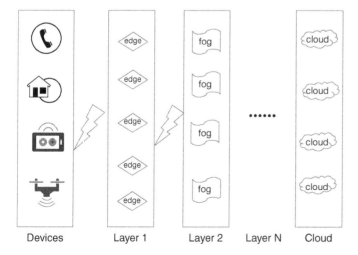

Figure 1.16 The architecture of intelligent IoT multi-tier computing.

resources at each layer also affects the quantity of offloading task. Finally, because of the influence of environmental factors, for example, the volume of resource is in real-time dynamic change, the unloading of tasks and allocation of resources present a great challenge. The heterogeneous mobile edge computing (Het-MEC) was proposed to integrally utilize the computing and transmission resources throughout the entire network, which is a reinforcement learning (RL)-based framework for Het-MEC (Zhang et al., 2020) (Wang et al., 2019c). Here, we explain the multi-tier computing architecture in the internet of intelligent things (IoIT). Different IoT applications have different levels of intelligence and efficiency in processing data. Multi-tier computing, which integrates cloud, fog, and edge computing technologies, will be required in order to deliver future IoT services (Yang, 2019).

1.3.2.2 Edge and Fog Computing

With the advent of a new round of the internet and industrial revolution, the internet will carry the convergence of the heterogeneous network and the dynamic reconfiguration of IoT equipment and personal user devices. To provide the higher performance of internet services, the IoT calculation capability is facing an unprecedented challenge from the actual needs of user services and IoT equipment's resource requirements. 5G (Yu, 2016) is about to be commercialized on a large scale mobile communication technology that will be integrated into the IoT. MEC can decentralize computing and storage services at the edge of the network, near the terminal equipment. Put another way, the MEC is an implementation of the edge computing paradigm that brings cloud computing capabilities to the edge of the mobile network, inside the radio access network (RAN) (He et al., 2018, Mach and Becvar, 2017). MEC nodes are generally located with the radio network controller or with a large base radio station (Dolui and Datta, 2017). Its biggest feature is to sink the core business to the edge of the mobile network, thus enhancing the various types of service quality.

The aforementioned implementations of edge computing share some features. First of all, they have the same aim: to extend cloud capabilities to the edge of the network. Also, they rely upon a decentralized infrastructure, even though, it is accessible through different types of networks (e.g. wireless, mobile, Bluetooth) and are composed of diverse devices. In addition, all edge implementations provide a set of benefits, mainly originated from the proximity to the edge of the networks: low latency, context and location awareness, high scalability and availability, and support to mobility. Undoubtedly, even if these implementations share the same goal and a number of features, they present some differences. They can be deployed in different ways, both in terms of the type of devices and proximity to end users. For instance, the deployment of MEC nodes is linked to the mobile network infrastructure, while mobile cloud computing (MCC) has a wider scope. There are also differences in terms of entities eligible to own these infrastructures. For example, since MEC nodes are bound to the edge of the mobile network infrastructure, only telecommunication companies can provide MEC services, while any entity can deploy an MCC infrastructure.

As a vital part of 5G mobile communication network technology, MEC has a wide range of application scenarios due to its high efficiency and low latency, thus applying it to the industrial internet (Chen et al., 2018b, Sun et al., 2018) also has significant advantages. At present, demand from applications with cross-platform and cross-sector is increasing, and the traditional ethernet has been unable to process the growing data in industry under new requirements of network latency.

According to the characteristics of the IoT, the edge computing platform can be introduced between the core network and the factory 5G wireless base station (Yu, 2016, Feng et al., 2017). The data in the terminal equipment is aggregated to the MEC server after passing through the base station and the internal gateway, then passes to the internet. The MEC server filters the uploaded data, caches part of the data to the edge data center (EDC), and processes and analyzes the data using the idle edge network resources inside the EDC.

Fog computing emerges from the crowd representing the highest evolution of the edge computing principles. In general, the goal of fog computing is to represent a complete architecture that allocates resources horizontally and vertically along the cloud-to-things continuum. Hence, it is not just a trivial extension of the cloud, but a new player in interacting with the cloud and the IoT to help enhance their interactions. However, the research on fog computing is still in its infancy and there are new differences.

Fog computing is often considered as an implementation of edge computing. However, fog computing provides distributed computing, storage, control, and networking capabilities closer to the user (Chiang et al., 2017). Fog computing is not limited to only the edge of the network, but it incorporates the edge computing concept. Fog computing provides a structured intermediate layer that fully bridges the gap between IoT and cloud computing. In fact, fog nodes can be located anywhere between end devices and the cloud; thus, they are not always directly connected to end devices. Moreover, fog computing does not only focus on the "things" side, but it also provides its services to the cloud. The N-tier architecture proposed by the OpenFog Consortium (Martin et al., 2017) is mainly aimed at giving an inner structure to the fog layer of the three-layer architecture, driving the stakeholders when it comes to deploying fog computing in a specific scenario. Indeed, although the deployment of fog software and fog systems is scenario specific, the key features of the fog architecture remain evident in any fog deployment.

In this vision, fog computing is not only an extension of the cloud to the edge of the network, nor a replacement for the cloud itself, rather a new entity working between cloud and the IoT to fully support and improve their interaction, integrating the IoT, edge, and cloud computing.

1.3.3 Intelligent Algorithms

1.3.3.1 Big Data

Big data defines huge, diverse, and fast growing data that requires new technologies to handle. With the rapid growth of data, big data has been brought to the attention of researchers to use it in the most prominent way for decision making in various emerging applications. Big data has always been defined by five most common characteristics: variety (Aftab and Siddiqui, 2018, Li et al., 2013, Batyuk and Voityshyn, 2016, Yadranjiaghdam et al., 2016), volume (Aftab and Siddiqui, 2018, Jain and Kumar, 2015), value (Demchenko et al., 2014, Gürcan and Berigel, 2018), veracity (Gürcan and Berigel, 2018, Benjelloun et al., 2015, Londhe and Rao, 2017), and velocity (Yadranjiaghdam et al., 2016, Kaisler et al., 2013, Yu et al., 2017). Variety indicates that the data is of multiple categories such as raw, structured, semi-structured, and unstructured data from various sources such as websites, social media sites, emails, and documents. The volume indicates very large quantities of generated data. The velocity concept deals with the speed of the data coming from various sources.

Value is the process of extracting valuable information from a huge set of data. It is important as it generates knowledge for people and business. And veracity refers to the accuracy of collected information. Data quality with its privacy is important for correct analysis.

Up to now, big data architecture patterns have emerged. Big data architecture pertains to the basic design and components required for the storage, management, and retrieval of big data in an IT domain. The big data architecture blueprint is categorized as the logical and physical structure and consists of four different layers: big data sources, storage, analysis and utilization. The sources layer includes a server or any data sources such as sensors, social media, and data warehouse. The storage layer collects all the structured data and stores it in the relational database management system or the unstructured data are placed into the Hadoop Distributed File System. Analysis is pivotal for the improvement of the business and the utilization of the outcome to counter complications.

Big data analytics, which is the bridge between big data and the IoT, is a transpiring field. IoT and big data both are two-fold, which can be considered as two sides for the same coin. Analytics of big data is to require to examine the massive sets of data which can contain different types. The social competence of big data and IoT is the emerging key to escalate the decision-making. As we experience the expansion of sensors over every organization/industry, IoT has most eminent attributes to analyze the information about things connected. Analysis of big data in the IoT demands a copious amount of space to store data. The current status of IoT is deficient without the big data. Currently, the IoT is already applied in wide-ranging domains and has been integrated into predicting natural calamities, regulating traffic, adjusting lighting at home, and also in agro-based industries. In the progressive world, the IoT would be much more in demand and have more prominence. This would result in various companies investing more in data centers to store and analyze all the huge amount of data generated from IoTs. Business can multiply their profits by integrating and adapting to personalized marketing, capitalizing on the customers' preference hiking the revenue. Both big data and IoT can be integrated into a larger scale with government and implement smart cities throughout the country to ensure enhanced lifestyle along with reducing time consumption, finances, and exploitation of energy. Agriculture industry could also flourish by evaluating the soil and determining the prerequisites for a better yield and profit.

Big data brings many attractive opportunities and application with a lot of challenges to handle. However, there are still some important challenges of big data to be mentioned. For instance, data does not come from a vacuum. It comes from many sources by underlying activities, examples like web logs, social networks, sensors, scientific research, and experiments that produce huge amount of data (Yu et al., 2017). Raw data generated from a source is too gigantic and everything is not always useful. These collected data have various structures that need to be further processed. So, initially, all data needs to be stored for pre-processing. Apart from these captured data, more data is automatically generated by the system called metadata that defines which type of data will be stored. The available storage is not sufficient enough to store such massive data. One solution is to upload it to the cloud. However, uploading terabytes and zetabytes of data will take a lot of time. Also, due to the rapid nature of data, it is not possible to use the cloud for real-time data processing (Behera et al., 2017). Fortunately, edge computing and fog computing can share a portion of the storage.

In addition, the data generated by users is heterogeneous in nature whereas data analysis algorithms expect homogeneous data for better processing and analysis. Data must be properly structured at the beginning of analysis. Structured data is well organized and manageable. Unstructured data represents all kinds of social network data, documents, and reviews (Behera et al., 2017). Unstructured data is costly to work with and also it is not feasible to convert all unstructured data to structured data. This diversity in data is challenging to handle and process (Katal et al., 2013).

For real-time applications, timeliness must be of top-most priority (Yin and Zhao, 2015). It is difficult to generate a timely response when the volume of data is very huge. Considering an example, it is important to analyze early detection of disease otherwise it is of no help to the patient. To find common patterns, first time is required to scan the whole dataset. And privacy of data is one of the foremost concerns in big data. Preserving the privcacy of a patient's data is essential as there is fear of inappropriate use of personal data which might get revealed when integrating such data from several other sources. Privacy of data is not only a technique but also a sociological problem (Katal et al., 2013). Each day, massive data is generated on social media where a user shares their private information, location, pictures, and so on, which not only reveal the identity of a user but can also be used for criminal activities and fraud.

Now we are in a digital age, which produces huge amounts of data every day. Properly dealing with the shortcomings of big data and making full use of its strengths and integrating innovation with the IoT, will improve the productivity and decision-making ability of society and enterprises.

1.3.3.2 Artificial Intelligence

The IoT is getting smarter. AI is being incorporated – in particular, machine learning – into IoT applications, with capabilities growing, including improving operational efficiency and helping avoid unplanned downtime. The key to such applications is finding insights in data. AI is playing an increasing role in IoT applications and deployments (Holdowsky et al., 2015, Schatsky et al., 2014). There is an apparent shift in the behavior of companies operating in this area. Venture capital investments in IoT start-ups with AI are rising sharply. Companies have made much progress at the intersection of AI and the IoT in recent years. And major vendors of IoT platform software are now offering integrated AI capabilities such as machine learning-based analytics.

AI is playing a starring role in IoT because of its ability to quickly wring insights from data. Machine learning, an AI technology, brings the ability to automatically identify patterns and detect anomalies in the data that smart sensors and devices generate, such as temperature, pressure, humidity, air quality, vibration, and sound. Companies are finding that machine learning can have significant advantages over traditional business intelligence tools for analyzing IoT data, including being able to make operational predictions up to 20 times earlier and with greater accuracy than threshold-based monitoring systems. And other AI technologies such as speech recognition and computer vision can help extract insight from data that used to require human review.

The powerful combination of AI and IoT technology is helping companies avoid unplanned downtime, increase operating efficiency, enable new products and services, and enhance risk management.

1.4 Typical Applications

1.4.1 Environmental Monitoring

Environmental monitoring is an important application area of the IoT. The automatic and intelligent characteristics of the IoT are very suitable for monitoring environmental information. Generally speaking, the structure of the environmental monitoring based on IoT includes the following parts. The perception layer provides the main function to obtain environmental monitoring information, such as temperature, humidity, and illumination, through sensor nodes and other sensing devices. (Lazarescu, 2013) introduces an application of the IoT in birdhouse monitoring. Because environmental monitoring needs to be perceived in a wide geographical range and contains a large amount of information, the devices in this layer need to be formed into an autonomous network through wireless sensor network technology, extract useful information employing collaborative work, and realize resource sharing and communication through access devices with other devices in the internet. The accessing layer transfers information from the sensing layer to the internet through the wireless communication network (such as wired internet network, Zigbee, LPWAN, WLAN network, GSM network, TDSCDMA network), satellite network, and other infrastructures. The network layer integrates the information resources within the network into a large intelligent network that can be interconnected to establish an efficient, reliable, and trusted infrastructure platform for the upper layer of service management and large-scale environmental monitoring applications. The service management layer conducts real-time management and control of the massive amount of information obtained by environmental monitoring within the network through a large central computing platform (such as high-performance parallel computing platform), and provide a good user interface for the upper application. The application layer integrates the functions of the bottom layer of the system and builds the practical application of the industry oriented to environmental monitoring, such as real-time monitoring of ecological environment and natural disasters, trend prediction, early warning, emergency linkage, etc. Through the above parts, environmental monitoring based on the IoT can realize collaborative perception of environmental information, conduct situation analysis, and predict development trends.

1.4.2 Public Safety Surveillance

Public safety surveillance based on the IoT with the characteristics of wide coverage of public security monitoring, multiple monitoring indicators, high continuity requirements, unsuitable environment for manual monitoring, and close correlation between perceived information content and people's lives, employs the technology of the IoT, in particular the technology of a sensor network to construct an information system engineering composed of a perception layer, network layer, and application layer, which mainly includes monitoring to ensure the safety of all kinds of production scenarios, the monitoring of producer safety, the monitoring of the safety of specific items, the monitoring of densely populated places, the monitoring of important equipment and facilities, and information collection of scenes, personnel, and items during emergency treatment in accidents. Public security is the cornerstone of national security and social stability. In order to effectively withstand all

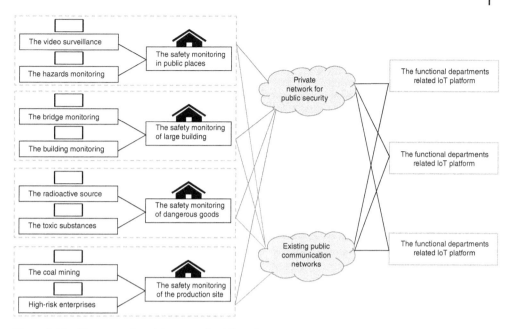

Figure 1.17 The network architecture of public safety surveillance.

kinds of man-made or natural disasters, countries will strengthen public security measures as the focus of government work. The IoT for public safety monitoring provides a new way to solve the problems facing public safety surveillance at present. The establishment of a complete public safety surveillance based on IoT will provide effective prevention mechanism for existing safety problems such as bridge tunnel collapse, hazardous material leakage, etc. The nationwide public safety surveillance based on IoT enables the timely, powerful, and transparent resolution of major safety incidents. Therefore, public safety surveillance based on the IoT should be given priority by the whole of society. The Figure 1.17 describes the network architecture of public safety surveillance based on the IoT, which is similar to the whole architecture of the IoT and consists of three parts: perception layer, network layer and application layer. However, due to the particular needs of public safety surveillance based on the IoT, there are some technical characteristics that other IoT applications do not have, which are summarized as follows.

In the perception layer, the types of perceived information are diverse, and the real-time requirement is high. The monitoring of most information (such as the safety of bridge buildings, monitoring of dangerous goods, etc.) requires high accuracy and is difficult to detect by manual means. Because of the high uncertainty of the information type of potential security hazards, a large number of different types of sensors should be deployed in densely staffed or high-risk production sites for a long time. Higher requirements are put forward for the networking strategy of perception layer, energy management, transmission efficiency, quality of service (QoS), sensor coding, address, frequency, and electromagnetic interference. These problems are also the key to the mature application of public safety surveillance based on the IoT. Because the information perceived by public safety surveillance based on the IoT involves the national key industries and the daily life of the people, once the information is leaked or improperly used, it may endanger national security, social stability, and people's privacy. Therefore, it is necessary for the information content of public safety surveillance based on IoT to be transmitted through a private network or 4G mobile network after taking security precautions to ensure the security, authenticity, and integrity of the information. It is necessary to establish a proprietary platform for public safety surveillance based on the IoT with different levels in view of the massive amount of data information and the serious harm that hidden dangers to security may bring. The service platform not only has the strong ability of information processing and integration, but also has timely links to relevant functional departments to deal with emergencies in case of public safety emergencies, so as to minimize losses and impact. In addition, the interconnection of public security IoT platforms of different levels is conducive to the maximum allocation of resources according to the hazard level of security incidents, so as to facilitate the timely, effective, and transparent resolution of public security incidents.

In public safety surveillance and environmental monitoring, massive sensor devices, such as for water level, temperature, cameras and microphones, obtain a large amount of data from the environment and transform it into more effective information through multi-level processing and analysis. In this process, in order to protect privacy, the storage, transmission and process of data must be strictly managed. Throughout the system, the transmission of data at each processing stage usually uses encryption technology to ensure the confidentiality of the data. With "smart" becoming the new default on devices, privacy risks are not always clear. Therefore, all IoT devices should adopt basic technologies for privacy protection, such as encrypted data and strong authentication. In terms of storage, we can only store the necessary and important information to protect information privacy. Information, such as video and audio, is transported only in the case of "need-to-know". Anonymization could be used to disguise the identity of the stored information. Sensitive data must be processed in a suitable manner and for the processing aim only. The admission and authentication of the data owner need to be acquired before exposing data information to third parties.

1.4.3 Military Communication

An interesting application scenario that is receiving great attention consists of a collaboration between sensors and mobile agents, typically unmanned air vehicles (UAVs) that are used for retrieving data from the sensor network, to clarify observations and to provide efficient information for mission-critical planning purposes. Such an architecture is illustrated in Figure 1.18, with the main advantage of the UAV is that it provides reach back to a remote command center. The sink is responsible for completing the request from the UAV to obtain information about the observation. Figure 1.18 shows the need for the sensor network to obtain and deliver critical information rapidly and dependably to support the overall mission operation and objectives. To support this, getting to the designated specified region of interest is a critical first step to the overall mission objective. A new routing protocol, called Swarm Intelligent Odour Based Routing (SWOB), had been proposed to solve this critical first step (Ghataoura et al., 2009). SWOB uses network topology and geographic location information to effectively coordinate the routing tasks for information agents to traverse the network to the region of interest. SWOB itself takes its inspiration on

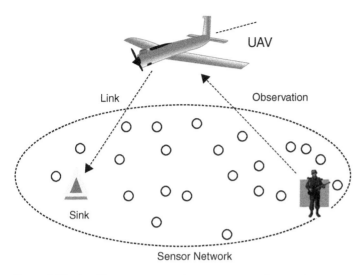

Figure 1.18 Intelligent transportation system scenario overview.

the basic principles and examples provided by social insects in odor localization and tracking. A wide variety of insects use plumes of pheromones (odor) to locate prey, mates, and other sources of particular interest. Insects themselves follow a route employing olfactory sensing to regions of higher pheromone concentration, since this represents a higher order of relevance in finding the required designated odor source. In this study, a virtual Gaussian odor plume is used to conceptually describe the odor dispersion effects found in nature and to establish the level of virtual odor concentration found in a sensor nodes environment, downwind from the odor source (region of interest). Using a virtual Gaussian odor plume model allows information agents to be controlled by an odor plume, with the eventual aim of guiding the designated region of interest. In this case, agents are forwarded in a unicast fashion to nodes that represent higher levels of odor concentration, dictated by the guidance of the virtual odor plume. As a result, the SWOB mechanism has advantages in energy and bandwidth efficiency, and distributed task management and scalability. So that nodes outside the Gaussian plume do not contribute to forwarding tasks and fewer nodes will be competing for bandwidth resource.

However, due to the limited resources of IoT devices in military communication, the existing data security protection method is not fully applicable to high security military communication. Moreover, the highly dynamic environment of the IoT network also makes the military communication more vulnerable and difficult to protect. Therefore, the security goals of confidentiality, authenticity, integrity and availability are still the pursuit of data proposed by the military communication system. Confidentiality means information transmission between objects must be protected from attackers. Access to the system and sensitive information is allowed for legal users only. Integrity is required to ensure data accuracy and completeness and keep it from being tampered with. In the end, to avoid any possible operational interruptions or failures, the availability of the security service must be increased.

1.4.4 Intelligent Manufacturing and Interactive Design

The IoT has been a hyped topic for nearly a decade now. Ever increasing, millions of devices get direct access to the internet, providing a plethora of applications, e.g. smart homes or mobile health management. This trend can also be found in industry where IoT components hardened for these environments have been introduced, called industrial IoT (IIoT) devices, which can be either sensors or actors, as well as mobile equipment such as smartphones, tablets, and smart glasses. Hence, mobile communication has become ubiquitous in smart factories. IIoT devices provide massive data on temperature, pressure, machine states, etc. In comparison to conventional data acquisition in machines, the amount of data will increase significantly and the interaction between machines on the edge as well as between distributed machines and cloud services (Munoz et al., 2016) will rise.

Digitalization, Industry 4.0 or Advanced Manufacturing are programs which are pushing the progress described before. Additionally, they are bringing the flexibility of production sites and systems in the digital era to a new level which has led to new constraints and requirements regarding the communication networks (Chen et al., 2018a). Although the basic concepts have been used for many decades now, production sites and systems are optimized using methods like the Toyota Production System, the Lean concept, and many other techniques. These require a continuous improvement of the whole setup of manufacturers, meaning that machines and components are physically moved around the site to increase the flow of produced parts and products (Ramprasad et al., 1999). Additionally, the demand for individual personalized products is increasing, not only for consumer products, but also for professional products. IIoT components in modern factories on communication networks are neglected and will lead to huge issues for network architecture, network performance engineering, and IT security. This causes flexible requirements on the autonomous configuration of the network and its elements from the component at the edge through the backbone to the source or sink of communication (Ma et al., 2017). Network slices are often used in the backbone parts of a network to allow for a differentiation of the various types of communication and data streams, especially in the context of 5G technology (Rost et al., 2017). This concept solely is neither able to cope with physical flexibility at the edge, nor with the demand of production systems that require end-to-end guarantees (Chien et al., 2019), which have been known in the past for performance engineering based on integrated services. These concepts focus on performance, but not on flexible architecture or IT security which are also required for flexible production systems beginning at the edge (Vilalta et al., 2016). One solution to overcome these challenges are software designed networks (SDNs), allowing for an agile and flexible network configuration. Virtual network functions (VNFs) are dividing the issue into smaller and more manageable elements. The network service, assembling the services of multiple VNFs, enables the machine park operator to connect existing machines to the cloud-based services such as data analytics. Subsequently, the data is processed to provide additional knowledge and services. Data analytics is supporting machine park operators to identify maintenance intervals, incidents with machines or potential increase in performance. SDNs will offer the flexibility for production systems as they are required (Choo et al., 2018).

Data analytics can be used to support machine park operators, but for data analytics machine data is required. Therefore, in this section the focus is on machine data acquisition in factories. Machines generate constantly data such as operational data, machine data, and process data (O/M/P data). Often, most of the data is only used within the machine, but due to topics such as big data and data analytics, the interest in using these data and further data is growing. As a result, the requirements to factory networks increase dramatically.The machine is controlled by a programmable logic controller (PLC) which reads sensor input data ($S_1 \dots S_n$) and generates output data for actuators ($A_1 \dots A_n$) across analogue and digital input and output (I/O) modules. The machine's MPC provides data directly via standardized high-level interfaces for data acquisition by central computers or manufacturing execution systems (MESs).

Furthermore, additional IIoT sensors are becoming increasingly important; for measuring specific temperatures and pressures inside a machine, or for considering ambient conditions such as temperature and humidity. With these retrofit procedures, existing older machines can be integrated into novel data analysis systems as well. Upcoming 5G technology for mobile radio edge communication will pave the way for this. These general conditions indicate that flexible SDN/NFV-based network components are useful for the digitalization of factories.

In addition, network services for intelligent manufacturing must fulfill manifold requirements and use cases as flexible and agile networks will become mandatory in the future.

1.4.5 Autonomous Driving and Vehicular Networks

Autonomous driving includes video cameras, radar sensors(ultrasonic radar, millimeter-wave radar) to sense the traffic or conditions around them and to navigate the road by integrating GPS (global positioning system), IMU (inertial measurement unit), and other information. Perception, planning, and control are the three core functions of existing autonomous driving. The perception layer obtains sensor and other shared communication information as input and transmits it to the planning layer. The planning layer will receive feedback information from the control layer in addition to the perception layer information, and the control layer will realize specific vehicle control.

In order to realize full-scene automatic driving, in addition to dealing with the basic driving environment, the autonomous vehicles also need to overcome the limitations of a rainy day, fog, night, and other harsh environments. At this time, the sensor of the autonomous vehicles alone is far from enough to meet the requirements.This requires the cars have the ability to "telepathically" communicate with participants in traffic scenarios such as roads, traffic signs, and other vehicles over long distances, and allows the car to perceive the complexity of the road in advance, from the distance beyond its line of sight. At this point, network and communication technology can realize autonomous driving more reliably and efficiently by obtaining real-time information shared by a large number of vehicles.

First, the intra-vehicle interconnection communication includes wired and wireless connections, which is further divided into a point-to-point communication mode and data bus communication mode. The wired connection technologies of intra-vehicle interconnection communication include CAN, LIN, FlexRay, MOST (media oriented system transport), idb-1394 (intelligent transport system data bus), D2B (digital data bus), LVDS (low differential signaling), ethernet, PLC (power-line communication). The wireless connection technologies of intra-vehicle interconnection communication include Bluetooth 5.0, ZigBee, UWB (ultra wideband), wireless fidelity, etc(Wang et al., 2019a).

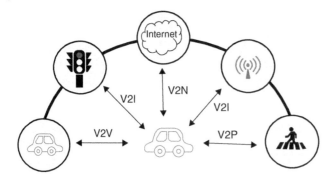

Figure 1.19   Description of VANET for autonomous vehicles.

Secondly, V2X technology is used for inter-vehicle communication (Dietzel et al., 2014). VANET (vehicular ad-hoc network) is the application of traditional MANET (mobile ad-hoc network) on a trafficked road. V2X technologies have V2V (vehicle-to-vehicle), V2I (vehicle-to-infrastructure), V2P (vehicle-to-pedestrian) and V2N (vehicle-to-network), which is shown in Figure 1.19. VANET adopts ICN (information-centric networking) architecture, which can meet the requirements of high scalability and low delay of autonomous driving. It can address data directly by changing the addressing scheme, rather than using the location of the data. In addition, the clustering can improve the scalability and reliability of VANET to meet the requirements of the autonomous vehicle networks (Cooper et al., 2017). The potential inter-vehicle communication technologies in autonomous driving include low power technologies (Bluetooth, ZigBee), IEEE 802.11 family technologies [wifi, DSRC (dedicated short range communications (Kenney, 2011))], base station driven technologies [WiMAX (worldwide interoperability for microwave access), LTE-V (long term evolution for vehicles)] (Chen et al., 2016) and some auxiliary technologies [HetVNET (heterogeneous vehicle network), SDN].

Current Cellular Vehicle-to-everything (C-V2X) technology, which is based on cellular communications, allows cars to have this capability. For example, communication with traffic lights can let the car know the status of traffic lights in advance, and thus it can slow down in advance. By communicating with other vehicles, we can inform other vehicles of emergency braking, lane changes, and turning status in time, so that other vehicles will have enough time to predict and deal with it. V2X communication technology can not only enable vehicle-to-vehicle communication, but also enable the big data platform in the cloud to communicate with the vehicle. In addition to enabling the vehicle to enjoy a variety of colorful network services, it can also enable the cloud platform to conduct unified scheduling and integration of all traffic participants, so as to achieve truly intelligent traffic without congestion.

1.5   Requirements and Challenges for Intelligent IoT Services

1.5.1   A Generic and Flexible Multi-tier Intelligence IoT Architecture

Sensing, communication, computing and networking technologies continue to generate more and more data, and this trend is set to continue. Future IoT devices could, in particular, be widely deployed for tasks such as environmental monitoring, city management, and medicine and healthcare, requiring data processing, information extraction, and real-time decision making. Cloud computing alone cannot support such ubiquitous deployments and applications because of infrastructure shortcomings such as limited communication bandwidth, intermittent network connectivity, and strict delay constraints.

To address this challenge, and ensure timely data processing and flexible services, multi-tier computing resources are required, which can be deployed and shared along the continuum from the cloud to things. Cloud computing technologies are usually centralized and used for global tasks. Multi-tier computing involves collaborations between cloud computing and fog computing, edge computing and sea computing technologies, which have been developed for regional, local and device levels, respectively (Figure 1.1). The integration of these different computing resources is vital for the development of intelligent IoT services.

Thus, it is essential to have effective communication and collaboration mechanisms across different levels and units. Similarly, interaction and collaboration between cloud, fog, edge and sea computing are vital in order to create an intelligent collaborative service architecture. As a result, this architecture actively connects the shared computing, communication and storage resources of all the nodes in the IoT network, and fully utilizes their capabilities at different locations and levels, in order to provide intelligent, timely and efficient services according to user requirements. Since most applications and their data do not require superior computing power, this architecture can significantly improve service quality and user experience while saving time, resources and costs.

1.5.2 Lightweight Data Privacy Management in IoT Networks

The IoT is composed of physical objects embedded with electronics, software, and sensors, which allows objects to be sensed and controlled remotely across the existing network infrastructure, and facilitates direct integration between the physical world and computer communication networks. The advancement and wide deployment of the IoT have revolutionized our lifestyle greatly by providing the most convenience and flexibility in our various daily applications. The typical applications of IoT include smart grid, healthcare, and smart city. The IoT has been widely applied in various applications such as environment monitoring, energy management, medical healthcare systems, building automation, and transportation. With the rapid deployment of IoT technologies and the variety of IoT applications, there is a need for new tools to ensure individual privacy requirements when collecting and using personal data (Dorri et al., 2017).

Unfortunately, due to the constraints of communication and computation resource of IoT devices, highly complex computation is delivered to the energy abundant cloud for considerably enhanced efficiency. However, due to the wide-area data transmission and resource constraints of the devices, IoT security and privacy remain a major challenge, and new efficient privacy-preserving solutions for intelligent IoT applications are required.

In recent years, fog computing-enhanced IoT has recently received considerable attention, as the fog devices deployed at the network edge can not only provide low latency and location awareness but also improved real-time and quality of services in IoT application scenarios (Lu et al., 2017, Mukherjee et al., 2017). However, existing security and privacy measurements for cloud-based IoT cannot be directly applied to the fog computing-enhanced IoT due to its features, such as mobility, heterogeneity, and large-scale geo-distribution. For example, all devices in fog networks have a certain level of reliance on one another. Authentication plays a major role in establishing an initial set of relations between IoT devices and fog nodes in the network. However, this is not sufficient as devices can always malfunction or are also susceptible to malicious attacks. In such a scenario, trust plays a major role in fostering relations based on previous interactions. Trust should play a two-way role in a fog network. That is, the fog nodes that offer services to IoT devices should be able to validate whether the devices requesting services are genuine. On the other hand, the IoT devices that send data and other valued processing requests should be able to verify whether the intended fog nodes are indeed secure. This requires a robust trust model in place to ensure reliability and security in the fog network.

1.5.3   Cross-domain Resource Management for Intelligent IoT Services

With the wide adoption of the IoT across the world, the IoT devices are facing more and more intensive computation task nowadays. However, with limited battery lifetime computation, bandwidth, and memory resources, the IoT devices are usually limited by their computing capability, latency, and energy, and IoT devices such as sensors, cameras and wearable devices have a computation bottleneck to limit the support for advanced applications such as real-time image processing and low latency online gaming. Fog and edge computing provide new opportunities for developments of IoT since fog and edge computing servers which are close to devices can provide more powerful computing resources. The IoT devices can offload the intensive computing tasks to the servers, while saving their computing resources and reducing energy consumption (Cui et al., 2019). However, the benefits come at the cost of higher latency, mainly due to additional transmission time, and it may be unacceptable for many IoT applications. One critical problem in the computation offloading is the selection of the fog and edge device from all the potential radio device candidates within the radio coverage area. This is further complicated by the determination of offloading rate, i.e. the number of computation tasks to offload to the edge device. An IoT device often requires a longer period to transmit the offloading data and receive the computation result compared with the local computation within the IoT device. This is further complicated by the chosen edge device carrying out heavy workloads and experiencing degraded radio channel fading and interference. It is, therefore, challenging for an IoT device to optimize the offloading policy within the dynamic network with time-variant radio link transmission rates, especially with an unknown amount of the renewable energy within a given time duration. Thus, it is a challenge to find a trade-off between various cross-domain resources, such as computing power, energy consumption and latency. IoT networks require effective and efficient resource management solutions.

1.5.4   Optimization of Service Function Placement, QoS, and Multi-operator Network Sharing for Intelligent IoT Services

IoT services are mostly deployed and executed in a distributed environment. Moving a composite application to a particular environment requires a distribution of contained services before deployment. Hence, application owners want to optimize the distribution of IoT services to improve network efficiency and user experience. A dynamic service provisioning, i.e. dynamic resource allocation, and dynamic deployment of services in a distributed environment enables high flexibility and optimized usage of the infrastructure. In particular, optimized distributions of services can be created by taking IoT network dependencies between services into account. Correlating the service dependencies and a particular state of infrastructure at deployment time leads to a best-qualified region in the infrastructure for the deployment (Gorlach and Leymann, 2012).

IoT services will have strict QoS requirements, including remote surgery via tactile internet, remote patient monitoring, drone delivery and surveillance. In these services QoS will play a vital role in determining user experience. For example, if a patient is monitored/operated on remotely or a drone is distantly maneuvered, the user experience will determine the amount of resources required to be allocated to reach user satisfaction. Hence, more adaptive and dynamic multimedia delivery methods are needed (Aazam and Harras, 2019).

Nowadays, the mobile and IoT networking world has undergone a rapid evolution. Driven by increasing demand and a highly competitive market, the choice of network providers and access technologies has significantly increased. Service offers need to provide dynamic services for mobile and IoT users at the same time. The services should support multiple services with high capacity using mobile networks, as well as low data rate, and long coverage using IoT networks (Hanafiah et al., 2012). This new demand will impact the requirements of multiple network coexistence, which becomes a challenge for researchers.

1.5.5 Data Time stamping and Clock Synchronization Services for Wide-area IoT Systems

Getting all devices in an IoT system to have a common notion of time is important to applications running on the system. This chapter discusses two basic system services achieving the common notion of time among the devices – data time stamping and clock synchronization. Data time stamping is to record the time of interest in terms of the wall clock; clock synchronization is to ensure the clocks of the devices in the system have the same value at all times (Yan et al., 2017). However, as the devices deeply embedded in the physical world face many uncertainties such as time-varying network connectivity and clock faults, ensuring accurate data time stamping and clock synchronization is challenging in practice. Atomic clocks, GPS, and clock synchronization and calibration protocols represent principal means to achieve data time stamping and clock synchronization. For massive deployments, chip-scale atomic clocks are still uneconomical solutions. Although GPS receivers can provide global time with μs accuracy, they generally do not work in indoor environments. Increased heterogeneity and limited resources in both hardware and software platforms are the key characteristics of IoT (Li et al., 2018). Since there are a large number of heterogeneous devices under the IoT access system, a number of them require different precision of clock synchronization for many situations such as data collection with space-time consistency. Therefore, how to realize the clock synchronization on IoT access system is needed urgently. On one hand, the way to access the reliable clock source according to different demands is one of the major issues. On the other hand, how to realize the clock synchronization for the heterogeneous underlying devices with different clock precision is another major issue.

1.6 Conclusion

This chapter has reviewed some well-known IoT technologies and related standards, including RFID, NFC, ZigBee, LoRa, Sigfox, and NB-IoT, as well as emerging network technologies. Their system architectures and technical advantages have been briefly discussed. Then, this chapter overviews a multiple-tier user-centered IoT network architecture, including new data collection technologies, computing power networks, and intelligent algorithms. In addition, key technologies in IoIT are introduced, such as mmWave, SDN, AI, and so on. Some applications relating to traditional IoT and IoIT are analyzed. With more and more IoT systems being deployed for different industrial sectors, it is very challenging to overcome the vertical barriers and mitigate the fragmentation problem across multiple application domains. It is also very difficult to guarantee system security and customer privacy while connecting and integrating several enterprise-level IoT platforms with heterogeneous data structures. At the end of this chapter, the requirements and challenges for intelligent IoT services are introduced.

References

Narrowband Internet of Things (NB-IoT); Technical Report for BS and UE radio transmission and reception (Release 13). https://www.3gpp.org/dynareport/36802.htm, 2017.

3GPP. Service Requirements for Machine-Type Communications (MTC). http://www.3gpp.org, 2014.

3GPP. Cellular system support for ultra-low complexity and low throughput Internet of Things (CIoT). https://portal.3gpp.org/desktopmodules/Specifications/SpecificationDetails.aspx?specificationId=2719, 2016a.

3GPP. Standardization of NB-IOT completed. https://www.3gpp.org/news-events/1785-nb_iot_complete, 2016b.

3GPP. Study on the security aspects of the next generation system. 2017.

Mohammad Aazam and Khaled A. Harras. Mapping QoE with resource estimation in IoT. In *2019 IEEE 5th World Forum on Internet of Things (WF-IoT)*. IEEE, apr 2019. doi: 10.1109/wf-iot.2019.8767254.

Nyoman Adhiarna and Jae-Jeung Rho. Standardization and global adoption of radio frequency identification (RFID): Strategic issues for developing countries. In *2009 Fourth International Conference on Computer Sciences and Convergence Information Technology*. IEEE, 2009. doi: 10.1109/iccit.2009.300.

U. Aftab and G. F. Siddiqui. Big data augmentation with data warehouse: A survey. In *2018 IEEE International Conference on Big Data (Big Data)*, pages 2785–2794, Dec 2018. doi: 10.1109/BigData.2018.8622206.

S. K. Agrawal and K. Sharma. 5g millimeter wave (mmwave) communications. In *2016 3rd International Conference on Computing for Sustainable Global Development (INDIACom)*, pages 3630–3634, March 2016.

N. Ahmed, H. Rahman, and Md.I. Hussain. A comparison of 802.11ah and 802.15.4 for IoT. *ICT Express*, 2(3):100–102, sep 2016. doi: 10.1016/j.icte.2016.07.003.

R. R. Al-Dahhan, Q. Shi, G. M. Lee, and K. Kifayat. Revocable, decentralized multi-authority access control system. In *2018 IEEE/ACM International Conference on Utility and Cloud Computing Companion (UCC Companion)*, pages 220–225, Dec 2018. doi: 10.1109/UCC-Companion.2018.00088.

Shadi Al-Sarawi, Mohammed Anbar, Kamal Alieyan, and Mahmood Alzubaidi. Internet of things (IoT) communication protocols: Review. In *2017 8th International Conference on Information Technology (ICIT)*. IEEE, may 2017. doi: 10.1109/icitech.2017.8079928.

Zigbee Alliance. Zigbee specification. http://www.zigbee.org/wp-content/uploads/2014/11/docs-05-3474-20-0csg-zigbee-specification.pdf, 2012.

M. M. Alsulami and N. Akkari. The role of 5g wireless networks in the internet-of- things (iot). In *2018 1st International Conference on Computer Applications Information Security (ICCAIS)*, pages 1–8, April 2018. doi: 10.1109/CAIS.2018.8471687.

Olumuyiwa Oludare FAGBOHUN and. Comparative studies on 3g,4g and 5g wireless technology. *IOSR Journal of Electronics and Communication Engineering*, 9(2):133–139, 2014. doi: 10.9790/2834-0925133139.

Michael Armbrust, Armando Fox, Rean Griffith, Anthony Joseph, Randy Katz, Andy Konwinski, Gunho Lee, David Patterson, Ariel Rabkin, Ion Stoica, and Matei Zaharia. A view of cloud computing. *Commun. ACM*, 53:50–58, 04 2010. doi: 10.1145/1721654.1721672.

Alos Augustin, Jiazi Yi, Thomas Clausen, and William Townsley. A study of LoRa: Long range & low power networks for the internet of things. *Sensors*, 16(9):1466, sep 2016. doi: 10.3390/s16091466.

A. Batyuk and V. Voityshyn. Apache storm based on topology for real-time processing of streaming data from social networks. In *2016 IEEE First International Conference on Data Stream Mining Processing (DSMP)*, pages 345–349, Aug 2016. doi: 10.1109/DSMP.2016.7583573.

R. K. Behera, A. K. Sahoo, and C. Pradhan. Big data analytics in real time - technical challenges and its solutions. In *2017 International Conference on Information Technology (ICIT)*, pages 30–35, Dec 2017. doi: 10.1109/ICIT.2017.39.

Behrtech. MIOTY: The only LPWAN solution for Industrial IoT standardized by ETSI. https://behrtech.com/mioty/, 2020.

F. Benjelloun, A. A. Lahcen, and S. Belfkih. An overview of big data opportunities, applications and tools. In *2015 Intelligent Systems and Computer Vision (ISCV)*, pages 1–6, March 2015. doi: 10.1109/ISACV.2015.7105553.

Mudit Ratana Bhalla and Anand Vardhan Bhalla. Generations of mobile wireless technology: A survey. *International Journal of Computer Applications*, 5(4): 26–32, aug 2010. doi: 10.5120/905-1282.

B. Chen, J. Wan, L. Shu, P. Li, M. Mukherjee, and B. Yin. Smart factory of industry 4.0: Key technologies, application case, and challenges. *IEEE Access*,6:6505–6519, 2018a. ISSN 2169-3536. doi: 10.1109/ACCESS.2017.2783682.

C. Chen, M. Lin, and C. Liu. Edge computing gateway of the industrial internet of things using multiple collaborative microcontrollers. *IEEE Network*, 32(1): 24–32, Jan 2018b. ISSN 1558-156X. doi: 10.1109/MNET.2018.1700146.

S. Chen, J. Hu, Y. Shi, and L. Zhao. Lte-v: A td-lte-based v2x solution for future vehicular network. *IEEE Internet of Things Journal*, 3(6):997–1005, Dec 2016. ISSN 2372-2541. doi: 10.1109/JIOT.2016.2611605.

M. Chiang, S. Ha, C. I, F. Risso, and T. Zhang. Clarifying fog computing and networking: 10 questions and answers. *IEEE Communications Magazine*, 55 (4):18–20, April 2017. ISSN 1558-1896. doi: 10.1109/MCOM.2017.7901470.

H. Chien, Y. Lin, C. Lai, and C. Wang. End-to-end slicing as a service with computing and communication resource allocation for multi-tenant 5g systems. *IEEE Wireless Communications*, 26(5):104–112, October 2019. ISSN 1558-0687. doi: 10.1109/MWC.2019.1800466.

K. R. Choo, S. Gritzalis, and J. H. Park. Cryptographic solutions for industrial internet-of-things: Research challenges and opportunities. *IEEE Transactions on Industrial Informatics*, 14(8):3567–3569, Aug 2018. ISSN 1941-0050. doi: 10.1109/TII.2018.2841049.

C. Cooper, D. Franklin, M. Ros, F. Safaei, and M. Abolhasan. A comparative survey of vanet clustering techniques. *IEEE Communications Surveys Tutorials*, 19(1):657–681, Firstquarter 2017. ISSN 2373-745X. doi: 10.1109/COMST.2016.2611524.

Vedat Coskun, Busra Ozdenizci, and Kerem Ok. The survey on near field communication. *Sensors*, 15(6):13348–13405, jun 2015. doi: 10.3390/s150613348.

Laizhong Cui, Chong Xu, Shu Yang, Joshua Zhexue Huang, Jianqiang Li, Xizhao Wang, Zhong Ming, and Nan Lu. Joint optimization of energy consumption and latency in mobile edge computing for internet of things. *IEEE Internet of Things Journal*, 6(3):4791–4803, jun 2019. doi: 10.1109/jiot.2018.2869226.

P. Datta and B. Sharma. A survey on iot architectures, protocols, security and smart city based applications. In *2017 8th International Conference on Computing, Communication and Networking Technologies (ICCCNT)*, pages 1–5, July 2017. doi: 10.1109/ICCCNT.2017.8203943.

Y. Demchenko, C. de Laat, and P. Membrey. Defining architecture components of the big data ecosystem. In *2014 International Conference on Collaboration Technologies and Systems (CTS)*, pages 104–112, May 2014. doi: 10.1109/CTS.2014.6867550.

S. Dietzel, J. Petit, F. Kargl, and B. Scheuermann. In-network aggregation for vehicular ad hoc networks. *IEEE Communications Surveys Tutorials*, 16(4):1909–1932, Fourthquarter 2014. ISSN 2373-745X. doi: 10.1109/COMST.2014.2320091.

K. Dolui and S. K. Datta. Comparison of edge computing implementations: Fog computing, cloudlet and mobile edge computing. In *2017 Global Internet of Things Summit (GIoTS)*, pages 1–6, June 2017. doi: 10.1109/GIOTS.2017.8016213.

Ali Dorri, Salil S. Kanhere, Raja Jurdak, and Praveen Gauravaram. Blockchain for IoT security and privacy: The case study of a smart home. In *2017 IEEE International Conference on Pervasive Computing and Communications Workshops (PerCom Workshops)*. IEEE, mar 2017. doi: 10.1109/percomw.2017.7917634.

Shahin Farahani. *ZigBee Wireless Networks and Transceivers*. Newnes, 2011.

Laith Farhan, Sinan T. Shukur, Ali E. Alissa, Mohmad Alrweg, Umar Raza, and Rupak Kharel. A survey on the challenges and opportunities of the internet of things (IoT). In *2017 Eleventh International Conference on Sensing Technology (ICST)*. IEEE, dec 2017. doi: 10.1109/icsenst.2017.8304465.

Szymon Fedor and Martin Collier. On the problem of energy efficiency of multi-hop vs one-hop routing in wireless sensor networks. In *21st International Conference on Advanced Information Networking and Applications Workshops (AINAW'07)*, volume 2, pages 380–385. IEEE, 2007.

M. Feng, S. Mao, and T. Jiang. Base station on-off switching in 5g wireless networks: Approaches and challenges. *IEEE Wireless Communications*, 24(4): 46–54, Aug 2017. ISSN 1558-0687. doi: 10.1109/MWC.2017.1600353.

Z. Feng and Y. Zhu. A survey on trajectory data mining: Techniques and applications. *IEEE Access*, 4:2056–2067, 2016. ISSN 2169-3536. doi: 10.1109/ACCESS.2016.2553681.

Klaus Finkenzeller. *RFID Handbook: Fundamentals and Applications in Contactless Smart Cards, Radio Frequency Identification and Near-field Communication*. John Wiley & Sons, 2010.

X. Foukas, G. Patounas, A. Elmokashfi, and M. K. Marina. Network slicing in 5g: Survey and challenges. *IEEE Communications Magazine*, 55(5):94–100, May 2017. ISSN 1558-1896. doi: 10.1109/MCOM.2017.1600951.

G. Gampala and C. J. Reddy. Massive mimo – beyond 4g and a basis for 5g. *International Applied Computational Electromagnetics Society Symposium (ACES)*, pages 1–2, 2018.

Darminder Singh Ghataoura, Yang Yang, George Matich, and Selex Galileo. SWOB: Swarm intelligent odour based routing for geographic wireless sensor network applications. In *MILCOM 2009 - 2009 IEEE Military Communications Conference*. IEEE, oct 2009. doi: 10.1109/milcom.2009.5380107.

A. Ghosh, A. Maeder, M. Baker, and D. Chandramouli. 5g evolution: A view on 5g cellular technology beyond 3gpp release 15. *IEEE Access*, 7:127639–127651, 2019. ISSN 2169-3536. doi: 10.1109/ACCESS.2019.2939938.

Katharina Gorlach and Frank Leymann. Dynamic service provisioning for the cloud. In *2012 IEEE Ninth International Conference on Services Computing*. IEEE, jun 2012. doi: 10.1109/scc.2012.30.

F. Gürcan and M. Berigel. Real-time processing of big data streams: Lifecycle, tools, tasks, and challenges. In *2018 2nd International Symposium on Multidisciplinary Studies and Innovative Technologies (ISMSIT)*, pages 1–6, Oct 2018. doi: 10.1109/ISMSIT.2018.8567061.

H. Guo, J. Liu, J. Zhang,W. Sun, and N. Kato. Mobile-edge computation offloading for ultradense iot networks. *IEEE Internet of Things Journal*, 5(6): 4977–4988, Dec 2018. ISSN 2372-2541. doi: 10.1109/JIOT.2018.2838584.

Syazalina Mohd Ali Hanafiah, Azita Laily Yusof, Norsuzila Ya'acob, and Mohd Tarmizi Ali. Performance studies on multi-operator sharing algorithm for cellular wireless network. In *2012 International Conference on ICT Convergence (ICTC)*. IEEE, oct 2012. doi: 10.1109/ictc.2012.6387153.

Ali Hazmi, Jukka Rinne, and Mikko Valkama. Feasibility study of 802.11ah radio technology for IoT and m2m use cases. In *2012 IEEE Globecom Workshops*. IEEE, dec 2012. doi: 10.1109/glocomw.2012.6477839.

L. He, Z. Yan, and M. Atiquzzaman. Lte/lte-a network security data collection and analysis for security measurement: A survey. *IEEE Access*, 6:4220–4242, 2018. ISSN 2169-3536. doi: 10.1109/ACCESS.2018.2792534.

Jonathan Holdowsky, M Mahto, M Raynor, and M Cotteleer. A primer on the technologies building the iot, 2015.

W. Hong, K. Baek, and S. Ko. Millimeter-wave 5g antennas for smartphones: Overview and experimental demonstration. *IEEE Transactions on Antennas and Propagation*, 65(12):6250–6261, Dec 2017. ISSN 1558-2221. doi: 10.1109/TAP.2017.2740963.

Huawei. NB-IOT Enabling New Business Opportunities whitepaper. https://e.huawei.com/en/material/his/iot/6ba6590551ed4ad8b7bbe3c751fe8ea4, 2016a.

Huawei. NB-IoT White Paper. http://carrier.huawei.com/en/technical-topics/wireless-network/NB-IoT/NB-IoT-White-Paper, 2016b.

International Telecommunication Union (ITU). ITU Internet Reports 2005: The Internet of Things-Executive Summary. Nov 2005.

Iotivity. IoTivity Architecture. https://iotivity.org/about/iotivity-architecture, 2020.

V. K. Jain and S. Kumar. Big data analytic using cloud computing. In *2015 Second International Conference on Advances in Computing and Communication Engineering*, pages 667–672, May 2015. doi: 10.1109/ICACCE.2015.112.

Xiaolin Jia, Quanyuan Feng, Taihua Fan, and Quanshui Lei. RFID technology and its applications in internet of things (IoT). In *2012 2nd International Conference on Consumer Electronics, Communications and Networks (CECNet)*. IEEE, apr 2012. doi: 10.1109/cecnet.2012.6201508.

S. Kaisler, F. Armour, J. A. Espinosa, and W. Money. Big data: Issues and challenges moving forward. In *2013 46th Hawaii International Conference on System Sciences*, pages 995–1004, Jan 2013. doi: 10.1109/HICSS.2013.645.

A. Katal, M. Wazid, and R. H. Goudar. Big data: Issues, challenges, tools and good practices. In *2013 Sixth International Conference on Contemporary Computing (IC3)*, pages 404–409, Aug 2013. doi: 10.1109/IC3.2013.6612229.

J. B. Kenney. Dedicated short-range communications (dsrc) standards in the united states. *Proceedings of the IEEE*, 99(7):1162–1182, July 2011. ISSN 1558-2256. doi: 10.1109/JPROC.2011.2132790.

L. U. Khan, I. Yaqoob, N. H. Tran,Z. Han, and C. S. Hong. Network slicing: Recent advances, taxonomy, requirements, and open research challenges. *IEEE Access*, pages 1–1, 2020. ISSN 2169-3536. doi: 10.1109/ACCESS.2020.2975072.

J. M. Khurpade, D. Rao, and P. D. Sanghavi. A survey on iot and 5g network. In *2018 International Conference on Smart City and Emerging Technology (ICSCET)*, pages 1–3, Jan 2018. doi: 10.1109/ICSCET.2018.8537340.

Mihai T Lazarescu. Design of a wsn platform for long-term environmental monitoring for iot applications. *IEEE Journal on emerging and selected topics in circuits and systems*, 3(1):45–54, 2013.

Il-Gu Lee and Myungchul Kim. Interference-aware self-optimizing wi-fi for high efficiency internet of things in dense networks. *Computer Communications,*89-90:60–74, sep 2016. doi: 10.1016/j.comcom.2016.03.008.

L. Li, S. Bagheri, H. Goote, A. Hasan, and G. Hazard. Risk adjustment of patient expenditures: A big data analytics approach. In *2013 IEEE International Conference on Big Data*, pages 12–14, Oct 2013. doi: 10.1109/BigData.2013.6691790.

Li Li, Hu Xiaoguang, Chen Ke, and He Ketai. The applications of WiFi-based wireless sensor network in internet of things and smart grid. In *2011 6th IEEE Conference on Industrial Electronics and Applications*. IEEE, jun 2011. doi: 10.1109/iciea.2011.5975693.

Yang Li, Rui Tan, and David K. Y. Yau. Natural timestamps in powerline electromagnetic radiation. *ACM Transactions on Sensor Networks*, 14(2): 1–30, jul 2018. doi: 10.1145/3199676.

Z. Liu, Z. L. Jiang, X. Wang, Y. Wu, and S. M. Yiu. Multi-authority ciphertext policy attribute-based encryption scheme on ideal lattices. In *2018 IEEE Intl Conf on Parallel*

Distributed Processing with Applications, Ubiquitous Computing Communications, Big Data Cloud Computing, Social Computing Networking, Sustainable Computing Communications (ISPA/IUCC/BDCloud/SocialCom/SustainCom), pages 1003–1008, Dec 2018. doi: 10.1109/BDCloud.2018.00146.

A. Londhe and P. P. Rao. Platforms for big data analytics: Trend towards hybrid era. In *2017 International Conference on Energy, Communication, Data Analytics and Soft Computing (ICECDS)*, pages 3235–3238, Aug 2017. doi: 10.1109/ICECDS.2017.8390056.

Rongxing Lu, Kevin Heung, Arash Habibi Lashkari, and Ali A. Ghorbani. A lightweight privacy-preserving data aggregation scheme for fog computing-enhanced IoT. *IEEE Access*, 5: 3302–3312, 2017. doi: 10.1109/access.2017.2677520.

Y. Ma, Y. Chen, and J. Chen. Sdn-enabled network virtualization for industry 4.0 based on iots and cloud computing. In *2017 19th International Conference on Advanced Communication Technology (ICACT)*, pages 199–202, Feb 2017. doi: 10.23919/ICACT.2017.7890083.

P. Mach and Z. Becvar. Mobile edge computing: A survey on architecture and computation offloading. *IEEE Communications Surveys Tutorials*, 19(3):1628–1656, thirdquarter 2017. ISSN 2373-745X. doi: 10.1109/COMST.2017.2682318.

Somayya Madakam, R. Ramaswamy, and Siddharth Tripathi. Internet of things (IoT): A literature review. *Journal of Computer and Communications*, 03 (05):164–173, 2015. doi: 10.4236/jcc.2015.35021.

B. A. Martin,F. Michaud, D. Banks, A. Mosenia,R. Zolfonoon, S. Irwan, S. Schrecker, and J. K. Zao. Openfog security requirements and approaches. In *2017 IEEE Fog World Congress (FWC)*, pages 1–6, Oct 2017. doi: 10.1109/FWC.2017.8368537.

Kais Mekki, Eddy Bajic, Frederic Chaxel, and Fernand Meyer. A comparative study of LPWAN technologies for large-scale IoT deployment. *ICT Express*, 5(1):1–7, mar 2019. doi: 10.1016/j.icte.2017.12.005.

Mithun Mukherjee, Rakesh Matam, Lei Shu, Leandros Maglaras, Mohamed Amine Ferrag, Nikumani Choudhury, and Vikas Kumar. Security and privacy in fog computing: Challenges. *IEEE Access*, 5:19293–19304, 2017. doi: 10.1109/access.2017.2749422.

Geoff Mulligan. The 6lowpan architecture. In *Proceedings of the 4th workshop on Embedded networked sensors - EmNets '07*. ACM Press, 2007. doi: 10.1145/1278972.1278992.

R. Munoz, J. Mangues-Bafalluy, R. Vilalta, C. Verikoukis, J. Alonso-Zarate, N. Bartzoudis, A. Georgiadis, M. Payaro, A. Perez-Neira, R. Casellas, R. Martinez, J. Nunez-Martinez, M. Requena Esteso, D. Pubill, O. Font-Bach, P. Henarejos, J. Serra, and F. Vazquez-Gallego. The cttc 5g end-to-end experimental platform: Integrating heterogeneous wireless/optical networks, distributed cloud, and iot devices. *IEEE Vehicular Technology Magazine*, 11 (1):50–63, March 2016. ISSN 1556-6080. doi: 10.1109/MVT.2015.2508320.

C Siva Ram Murthy and BS Manoj. *Ad hoc wireless networks: Architectures and protocols, portable documents*. Pearson education, 2004.

NGMN. Description of network slicing concept. 2016.

Gianni Pasolini, Chiara Buratti, Luca Feltrin, Flavio Zabini, Cristina De Castro, Roberto Verdone, and Oreste Andrisano. Smart city pilot projects using LoRa and IEEE802.15.4 technologies. *Sensors*, 18(4):1118, apr 2018. doi: 10.3390/s18041118.

Uroš M Pešović, Jože J Mohorko, Karl Benkič, and žarko F Čučej. Single-hop vs. multi-hop–energy efficiency analysis in wireless sensor networks. In *18th Telecommunications Forum, TELFOR*, 2010.

S. Ramprasad, N. R. Shanbhag, and I. N. Hajj. Decorrelating (decor) transformations for low-power digital filters. *IEEE Transactions on Circuits and Systems II: Analog and Digital Signal Processing*, 46(6):776–788, June 1999. ISSN 1558-125X. doi: 10.1109/82.769785.

P.P. Ray. A survey on internet of things architectures. *Journal of King Saud University - Computer and Information Sciences*, 30(3):291–319, jul 2018. doi: 10.1016/j.jksuci.2016.10.003.

P. Rost, C. Mannweiler,D. S. Michalopoulos, C. Sartori,V. Sciancalepore,N. Sastry, O. Holland, S. Tayade, B. Han, D. Bega, D. Aziz, and H. Bakker. Network slicing to enable scalability and flexibility in 5g mobile networks. *IEEE Communications Magazine*, 55(5):72–79, May 2017. ISSN 1558-1896. doi: 10.1109/MCOM.2017.1600920.

David Schatsky, Craig Muraskin, and Ragu Gurumurthy. Demystifying artificial intelligence: what business leaders need to know about cognitive technologies. *A Deloitte Series on Cognitive Technologies*, 2014.

J Schlienz and D Raddino. Narrowband internet of things whitepaper. 2016.

Sigfox. Sigfox Device ARIB Mode White Paper. https://support.sigfox.com/docs/sigfox-device-arib-mode-white-paper, 2018a.

Sigfox. Sigfox Device ETSI Mode White Paper. https://support.sigfox.com/docs/sigfox-device-etsi-mode-white-paper, 2018b.

W. Sun, J. Liu, Y. Yue, and H. Zhang. Double auction-based resource allocation for mobile edge computing in industrial internet of things. *IEEE Transactions on Industrial Informatics*, 14(10):4692–4701, Oct 2018. ISSN 1941-0050. doi: 10.1109/TII.2018.2855746.

C. K. Toh. *Wireless ATM and ad-hoc networks*. Kluwer Academic Press, 1997.

Chai K Toh. *Ad hoc mobile wireless networks: protocols and systems*. Pearson Education, 2001.

Riccardo Trivisonno, Xueli An, and Qing Wei. Network slicing for 5g systems: A review from an architecture and standardization perspective. pages 36–41, 09 2017. doi: 10.1109/CSCN.2017.8088595.

International Telecommunication Union. Overview of the Internet of things. http://handle.itu.int/11.1002/1000/11559, 2016.

R. van Kranenburg and S. Dodson. *The Internet of Things: A Critique of Ambient Technology and the All-seeing Network of RFID*. Network notebooks. Institute of Network Cultures, 2008. ISBN 9789078146063. URL https://books.google.com/books?id=PilgkgEACAAJ.

C Vedat, Kerem Ok, and O Busra. Near field communication: From theory to practice. *Istanbuh NFC Lab-lstanbul, ISIK University*, pages 82–94, 2012.

R. Vilalta,A. Mayoral,D. Pubill, R. Casellas, R. Martínez, J. Serra, C. Verikoukis, and R. Muñoz. End-to-end sdn orchestration of iot services using an sdn/nfv-enabled edge node. In *2016 Optical Fiber Communications Conference and Exhibition (OFC)*, pages 1–3, March 2016.

J. Wang,J. Liu, and N. Kato. Networking and communications in autonomous driving: A survey. *IEEE Communications Surveys Tutorials*, 21(2):1243–1274, Secondquarter 2019a. ISSN 2373-745X. doi: 10.1109/COMST.2018.2888904.

P. Wang, C. Yao,Z. Zheng, G. Sun, and L. Song. Joint task assignment, transmission, and computing resource allocation in multilayer mobile edge computing systems. *IEEE Internet of Things Journal*, 6(2):2872–2884, April 2019b. ISSN 2372-2541. doi: 10.1109/JIOT.2018.2876198.

P. Wang, Z. Zheng, B. Di, and L. Song. Hetmec: Latency-optimal task assignment and resource allocation for heterogeneous multi-layer mobile edge computing. *IEEE Transactions on Wireless Communications*, 18(10):4942–4956, Oct 2019c. ISSN 1558-2248. doi: 10.1109/TWC.2019.2931315.

B. Yadranjiaghdam, N. Pool, and N. Tabrizi. A survey on real-time big data analytics: Applications and tools. In *2016 International Conference on Computational Science and Computational Intelligence (CSCI)*, pages 404–409, Dec 2016. doi: 10.1109/CSCI.2016.0083.

Zhenyu Yan, Yang Li, Rui Tan, and Jun Huang. Application-layer clock synchronization for wearables using skin electric potentials induced by powerline radiation. In *Proceedings of the 15th ACM Conference on Embedded Network Sensor Systems*. ACM, nov 2017. doi: 10.1145/3131672.3131681.

Yang Yang. Multi-tier computing networks for intelligent iot. *Nature Electronics*, 2, 01 2019. doi: 10.1038/s41928-018-0195-9.

Yang Yang, Hui hai Wu, and Hsiao hwa Chen. SHORT: Shortest hop routing tree for wireless sensor networks. In *2006 IEEE International Conference on Communications*. IEEE, 2006a. doi: 10.1109/icc.2006.255606.

Yang Yang, Feiyi Huang, Xuanye Gu, Mohsen Guizani, and Hsiao-Hwa Chen. Double sense multiple access for wireless ad hoc networks. In *Proceedings of the 3rd international conference on Quality of service in heterogeneous wired/wireless networks - QShine 06*. ACM Press, 2006b. doi: 10.1145/1185373.1185386.

Yang Yang, Jing Xu, Guang Shi, and Cheng-Xiang Wang. *5G Wireless Systems*. Springer International Publishing, 2018. doi: 10.1007/978-3-319-61869-2.

C. Yao,X. Wang,Z. Zheng, G. Sun, and L. Song. Edgeflow: Open-source multi-layer data flow processing in edge computing for 5g and beyond. *IEEE Network*, 33(2):166–173, March 2019. ISSN 1558-156X. doi: 10.1109/MNET.2018.1800001.

J. Yin and D. Zhao. Data confidentiality challenges in big data applications. In *2015 IEEE International Conference on Big Data (Big Data)*, pages 2886–2888, Oct 2015. doi: 10.1109/BigData.2015.7364111.

S. Yu, M. Liu, W. Dou, X. Liu, and S. Zhou. Networking for big data: A survey. *IEEE Communications Surveys Tutorials*, 19(1):531–549, Firstquarter 2017. ISSN 2373-745X. doi: 10.1109/COMST.2016.2610963.

Y. Yu. Mobile edge computing towards 5g: Vision, recent progress, and open challenges. *China Communications*, 13(Supplement2):89–99, N 2016. ISSN 1673-5447. doi: 10.1109/CC.2016.7833463.

Morteza M Zanjireh and Hadi Larijani. A survey on centralised and distributed clustering routing algorithms for wsns. In *2015 IEEE 81st Vehicular Technology Conference (VTC Spring)*, pages 1–6. IEEE, 2015.

Morteza M Zanjireh, Ali Shahrabi, and Hadi Larijani. Anch: A new clustering algorithm for wireless sensor networks. In *2013 27th International Conference on Advanced Information Networking and Applications Workshops*, pages 450–455. IEEE, 2013.

S. Zhang. An overview of network slicing for 5g. *IEEE Wireless Communications*, 26(3):111–117, June 2019. ISSN 1558-0687. doi: 10.1109/MWC.2019.1800234.

Y. Zhang, B. Di,P. Wang, J. Lin, and L. Song. Hetmec: Heterogeneous multi-layer mobile edge computing in the 6g era. *IEEE Transactions on Vehicular Technology*, pages 1–1, 2020. ISSN 1939-9359. doi: 10.1109/TVT.2020.2975559.

Cheng Zhao, Wuxiong Zhang, Yang Yang, and Sha Yao. Treelet-based clustered compressive data aggregation for wireless sensor networks. *IEEE Transactions on Vehicular Technology*, 64(9):4257–4267, sep 2015. doi: 10.1109/tvt.2014.2361250.

C. Zhong, Z. Zhu, and R. Huang. Study on the iot architecture and gateway technology. In *2015 14th International Symposium on Distributed Computing and Applications for Business Engineering and Science (DCABES)*, pages 196–199, Aug 2015. doi: 10.1109/DCABES.2015.56.

2

Computing and Service Architecture for Intelligent IoT

2.1 Introduction

Due to more data and more powerful computing power and algorithms, IoT applications are becoming increasingly intelligent, which are shifting from simple data sensing, collection and representation tasks towards complex information extraction and analysis. Diverse intelligent IoT applications demand different levels of intelligence and efficiency in processing data. For example, intelligent agriculture applications care more about low-power sensors, while smart driving applications require powerful computing capabilities at device levels and low-latency communications with neighboring equipment. Traditional computing and service architecture based on cloud computing can not satisfy diverse demands of different intelligent IoT applications, and multi-tier computing network architecture coined by cloud, fog and edge computing will be required for intelligent IoT (Yang, 2019). In Section 2.1, the multi-tier computing network architecture for intelligent IoT applications is presented, along with two important frameworks, i.e. Cost Aware Task Scheduling (CATS) (Liu et al., 2020) and Fog as a Service Technology (FA^{2}ST) (Chen et al., 2018a). Multi-tier computing network architecture is a computing architecture, which advocates active collaborations between cloud, fog and edge computing technologies for intelligent and efficient data processing at different levels and locations. It comprises not only Computing, Communication and Caching (3C) resources but also a variety of embedded AI algorithms along the cloud-to-things continuum, thus supporting different intelligent IoT applications demanding different levels of intelligence and efficiency in processing data. To underpin such a multi-tier computing network architecture, two important frameworks, i.e. CATS and FA^{2}ST is developed. Specifically, CATS is an effective resource sharing framework that utilizes a practical incentive mechanism to motivate efficient collaboration and task scheduling across heterogeneous resources at multiple devices, edge/fog nodes and the cloud, which are probably owned by different individuals and operators. While FA^{2}ST is a flexible service provisioning frame work that is able to discover, orchestrate, and manage micro-services and cross-layer 3C resources at any time, anywhere close to end users, thus guaranteeing high-quality services under dynamic network conditions. In Section 2.2 and Section 2.3, two intelligent application scenarios and the corresponding technical solutions are described in detail, as illustrative case studies of the multi-tier computing network architecture. Firstly, in Section 2.2, an on-site cooperative Deep Neural Network (DNN) inference framework, which is based on edge computing, is proposed to execute DNN inference tasks

Intelligent IoT for the Digital World: Incorporating 5G Communications and Fog/Edge Computing Technologies,
First Edition. Yang Yang, Xu Chen, Rui Tan, and Yong Xiao.

with low latency and high accuracy for industrial IoT applications, thus meeting the strict requirements on service delay and reliability. Secondly, a three-tier collaborative computing and service framework, which is based on fog computing, is proposed to support dynamic task offloading and service composition in simultaneous localization and mapping(SLAM) for a robot swarm system, which requires timely data sharing and joint processing among multiple moving robots. Both cases are implemented and evaluated in real experiments, and a set of performance metrics demonstrates the effectiveness of the proposed multi-tier computing network and service architecture in supporting intelligence IoT applications in stationary and mobile scenarios.

2.2   Multi-tier Computing Networks and Service Architecture

Over the past decades, computing networks have evolved from distributed computing (e.g., grid computing) to centralized cloud computing. And now, the pendulum swings back. Cloud computing is an architecture with centralized resources and management in the cloud, referring to data centers, backbone IP networks, and cellular core networks (Mao et al., 2017). Due to centralized resources and management, it gives end devices and consumers elastic on-demand resource allocation, reduced management effort, flexible pricing models (pay-as-you-go), and easy application and service provisioning (Mouradian et al., 2017). However, the main advantage of cloud computing – consolidated resources and management – is also its main weakness. Centralized resources and management mean that functions and controls are located far from where tasks are generated. Due to the long physical distance, limited communication bandwidth, intermittent network connectivity, etc., cloud computing alone cannot meet the requirements of many delay-sensitive applications in 5G, such as automatic driving. Aimed at providing cloud computing capabilities at the edge of the network, in 2014, the Industry Specification Group (ISG) within the European Telecommunication Standards Institute (ETSI) proposed mobile edge computing (MEC). As defined by ETSI, MEC is a network architecture that provides IT and cloud-computing capabilities within radio access networks in close proximity to mobile subscribers (Hu et al., 2015). In March 2017, the ETSI expanded the scope of MEC and after that replaced the term "mobile" with "multi-access" (Mouradian et al., 2017). The edges of non-mobile networks are also being considered in multi-access edge computing (MEC). Fog computing, first initiated by Cisco, was further promoted by the OpenFog Consortium to extend and generalize edge computing (Chiang and Zhang, 2016). The OpenFog Consortium defines fog computing as "a system-level horizontal architecture that distributes resources and services of computing, storage, control and networking anywhere along the continuum from cloud to things" (OpenFog Consortium, 2017). A key difference between MEC and fog computing is that MEC functions only in stand-alone mode, while fog computing has multiple interconnected layers and could interact with the distant cloud and the network edge.

On the other hand, due to more data and more powerful computing power and algorithms, IoT applications are becoming increasingly intelligent, and are shifting from simple data sensing, collection, and representation tasks towards complex information extraction and analysis. In the future, different intelligent IoT applications will demand different levels of intelligence and efficiency in processing data. For example, smart cars and drones require low-latency communications for receiving monitoring data and control messages. Autonomous driving and three-dimensional virtual-reality games need powerful computing capabilities at device and edge levels, as well as broad communication bandwidths. Industrial IoT and smart city management systems demand extreme network reliability, data security and service availability. To support future intelligent IoT services, the multi-tier computing network (Yang, 2019), which integrates cloud, fog, and edge computing technologies, is proposed in this chapter. In addition, to underpin the multi-tier computing network infrastructure and architecture, CATS (cost aware task scheduling) task scheduling framework (Liu et al., 2020), which deals with efficient resource sharing and efficient task scheduling within multi-tier networks, and FA2ST (fog as a service technology) service architecture (Chen et al., 2018a), which supports dynamic service management, are developed as fundamental service architectures or functionalities.

2.2.1 Multi-tier Computing Network Architecture

A multi-tier computing network is a computing architecture involving collaborations between cloud computing, fog computing, edge computing, and sea computing (Jiang, 2010) technologies, which have been developed for regional, local, and device levels, respectively, as shown in Figure 2.1 (Yang et al., 2020a).

Cloud computing, edge computing, and fog computing are not competitive but interdependent. They complement each other to form a service continuum, and fog is the bridge that connects centralized clouds and distributed edges of the network. For example, together with the edge, fog ensures timely data processing, situation analysis, and decision making at the locations close to where the data is generated and should be used. Together with the cloud, fog supports more intelligent applications and sophisticated services in different industrial verticals and scenarios, such as cross-domain data analysis, pattern

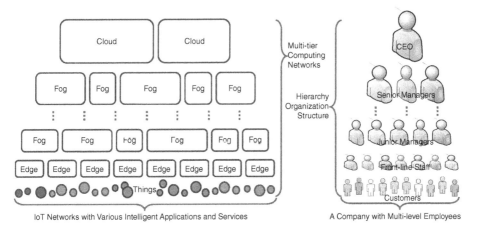

Figure 2.1 A multi-tier computing network architecture. The architecture integrates cloud, fog, edge, and sea computing technologies (Yang et al., 2020a).

recognition and behavior prediction. Thus, the integration of these different computing technologies and resources is vital for the development of intelligent IoT applications.

Multi-tier computing networks can be thought of as a large company with a top-down, multi-tier organization structure: managers and employees at different levels in the company have different resources, capabilities and responsibilities in terms of data access and processing, task assignment, customer development, and decision making, as shown in Figure 2.1. Cloud computing is equivalent to the top hierarchical level in the company, possessing the most information sources, the strongest analytical intelligence, the maximum storage space and the highest decision-making authority. As such, cloud computing is expected to handle challenging tasks at the global level, such as cross-domain data analysis and processing, abnormal behavior diagnosis and tracing, hidden problem prediction and searching, new knowledge discovery and creation, and long-term strategic planning and decisions. Edge computing, on the other hand, is equivalent to front-line staff, having the least resources and capabilities but being able to directly interact with customers in different application domains. Therefore, edge computing is good at handling delay-sensitive tasks at the local level, such as data collection, data compression, information extraction, and event monitoring.

Between the cloud and the edge within the network, there is fog computing, which is equivalent to mid-level management in the company. Like an efficient management system with many levels of resources, duties and responsibilities, fog computing is a hierarchy of shared computing, communication, and storage resources that can collaboratively handle complex and challenging tasks at the regional level, such as cross-domain data analysis, multi-source information processing and on-site decision making for large service coverage. Because user requirements are usually dynamic in terms of time and space, fog computing can provide a flexible approach to incorporating distributed resources at different geographical or logical locations in the network, thus offering timely and effective services to customers.

The devices, or things, of the IoT network are equivalent to the customers of the company that have numerous different requests and demands for intelligent applications and services. Each device has limited processing, communication, storage, and power resources. However, collectively, they contribute to the concept of sea computing at the device level, which supports data sensing, environment cognition, mobility control, and other basic functions for individual things in real-time.

To make the company a success, it is essential to have effective communication and collaboration mechanisms across different levels and units. Similarly, interaction and collaboration between cloud, fog, edge, and sea computing are vital in order to create an intelligent collaborative service architecture. By actively connecting the shared computing, communication, and storage resources of all the nodes in the IoT network and fully utilizing their capabilities at different locations and levels, the multi-tier computing network can intelligently and efficiently support a full-range of IoT applications and services in a timely way.

To this end, in the following sections, we first introduce the cost aware task scheduling framework of CATS as a fundamental service architecture built on the multi-tier computing network infrastructure. CATS not only supports efficient collaborative task scheduling across multi-tier resources, but also motivates the effective resource sharing within multi-tier computing networks. Furthermore, to combat the dynamics of multi-tier computing networks in nature, we also develop FA2ST service architecture for the cross-domain collaborative resources management under dynamic user requirements and network environments.

2.2.2 Cost Aware Task Scheduling Framework

Cost aware task scheduling (CATS) is a cost aware task scheduling framework customized for multi-tier computing networks, which can efficiently motivate resource sharing among multi-tier resources, and effectively schedule tasks from end user equipment (UE) to multi-tier resources to reduce cost (e.g., delay, energy, and payment) of processing tasks. It supports flexible cross-layer task scheduling along the cloud-to-things continuum, aiming at reducing the cost of processing tasks from the view of UE. The main features of CATS include (1) a computation load based hybrid payment model to motivate cloud and fog to share resources and to serve UE, (2) a weighted cost function mapping delay, energy and payment into one single unit, and (3) a distributed task scheduling algorithm with near-optimal performance guarantee (Liu et al., 2020).

As illustrated in Figure 2.2, a multi-tier computing network typically consists of multiple UE, multiple proximal fog nodes (FNs) and a remote cloud, from bottom to top. The cloud, together with the fog nodes, provides the cloud computing-like services to UEs and charges for this. The cloud is operated by the cloud service provider, such as Amazon, Microsoft, Alibaba, while the fog nodes are operated by the fog service providers, like telecom operators (telecom operators can provide storage, computing and other cloud computing-like services to UE by installing servers within or in the proximity of base stations or access points). It is worth noting that only the scenario of single cloud service provider and single fog service provider is supported in the present CATS framework; the more complicated scenario with multiple cloud service providers and/or multiple fog service providers, i.e. a multi-tenant problem, is for further study.

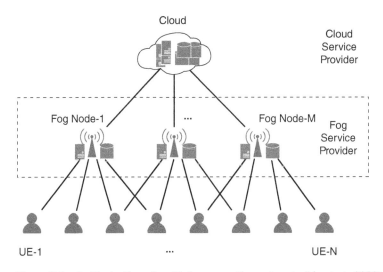

Figure 2.2 An illustration of multi-tier computing networks (Liu et al., 2020).

For the consideration of delay and/or energy consumption, UE can offload their computing tasks to fog nodes or cloud for processing, which depends on the quality of service (QoS) provided by them, say the delay and energy consumption, as well as the money paid to the cloud and fog nodes. Generally, the cloud possesses powerful computation capability, but is far away from the end UE, and thus the communication cost is high. While the fog nodes are typically close to end UE and thus the communication cost is comparatively low, the computation capability is limited. Therefore, how to effectively schedule tasks within multi-tier computing networks is a critical challenge, given different features of computing resources and diverse requirements of tasks. In addition, how to motivate the multi-tier resources is also a key problem, which is the premise of the formation of multi-tier computing networks. To solve these challenges, in this section, we propose CATS.

2.2.2.1 Hybrid Payment Model

Since the cloud and fog nodes will consume their own resources and energy when processing tasks from UEs, they need to be motivated to share resources and thus participate in the computation offloading. Generally, for nodes in the same layer, if they can offload tasks to each other, then a "tit-for-tat" incentive mechanism can be adopted, where one exploits more resources from the others only if it contributes more resources to them (Pu et al., 2016; Yang et al., 2018). For nodes across different layers, the price-based incentive mechanisms are usually adopted to advocate resource sharing and computation offloading among each other (Nguyen et al., 2018; Zhang and Han, 2017; Liu et al., 2017; Zhang et al., 2017), which may be the most straightforward way. Thus, in the CATS framework, the latter one is adopted, i.e. the UE will pay for offloading tasks to the cloud and fog nodes.

Generally speaking, the cost of cloud and fog nodes for processing tasks increases with the computation load of tasks, such as energy consumption. Thus, the money paid to the cloud and fog nodes should also increase. However, for a single fog node, the delay and energy consumption experienced by a single piece of UE would get worse if the UE served by the fog node got more. That's because the resources allocated to a single piece of UE will decrease with an increase in UE served by the same fog node, i.e. resource competition and service congestion. While for the cloud, since the resources of cloud are usually adequate, the QoS for every UE can be guaranteed according to the service level agreement (SLA). Therefore, to compensate for the UE, the fog nodes are willing to give a discount for the decreased QoS.

For the convenience of presentation and analysis, we denote the set of UE, the set of fog nodes and the cloud by $\mathcal{N} = \{1, \cdots, N\}$, $\mathcal{M} = \{1, \cdots, M\}$ and c, respectively. In the CATS framework, the money paid to fog node m by UE n is modeled as $C_{n,m} = c^{\mathrm{fog}}(1 - \alpha(n_m - 1))z_n\gamma_n$. c^{fog} is a constant, which means the price per unit computation amount (in CPU cycles) charged by the fog node or fog service provider. α is the discount ratio. n_m is the amount of UE served by fog node m. z_n is the data size of task n (in bits)[1], and γ_n is the is the processing density of task n, i.e. the CPU cycles required to process a unit bit of data. The product of z_n and γ_n is the computation amount of task n. Correspondingly, the money charged by the cloud can be presented as $C_{n,c} = c^{\mathrm{cloud}}z_n\gamma_n$. Similarly, c^{cloud} is the

---

1 We use task n and the task of UE n interchangeably.

price charged by the cloud or cloud service provider for unit computation amount. The payment models for the fog nodes and cloud are termed as the dynamic payment model and static payment model, respectively.

From the viewpoint of business, the different modeling above can also be explained as follows. The cloud service providers are generally large companies with mature markets and weak bargaining power of individual users, and thus their payment model is relatively fixed. While the fog service providers are typically small start-up companies, being the new market players, users have comparatively stronger bargaining power against them, and thus their payment model can be dynamically adjusted. In order to grab market share from the cloud service providers, the fog service providers are willing to adopt the price reduction strategy to attract more users so as to increase the final gross revenue.

2.2.2.2 Weighted Cost Function

CATS aims at not only motivating the resource sharing among multi-tier resources, but also reducing the delay and energy consumption of task processing by flexibly offloading them to fog nodes and/or cloud. Thus, except for the hybrid payment model aforementioned, delay model and energy consumption model also play a key role in CATS. However, due to the complicated physical world, such as network environments, electronic circuits and so on, it is quite difficult to exactly model the delay and energy consumption. Therefore, based on various theoretical models in the literature, we propose the average based models to approximate the delay and energy consumption, involving the communication aspect and computation aspect.

Delay model

The delay mainly consists of communication related delay and computation related delay. Generally, the communication delay can be calculated by dividing the data size by the data rate, and the computation delay can be approximated by dividing the computation amount by the computation capability. Specifically, if we denote the data rate by R (in bits per second) and computation capability by f (in CPU cycles per second, i.e. CPU clock frequency), the communication/computation delay of transmitting/computing a task with data size z and processing density γ can be represented as z/R and $z\gamma/f$, respectively.

However, for different task processing modes, more elaborate processing procedure and resource allocation need to be considered and handled. To be specific, for local computing, since no communication is necessary, the final delay of task processing equals to the local computation delay. Further, because the computation capability of local device can be monopolized by the local task, no resource allocation needs to be considered, i.e. $f = f_n$, where f is as defined above and f_n is the computation capability of UE n. Thus, for local computing, the delay model is as $T_{n,l} = z_n\gamma_n/f_n$.

For fog node computing, i.e. offloading tasks to fog nodes for processing, both communication and computation are involved. Since the fog node servers are typically colocated with BSs or APs, the final delay of task processing can be calculated by simply adding the two parts. However, due to the constraints of communication resources and computation capabilities of a single fog node, the resource allocation needs to be considered, among UEs offloading tasks to the same fog node. For simplicity, here, the average bandwidth sharing and the average computation resource allocation are adopted, which can be used to model

TDMA and OFDMA MAC protocols and queuing based task processing procedure (Jošilo and Dán, 2017, 2018). That is, $R = R_{n,m}/n_m$ and $f = f_m/n_m$, where R and f are as defined above. $R_{n,m}$ is the data rate between UE n and fog node m when the channel is monopolized by UE n, f_m is the computation capability of fog node m, and n_m is the amount of UE served by fog node m, as mentioned above. In conclusion, for fog node computing, the delay can be formulated as $T_{n,m} = n_m z_n/R_{n,m} + n_m z_n \gamma_n/f_m{}^2$.

For cloud computing, because UEs connect to the cloud servers via the wide area network and the cloud servers are typically with adequate computation capabilities, the data rate and allocated computation capability can be guaranteed for every UE/task according to the SLA and estimated by the average data rate $R_{n,c}$ and the average computation capability f_c in a time interval. In addition, since the cloud servers are generally far away from BSs or APs, the extra round trip delay t_n^{rt} between BSs/APs and cloud servers needs to be considered. As a result, for cloud computing, the delay is modeled as $T_{n,c} = z_n/R_{n,c} + z_n \gamma_n/f_c + t_n^{rt}$.

Energy consumption model

Once the delay models have been established, the energy consumption models can be formulated as the product of power and time (delay). To be specific, for communication energy consumption, we denote the transmitting power of UE n for the fog node computing and the cloud computing as p_n^{fog} and p_n^{cloud}, respectively. Then, the corresponding transmitting energy consumption is $E_{n,m}^{trans} = p_n^{fog} n_m z_n/R_{n,m}$ and $E_{n,c}^{trans} = p_n^{cloud} z_n/R_{n,c}$, respectively. For computation energy consumption, according to the circuit theory, the CPU power consumption can be divided into several factors including the dynamic, short circuit and leakage power consumption, where the dynamic power consumption dominates the others (Burd and Brodersen, 1996; Yuan and Nahrstedt, 2006; De Vogeleer et al., 2013; Zhang et al., 2013). In particular, it is shown that the dynamic power consumption is proportional to the product of the square of circuit supplied voltage and CPU clock frequency (De Vogeleer et al., 2013). It is further noticed that the clock frequency of the CPU chip is approximately linear proportional to the voltage supply when operating at the low voltage limits (Burd and Brodersen, 1996; Zhang et al., 2013). Thus, the power consumption of a CPU cycle is given by κf^3, where κ is a constant related to the hardware architecture. As a result, the local computation energy consumption of task n can be written as $E_{n,l}^{comp} = \kappa_n z_n \gamma_n f_n^2$ (Mao et al., 2017; Zhang et al., 2013).

Weighted cost function

Delay and energy consumption, together with the monetary payment, make up the gross cost of UEs for processing tasks. However, these metrics are with different units, and thus we need to map them into a unified unit. To this end, we construct a weighted cost function as a weighted linear combination of delay, energy consumption and monetary payment. To be specific, denote the weights of delay, energy consumption and monetary payment by λ_n^T, λ_n^E and λ_n^C, for every UE. $0 \leq \lambda_n^T, \lambda_n^E, \lambda_n^C \geq 1$ The gross cost of UE n is defined as $O_n = \lambda_n^T T_n + \lambda_n^E E_n + \lambda_n^C C_n$. Such a definition has a definite physical explanation and can flexibly apply to different application requirements. For example, if UE n is a delay-sensitive

---

2 In this formulation, the delay of transmitting results back to UEs is ignored, although it is supported by CATS framework. It is the same for the cloud computing.

user, then λ_n^{T} will be relatively larger. Similary, if UE n cares more about energy consumption, then λ_n^{E} will increase.

2.2.2.3 Distributed Task Scheduling Algorithm

Based on the hybrid payment model and weighted cost function above, we develop a distributed task scheduling algorithm to minimize the gross cost of every UE. To be specific, consider a slotted time structure, and every task scheduling time slot is further divided into multiple decision slots. To begin with, every UE chooses local computing, and the cloud and fog nodes broadcast the delay, energy consumption and payment related parameters to UEs. Then, at every decision slot:

Fog nodes broadcasting the workload status to UEs
The fog nodes broadcast their current workload status, i.e. n_m, to UEs.

UEs determining the best task processing mode
Based on the workload status information from fog nodes, the UE measures the channel conditions and then determine the best task processing mode, i.e. local computing, fog node computing and cloud computing.

UE contending for the update opportunity of task processing mode
The UE sends the request to the fog nodes or cloud to contend for the opportunity to update the task processing mode. The requests will be aggregated at a centralized controller, which is responsible for determining the UE winning the update opportunity.

UE updating the task processing mode
The centralized controller will broadcast the decision to the UEs. Only the UE permitted can update the task processing mode, while other UE will maintain their current task processing mode.

If no UE wants to update the current task processing mode, the centralized controller will broadcast the END message to all UE and the algorithm terminates (Liu et al., 2020).

2.2.3 Fog as a Service Technology

The FA2ST is a micro-service based service framework that enables cross-layer resource discovery, fine-grained resource management, and dynamic service provision for fog as-a-service in dynamic IoT environments. It supports a cloud-to-things service continuum from service providers' perspective by a fog choreography system that concordances different communication, computation, and storage resources to provide and manage IoT services for varying application domains. The main features of FA2ST include (1) a dynamic fog network that manages service deployment and localizes resources for an efficient transition process, and (2) an adaptable service management model to wrap diverse resources and enable fine-grained service provisioning across multiple resource and application domains.

As illustrated in Figure 2.3(right), the FA2ST system enables microservices on multiple shared autonomous fog nodes, denoted by transparent ellipses. Fog nodes can differ

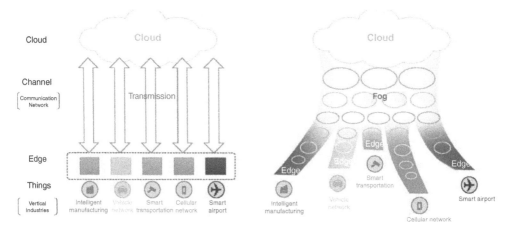

Figure 2.3 The cloud–edge–things system (left), the FA2ST system (right) (Chen et al., 2018a).

in their capacity and specificity, and resources on them are not restricted to any particular application domain. Such fog nodes are linked together to make a fog network. Authorized third-party fog nodes are allowed to join the fog network. FA2ST proposes fog choreography to manage microservices and provide value-added fog services through a decentralized service composition model. Service discovery and composition are enabled in the fog network. During the execution of a fog service, the participating fog nodes will collaborate with each other to identify feasible microservices for meeting the end users' specific requirements (denoted by colored branches). In this process, the fog nodes and the microservices on them, which are required to support an application, are composed on-demand and released for future reuse after current tasks.

FA2ST provides holistic end-to-end fog service provisioning, with a fine-grained allocation of control logic, functionalities, information, and resources based on their location, granularity, capability, and so on. Service integration in the FA2ST system can happen anywhere, in the cloud or on the fog nodes closer to the users. For example, an autonomous car will need local services from multiple application domains to make driving decisions, such as obtaining information about surrounding vehicles from onboard sensors and nearby vehicles; road and traffic conditions from the smart transportation systems; and weather conditions from the external weather applications. In this case, FA2ST allows services to be provided by different fog nodes, including in-vehicle fog nodes, road-side fog nodes, and weather station fog nodes. Therefore, FA2ST enables scalable, flexible, multi-level cross-domain resource and service management.

2.2.3.1 FA2ST Framework

FA2ST is realized as a framework that consists of three layers to meet user requirements on resource availability, architectural scalability, system interoperability, and service flexibility, which are FA2ST infrastructure, FA2ST platform, and FA2ST based software layer, as shown in Figure 2.4 (Chen et al., 2018a).

The FA2ST infrastructure manages different types of devices, maintaining network connectivity among them, and visualizing the computing, networking and storage resources.

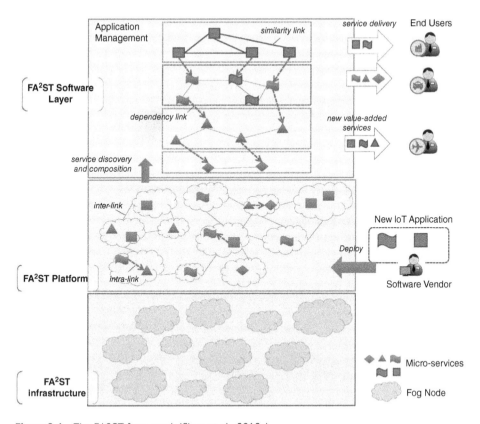

Figure 2.4   The FA2ST framework (Chen et al., 2018a).

Software and hardware resources on each device are abstracted as a fog node that is made available to the FA2ST platform users. The FA2ST platform organizes the fog nodes and maintains a network of these fog nodes ("fog network") to provide a horizontal computing platform across different networks, devices, and semantic domains. It allows software vendors to deploy IoT applications over the distributed FA2ST infrastructure, which includes middleware, operating systems, and a node management module to cope with heterogeneity problems to allow software vendors to concentrate on dynamic user demands and their application's functionality and behavior.

The FA2ST software layer is the core of the FA2ST framework. It deploys and manages microservices to meet IoT requirements. An IoT application is normally a composition of multiple functionalities that satisfy user demands and can be supported by a set of microservices. For dynamic user demands, the fog nodes will cooperatively resolve the functionalities on the fly. The FA2ST software layer allows each fog node to discover the needed microservices and manage their composition and invocation. It flexibly composes services in a user's local computing environment to enable and guarantee seamless application execution even when the computing environment changes. In other words, when an end user moves to a new environment or when a service becomes unavailable, the FA2ST software layer supports service re-composition to ensure application continuity.

Figure 2.4 shows how different layers in the FA2ST framework work together. The FA2ST infrastructure organizes fog nodes and opens their computing, networking, sensing, and storage resources to the IoT environment. IoT software vendors implement software functionalities and deploy microservices on appropriate fog nodes through the FA2ST platform, and the platform manages the computing and storage resources for those microservices. The FA2ST software layer hides an application's implementation details and provides service APIs (e.g., service description) to make the software functionalities accessible. Microservices are selected and composed at runtime to satisfy end users' software requirements. The FA2ST software layer also monitors the execution of the composed microservices and adapts the composition to any environmental change or users' change on their requirements that occurs during the execution.

2.2.3.2 FA2ST Application Deployment

FA2ST application deployment is a two-step process. It includes a vendor-driven deployment and an on-demand rc-deployment, as shown in Figure 2.5 (Chen et al., 2018a). In the vendor-driven deployment step (Figure 2.5 (a)), a software vendor uploads an application's microservices on the FA2ST platform and provides a profile of potential clients to the system. The profile indicates essential non-functional service requirements like the desired QoS, the service coverage areas, and the service cost, just to name a few. The node

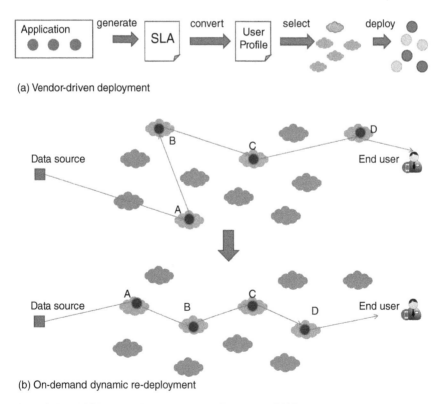

(a) Vendor-driven deployment

(b) On-demand dynamic re-deployment

Figure 2.5 FA2ST application deployment (Chen et al., 2018a).

management module on the FA2ST platform further selects a particular set of fog nodes according to the profile and deploys several copies of the microservices on the fog nodes and makes them easy to be accessed by target clients. Once the microservices are available for use, the system starts to adapt the microservice's host nodes according to the runtime service demands and the actual QoS to optimize the static deployment. In the on-demand re-deployment step [Figure 2.5(b)], when a user requests an application, a service request will be sent over the fog network and will be resolved cooperatively by the fog nodes. The QoS will be calculated and recorded by these fog nodes. The FA2ST system uses the historical data to check if there exists a microservice that has become the performance bottleneck and decides whether or not to re-deploy it. Once a re-deployment is necessary, the current hosting fog node selects a new host in its vicinity, deploys the microservice and releases its own resources that were previously allocated to the microservice.

As one application's microservices may not all be deployed on the same fog node, there exists heterogeneity in communications or local data types among the microservices' providers. To deal with the heterogeneity, the service invocation in the FA2ST system is not restricted to a specific application workflow. We invoke microservices for applications through dynamic service composition and compose appropriate services or data converters to deal with heterogeneity. This also enables new value-added services to be provided to end users.

2.2.3.3 FA2ST Application Management

In the FA2ST, applications are managed as a service overlay by the FA2ST software layer. We leverage a dynamic service discovery and composition model (Chen and Clarke, 2014) to efficiently discover services over the fog network and extend it to realize cross-application service management. As shown in Figure 2.4, we define similarity links and dependency links between services. The services that provide the same functionality and have physical connections are linked by a similarity link, and a dependency link will be built if a service's execution depends on another service's execution result. We further define inter-links and intra-links between services: an inter-link connects services from different fog nodes, and an intra-link connects services on the same fog node. If a microservice is re-deployed in another fog node, the links to the corresponding microservices will change accordingly. This model allows fog nodes to share their services, data or resources in their vicinity without registration. The admission control of a service discovery process migrates from one fog node to another through inter-links when providing services for an application, and the service discovery within a fog node uses intra-links to find out a series of microservices through their similarity or dependency relations. During execution, participating fog nodes select the remaining execution path and invoke each subsequent service. In other words, FA2ST merges the service selection and binding process with the service invocation process to select and bind a service only when it is going to be used immediately, which reduces interactions among fog nodes while keeping service composition flexible during the service execution process (Chen et al., 2016). This prevents the system from binding a service that may no longer be available when it is about to be invoked and ensures the system always selects the best service (a.k.a., a service with the best QoS) in the remaining execution.

In the following sections, two illustrative application case studies are given. One is the edge intelligence for industrial IoT, which deploys the deep learning algorithms on multi-tier computing networks and utilizes the task scheduling framework to realize the near real-time industrial IoT applications. The other one is collaborative SLAM of robot swarm, which implements the collaborative SLAM of moving robot swarm under the FA2ST service framework.

2.3 Edge-enabled Intelligence for Industrial IoT

2.3.1 Introduction and Background

2.3.1.1 Intelligent Industrial IoT

In the big data era, industrial companies are confronted with market pressures in managing both product quality and manufacturing productivity. Many industrial companies, however, are not prepared to deal with big data yet, even though they have understood the value of intelligent transformation already. For companies who lack intelligent assisted tools, the challenge lies not in weighing the benefit of industrial data but in employing executable methods on realistic data processing. Now, as the revolution of integration with innovative information and communication technology and industry, referred to as smart industry, the companies meet the opportunity to reorganize their manufacturing more effectively as well as efficiently.

Smart industry, also known as Industry 4.0, describes the industrial revolution of automation and data exchange in manufacturing technology, directing a tendency of combining artificial intelligence (AI) and traditional industrial manufacture (Bisio et al., 2018). A significant feature in smart industry is the tremendous amount of data generated by the widespread IoT devices, which is essential for industrial intelligence. The data is sensed by various kinds of sensors during the manufacturing process, such as product scanned images from product surface inspection and vibration signal from machinery defect diagnosis (Wang et al., 2018). Using these data, intelligent approaches are able to perform a propulsive role for the automation in the manufacturing process.

Among AI approaches driven by big data, deep learning (DL) is the most popular one, which has been applied in a broad spectrum of scenarios (Sze et al., 2017), varying from voice control in a smart home to cancer detection in an intelligent hospital. The superior performance of the DNN model comes from the ability to extract high-level features from raw data, over which a large scale of mathematic calculation is conducted to acquire the high-level representation of the input space. For this reason, while DNN-based approaches process a tremendous amount of raw data, they require a tremendous amount of computation. Today's IoT devices in smart factories, however, fail to afford large-scale computation within reasonable processing latency and energy consumption (Wang et al., 2016). Towards this problem, traditional wisdom resorts to powerful cloud data centers for heavy computation such as DNN inference. Specifically, raw sensory data generated in factories is sent to the cloud data center for processing, along with inference results being sent back to the factories once it finished. Although this cloud-centric approach releases IoT device loading, it may cause high end-to-end latency and heavy energy consumption because of the heavy transmission between the IoT devices and the remote cloud data center, which is a negative for efficiency pursuance (Chen et al., 2018b).

2.3.1.2 Edge Intelligence

To alleviate the latency and energy bottlenecks, the edge computing paradigm based on the multi-tier computing network can be introduced to smart industry. As Figure 2.6(a) shows, consider the scenario where a surface inspection camera is set on the automated assembly line. The products are flowing on the automated assembly line, waiting for inspection one by one. Once a product arrives at the right position, the surface of it will be captured by the inspection camera, an IoT device with limited computation capability. Then, the images will

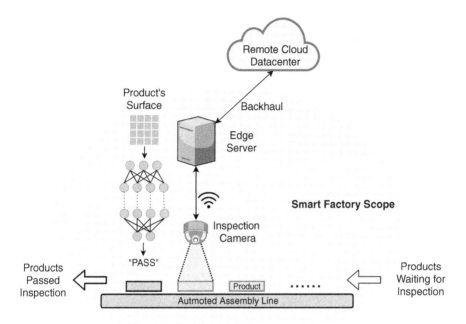

(a) Product surface inspection powered by edge intelligence

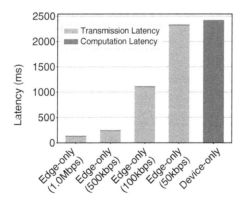

(b) AlexNet latency breakdown

Figure 2.6 Example scenario: edge intelligence enabled product surface inspection.

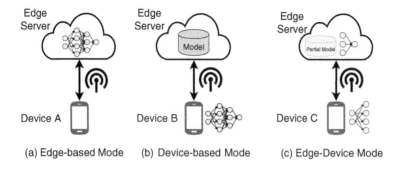

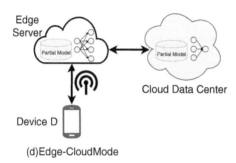

Figure 2.7 Major edge-centric inference modes: edge-based, device-based, edge-device and edge-cloud.

be uploaded to the edge servers for near real-time inference and analysis utilizing the DL model deployed on the edge servers. The images may also be transmitted to the cloud servers for further inference and analysis. Edge computing pushes the cloud capabilities from the remote network core to the network edges close to end devices, enabling a low-latency and energy-efficient inference task processing.

In general, edge-centric inference approaches can be classified into four modes, namely edge-based, device-based, edge-device and edge-cloud, as illustrated in Figure 2.7. Specifically, we demonstrate four different DNN model inference modes in Figure 2.7, and describe the main workflow of each mode as follows. It is worth noting that the four edge-centric inference modes can be adopted in a system simultaneously to carry out complex AI model inference tasks (e.g., cloud–edge–device hierarchy in the multi-tier computing network), by efficiently pooling heterogeneous resources across a multitude of end devices, edge nodes and cloud.

- **Edge-based**: In Figure 2.7(a), device A is in the edge-based mode, where the device receives the input data and then sends it to the edge server. After the DNN model inference is finished at the edge server, the prediction results will be returned to the device. Under this inference mode, since the DNN model is on the edge server, it is easy to implement the application on different mobile platforms. Nevertheless, one main disadvantage is that the inference performance depends on network bandwidth between the device and the edge server.

- **Device-based**: In Figure 2.7(b), device B is in the device-based mode, where the mobile device obtains the DNN model from the edge server and performs the model inference locally. During the inference process, the mobile device does not communicate with the edge server. To achieve reliable inference, a large amount of resources are required such as CPU, GPU, and RAM on the mobile device. The performance depends on the local device itself.
- **Edge-device**: In Figure 2.7(c), device C is in the edge-device mode, where the device first partitions the DNN model into multiple parts according to the current system environmental factors such as network bandwidth, device resource and edge server workload, and then executes the DNN model up to a specific layer with the intermediate data sent to the edge server. The edge server will next execute the remain layers and send the prediction results back to the device. Compared to the edge-based mode and the device-based mode, the edge-device mode is more reliable and flexible. It may also require a significant amount of resources on the mobile device because the convolution layers at the front position of a DNN model are generally computational-intensive.
- **Edge-cloud**: In Figure 2.7(d), device D is in the edge-cloud mode, which is similar to the edge-device mode and suitable for the case where the device is highly resource-constrained. In this mode, the device is responsible for input data collection and the DNN model is executed through edge-cloud synergy. The performance of this model heavily depends on the network connection quality.

2.3.1.3 Challenges

Edge intelligence has great potential in reducing the latency of industrial applications comparing to the cloud computing paradigm, but the latency may still fail to meet the strict manufacturing requirement due to the unpredictable bandwidth. Take the automated assembly line of Figure 2.6(a) as an example again. For a 24-hour continuous working pipeline, each product only shares a small piece of inspection time so as to improve the manufacturing efficiency. If each product is delayed more than presupposed during the inspection, the accumulated excess time will be considerable, resulting in a sharp decline in outcomes and huge loss of profits. Using traditional deep learning approaches, however, it is difficult for the resource-limited inspection camera to finish the inference tasks under the strict latency requirement.

As an illustration, we take a Raspberry Pi and a desktop PC to emulate the IoT device and the edge server respectively, processing image recognition inference with the classical AlexNet model (Krizhevsky et al., 2017) over the Cifar-10 dataset. The breakdown of the end-to-end latency is plotted in Figure 2.6(b), where the bandwidth between the IoT device and the edge server varies. As Figure 2.6(b) shows, it takes more than 2 s to finish the whole inference locally on the Raspberry Pi, which is too inefficient to be employed in practical manufacturing inspection. Then we apply edge intelligence for the scenario. By the wireless connection, the captured image of the surface can be transferred and thus the DNN computation is offloaded to the edge server. Nevertheless, if the whole DNN computation is executed only on the edge server, the latency requirement is still unsatisfied. The computation time of the edge server limits within 10 ms while the transmission time is highly sensitive to the available bandwidth, i.e. the latency now is dominated by the available bandwidth. Specifically, the latency of edge-only DNN inference increases rapidly as the available bandwidth drops from 1.0 Mbps to 50 kbps. Therefore, for real-time applications like product defect inspection, neither the edge-only nor the device-only approaches serve as a practical solution, which motivates us to design an adaptive edge intelligence framework for IIoT.

To obtain a better understanding of the latency bottleneck of DNN execution on IoT devices, we further explore the runtime and the output data size of each layer, as Figure 2.8(a) shows. Interestingly, the experimental result varies by layer but the first half consumes most of the layer runtime, indicating that most of the computation is executed at the convolution layers. Another inspiring fact is that the layer that consumes much runtime may not export a large size of data. Base on these observations, an intuitive idea is DNN partitioning. Specifically, the computation of the DNN model could be partitioned into two parts, with the computation-intensive part executed on the powerful edge server and the rest runs locally. For instance, select the position after the third convolution layer (conv3, the layer with dot mark in Figure 2.8(a)) as the partition point, i.e. offload the computation before the partition point to the edge server and remain the rest computation on the device. Then the question occurs naturally: how do we select the partition point

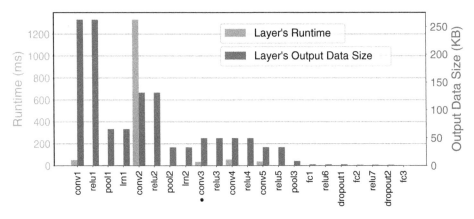

(a) The runtime of AlexNet layers on Raspberry Pi

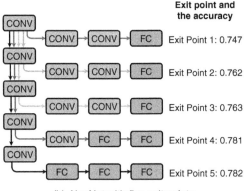

(b) AlexNet with five exit points

Figure 2.8 DNN partitioning and DNN right-sizing.

to minimize the whole latency? To answer this question, we need to jointly consider the computation latency and transmission latency.

While DNN partitioning tries to maximize the use of integrated computing power of the edge server and the IoT device, the total latency of the whole DNN inference is still constrained by the local execution part. To further reduce the latency, we combine the DNN partitioning method with DNN right-sizing. DNN right-sizing performs an early-exit mechanism during the DNN execution and trains a DNN model to obtain the accuracy of each exit point. Figure 2.8(b) illustrates a trained DNN model with five exit points, where each early-exit branch achieves the accuracy of 0.747, 0.762, 0.763, 0.781, and 0.782 respectively (only the convolution layers and the fully connected layers are drawn for ease of illustration). By training the DNN model with multiple exit points, we can obtain a series of branches with different size and then choose the branch with the maximum accuracy while meeting the latency requirement. Applying DNN right-sizing and DNN partitioning incurs the trade-off between accuracy and latency of the DNN inference since an early-exit leads to a shorter branch with lower accuracy as Figure 2.8(b) shows. Considering that some industrial applications require a strict deadline but can tolerate moderate accuracy loss, we can strike a balance on the trade-off in an on-demand manner (i.e. adaptive to the available bandwidth).

In conclusion, to achieve a real-time DNN-based industrial application in the edge computing paradigm, in this section, we propose an on-demand cooperative inference framework for IIoT, named Boomerang (Zeng et al., 2019). Under the requirement of industrial edge intelligence, Boomerang is designed with two keys. One is DNN right-sizing, which is employed to accelerate the execution of inference by early-exiting DNN execution at a selected intermediate DNN layer. The other is DNN partitioning, where the DNN computation is partitioned adaptively and distributed to the IoT devices and the edge server according to the available bandwidth. To make a reasonable choice, Boomerang jointly optimizes the two keys to satisfy the requirement for inference tasks on the manufacturing process. Moreover, to account for the massive number of heterogeneous IoT devices and reduce the manual overhead of model profiling at the install phase, we improve the basic Boomerang with a DRL model, achieving end-to-end automatic DNN right-sizing and DNN partitioning plan generation. Based on Raspberry Pi, the emulation evaluation results demonstrate the effectiveness of Boomerang (both the basic version and DRL version) for low-latency edge intelligence.

2.3.2 Boomerang Framework

In this section, we introduce Boomerang, an on-demand cooperative inference framework that automatically and intelligently selects the exit point and partition point to meet the requirements of industrial application.

2.3.2.1 Framework Overview

Figure 2.9 presents the workflow of the Boomerang framework, illustrated in three phases: install phase, optimization phase and runtime phase.

At the install phase, the employed DNN model is installed in advance on both the devices and the edge server. As the installation and configuration finish, Boomerang will run the

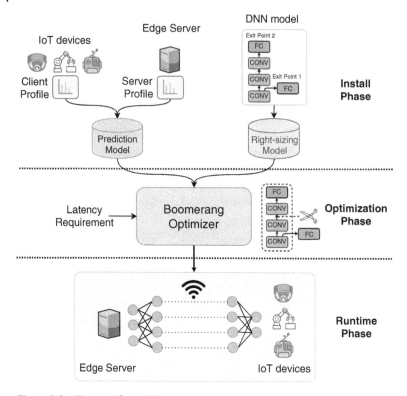

Figure 2.9 The workflow of Boomerang.

DNN model to profile the local DNN inference on both sides respectively. Specifically, during the profiling, the exact runtime of each layer is recorded. The profiling results are aggregated in the prediction model, where a series of regression models are trained to achieve the accurate prediction on each layer's runtime. Meanwhile, the installed DNN model feeds the right-sizing model, where the DNN model with multiple exit points is trained to obtain the exact accuracy of each branch. A further description of the prediction model and the right-sizing model will be detailed in Section 2.3.2.2. Since both the DNN profiling and the model training is infrastructure-dependent, given the exact employed devices and DNN model, the initialization on the devices will be executed only once during the whole application lifetime. In addition, this mechanism guarantees that devices of the same type share the same profiling result if they share the same DNN model. Therefore, for devices of the same type, it is feasible for the DNN model to be trained only on one device and then share the profiling result with the others. This will significantly reduce the profiling cost when there is a large number of IoT devices and thus save expensive industrial energy.

At the optimization phase, the trained prediction model and right-sizing model are fed into the Boomerang optimizer, together with the application latency requirement, to jointly optimize the DNN partitioning and DNN right-sizing. Once a DNN query (e.g., a scanned surface image) arrives, the Boomerang optimizer will record the current available bandwidth and run the points' selection algorithm (detailed in Section 2.3.2.3) to select the optimal exit point and partition point that maximize the inference accuracy while satisfying the application latency requirement.

At the runtime phase, the inference plan is generated according to the selected exit point and partition point, which guides the cooperative inference on the IoT device and the edge server. Specifically, the edge server executes the first half of computation while the IoT device finishes the remaining half. In the example scenario of product surface inspection, once a product surface image is captured, it will be transferred to the edge server, where the first half of the inference computation is performed and the intermediate processing result will be sent back, and then the client will finish the remaining inference computation locally.

2.3.2.2 Prediction Model and Right-sizing Model

It is observed that the latency of each type of layer is dominated by a variety of configurable parameters (Kang et al., 2017). For a typical convolution neural network (CNN) model, the types of layers cover convolution, relu, pooling, normalization, dropout and fully connected. And the dominated parameters include the size of convolution kernel, the convolution stride, the number of features and so on. Since these configurable parameters and the running latency results can be easily collected, an intuitive idea is to establish a regression model between them. Therefore, we vary the configurable parameters and measure the latency for each configuration on both the IoT device and the edge server. With these profiling results, we train a regression model to accurately predict the execution performance of each DNN layer, which will then be used in the Boomerang optimizer for point selection at the optimization phase.

The right-sizing model is responsible for providing the available early-exit branches and the corresponding accuracy. Specifically, the DNN model with multiple exit points is input to the right-sizing model, where each early-exit branch is executed to obtain the accuracy result. The trained right-sizing model is a series of exit points with accuracy attached, which are employed in the Boomerang optimizer for the point selection algorithm. In this work, we developed a right-sizing model based on BranchyNet (Teerapittayanon et al., 2016), which is an open source DNN training framework that supports the early-exit mechanism.

2.3.2.3 Boomerang Optimizer

The Boomerang optimizer runs the point selection algorithm at the optimization phase, requiring four inputs: (1) the regression functions from the prediction model, (2) the exit points from the right-sizing model, (3) the latency requirement from the application profile, and (4) the observed bandwidth. A typical iteration of the point selection algorithm is given in Figure 2.10. Since the DNN early-exit mechanism damages the accuracy of inference (Teerapittayanon et al., 2016), a relatively longer branch will gain a better inference performance. For this reason, the selection starts with the last exit point (exit point 2 in the figure) and then the partition point will be selected iteratively on this early-exit branch. For a selected pair of exit point and partition point, the regression functions are employed to estimate the inference latency of DNN layers on both sides while the transmission latency between them will be calculated according to the observed bandwidth, as illustrated in step (b). We obtain the whole runtime for the current pair of exit point and partition point by simple addition. After traversing all the partition points on the current early-exit

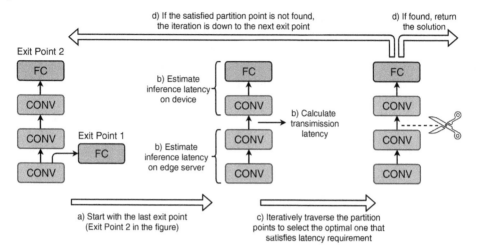

Figure 2.10 Point selection algorithm in the Boomerang optimizer.

branch, we select the optimal one with the minimum runtime and check whether it satisfies the latency requirement or not. If so, this pair will be returned as the selected points, otherwise the process will repeat to the next exit point until it finds the satisfied pairs, as step (d) shows. The selected pair of exit point and partition point will then be shared on both sides, where the exact inference plan is generated. If no solution is found after exhaustive traversing, there exists no solution for the given DNN model and latency requirement.

Since the prediction model and the right-sizing model is pre-trained at the install phase, the Boomerang optimizer mainly involves linear search operations that can be done at fast speed (no more than 1 ms in our experiments). This will save expensive industrial machine-hours and thus improve manufacturing productivity.

2.3.2.4 Boomerang with DRL

Although our purposed Boomerang based on a prediction model and a right-sizing model performs fast at the runtime phase, it still requires a lot of initialization work at the install phase. This may incur an enormous amount of energy consumption and time overhead in profiling and model training, especially when there are many types of IoT devices. Considering the tremendous amount of heterogeneous IoT devices in smart factories, the manual overhead for regression based model profiling can be significant at the install phase, which hurts the energy efficiency. Motivated by this, we further apply Deep Reinforcement Learning (DRL) (Wang et al., 2019) to generate a satisfactory inference plan for Boomerang in an autonomous manner. Deep reinforcement learning provides an end-to-end solution for sequential decision making problems, which is suitable for the selection of the exit point and the partition point. We employ the commonly adopted Deep Q-Network (DQN) as our DRL model, and Figure 2.11 shows a typical episode on the workflow of our DRL model. Essentially the selection of points is a trade-off between accuracy and latency, therefore we denote the state with (1) accuracy, (2) data size, (3) transmission latency, (4) device inference latency, and (5) edge inference latency. With the state input, the DQN works and outputs a series of probabilities on the inference plans,

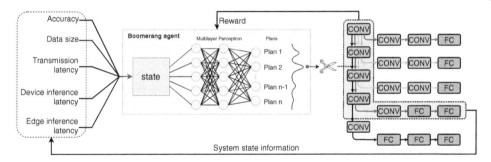

Figure 2.11 The workflow of the DRL model for point selection.

among which the plan with the highest probability is selected and applied to the DNN model right-sizing and partition. Then the inference will be executed according to the selected plan and the total latency and accuracy are recorded to obtain a reward according to the following reward function:

$$R_t = \begin{cases} e^{\tau a_t}, & \text{if } l_t \leq L, \\ 0, & \text{else,} \end{cases} \tag{2.1}$$

where e is the natural base, L is the application latency requirement, a_t and l_t are the accuracy and the latency of the current plan, respectively. Furthermore, a hyper-parameter τ is to tune the reward amplitude to optimize the overall performance. For the current selected plan, the obtained reward will be fed back to update the DQN and the system state information will be collected as the next input. During the whole training process the episodes will repeat until the plan selection reaches a convergence and the convergent result is the plan that satisfies the latency requirement. There exists no satisfying solution if there is no convergence. As the DRL model finishes training, the plan result will be sent to the devices as well as the edge server to guide their DNN inference.

The DRL model is trained at the install phase, requiring input as following: (1) the employed DNN model, (2) the application latency requirement, (3) the computational capability of the IoT device, and (4) the computational capability of the edge server. Except for the DNN model, all the other three required inputs are short technical parameters (which are usually given in the instruction manual) that can be easily collected and preset. Regardless of how many types of device there are, there is no need for the IoT devices to execute extra profiling, and only a list of computational capability parameters is required thus resulting in a saving of energy.

2.3.3 Performance Evaluation

In this section, we evaluate Boomerang in terms of latency requirement and accuracy.

2.3.3.1 Emulation Environment
Boomerang is a framework that is suitable for various kinds of DNN models but in our emulation we focus on the convolution neural network and set the inference task as image recognition. Specifically, we implement the classical AlexNet model and perform inference tasks over the large-scale Cifar-10 dataset. AlexNet has five exit points, as shown in

Figure 2.8(b), and each exit point corresponds to a branch that covers 12, 16, 19, 20 and 22 layers respectively.

The emulation involves a Raspberry Pi 3 as the IIoT device and a desktop PC as the edge server. The Raspberry Pi 3 has a quad-core ARM processor at 1.2 GHz with 1 GB RAM while the desktop PC is equipped with a quad-core i7-6700 Intel processor at 3.4 GHz with 8 GB RAM. The available bandwidth between the IoT device and the edge server is controlled by the WonderShaper tool. For Boomerang using the prediction model and right-sizing model (named regression-based Boomerang in Figure 2.12), the right-sizing model is implemented with Chainer (Tokui et al., 2015), which is an open source neural network framework allowing flexible structure control, and the prediction model manages independent variables with linear regression models. For Boomerang using the DRL model (named DRL-based Boomerang in Figure 2.12), we implement the DQN model with TensorFlow.

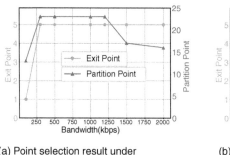

(a) Point selection result under different bandwidths (regression-based Boomerang)

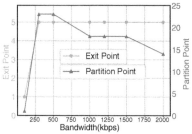

(b) Point selection result under different bandwidths (DRL-based Boomerang)

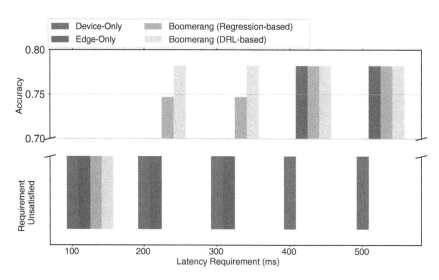

(c) Accuracy comparison under different latency requirements

Figure 2.12 Evaluation results under different bandwidths and different latency requirements.

2.3.3.2 Evaluation Results

In our experiment, we deploy Boomerang on the Raspberry Pi 3 and the desktop PC. Since both latency requirements and the available bandwidth play an essential role in Boomerang, we evaluate the performance and efficiency of Boomerang by varying these two factors.

First, we investigate the effect of bandwidth with the latency requirement fixed at 1000 ms and bandwidth varying from 50 kbps to 2.0 Mbps. Figure 2.12(a) shows the exit points and partition points selected by regression-based Boomerang under different bandwidths. As we can see, while the partition point fluctuates, the selected exit point gets later as the bandwidth increases, indicating that higher bandwidth leads to a larger size of DNN computation and thus better accuracy performance. Figure 2.12(b) shows the same experiment for DRL-based Boomerang. The trend of the DRL's selection result is similar with the regression-based Boomerang, while in DRL-based Boomerang the partition point drops earlier as the bandwidth climbs.

Figure 2.12(c) integrates the performance of four approaches, where the bandwidth is set to 400 kbps. The rectangle of each approach is located lower if the approach does not satisfy the latency requirement. As seen in Figure 2.12(c), all four approaches remain lower when the latency requirement is very strict (100 ms). As the latency requirement relaxes, inference by Boomerang works earlier than the other approaches, which is illustrated at 200 ms and 300 ms requirement. Notably, although both regression-based and DRL-based Boomerang works under these two requirements, the latter achieves higher accuracy. It is reasonable since regression-based Boomerang employs regression models as the prediction model, which still incurs moderate errors in predicting the layer's runtime. When the latency requirement is very stringent ($\leq$300 ms), these moderate errors may make a negative influence on point selection, for which Boomerang executes a sub-optimal inference plan. Nevertheless, DRL-based Boomerang performs end-to-end inference plan generation that avoids the estimation part and thus gains a more stable performance under the strict latency requirement. When the required latency spans to 400 ms and longer, all the approaches meet the requirement except for local inference due to the limited local computing resource. Considering an industrial manufacturing environment, this feature of Boomerang allows a lower latency for task processing, which helps achieve higher efficiency and profits.

2.4 Fog-enabled Collaborative SLAM of Robot Swarm

2.4.1 Introduction and Background

Robots are now popular and have entered people's lives and industries in many fields. They bring us much convenience in daily life, save a huge amount of manpower in factories, and complete tasks that are "mission impossible" for human beings. In the real world, robots are often needed to explore a previously unknown environment. For example, when people are buried in a collapsed building in an earthquake, and the space/conditions do not allow a rescue person/animal to enter, an advanced method is to send a robot with suitable size and shape to detect the location of the people, environment, and living conditions like oxygen levels, temperature, etc. Mapping information is critical to rescue planning. This requires the explorer robot to construct a map of the space and meanwhile know its own location and orientation. In robotics, performing concurrent construction of a map of the environment and estimation of the robot state is called simultaneous localization and mapping (SLAM).

Robot SLAM usually involves the following steps: collecting data by sensors, front-end odometry, back-end optimization, loop closure, and mapping. Sensors are used to collect environment information. Odometry is to estimate the movement of the robot based on the sensing data and optimization is to reduce noise and distortions in the first two steps to improve accuracy. Loop closure is to recognize the intersecting points of the movement then to estimate the real environment topology. In these steps, optimization has the heaviest computation.

Generally, in the rescue use case, robot SLAM needs to be low cost, low-power consumption, accurate, and speedy. However, these requirements mutually restrict each other. First, tens or hundreds of rescue robots may be used in one saving campaign so robot cost is a big concern. Popular SLAM sensors include laser radars and cameras; however, although generally laser radars perform better in accuracy, they are much more expensive than usual cameras, so the latter is more suitable for massive rescue robots. Second, accurate mapping and localization are very important to rescue path planning, and to achieve this a high-performance computing unit is needed, especially at the step of optimization that involves many advanced algorithms, and this of course, is contradictory to the low-cost requirement aforementioned. Third, rescue robots use battery; hence, battery life is a key consideration. Reducing the computing tasks of robots, e.g., using low-performance algorithms, is an effective way to save robot energy consumption; however, it may lead to inaccurate SLAM. Fourth, as we all know, time is critical in many rescue cases; hence, robots are required to move as quickly as possible and perform fast SLAM, imposing much pressure on the onboard computing unit. Similarly, low-complexity algorithms may be used to save SLAM time; however, a consequence is that the SLAM accuracy may be affected. Fifth, in a large area where multiple robots are used, collaboration between the robots in SLAM is required to merge the maps finally, thus a network is needed and one robot needs to be a leader in SLAM and to merge the maps. This lead robot, of course, will consume additional energy. However, the this network is not usually available in disaster scenarios.

An effective solution to robot SLAM issues is fog computing. OpenFog has identified eight pillars that a fog network needs to consider. They are also challenges and requirements faced by fog-enabled robot SLAM.

- **Security**: robot SLAM enabled by fog has privacy critical, mission critical, and even life critical aspects, particularly in rescue applications. Connecting robots to the fog network and cloud exposes both robots and the SLAM process to possible hacking and may suffer loss of map data and/or tampering with the SLAM process.
- **Scalability**: it needs to address the dynamic technical and business needs behind deployment of the fog network for robot SLAM, meaning the performance, capacity, security, reliability, hardware, and software of the fog network should be scalable. An FN may scale its internal capacity though the addition of hardware and software, and a fog network may scale up and out by adding a new node if computing tasks are too heavy.

- **Open**: openness is essential to the success of a ubiquitous fog network. It needs diverse vendors for consideration of cost, quality, and innovation. With interoperability and software technologies, fog computing can be deployed or defined anywhere. A robot can perform fog-enabled SLAM when it has connection to a network nearby with required software.
- **Autonomy**: means the fog network can assist SLAM when its external service or connection is lost. This eliminates the need for centralized processing, e.g., at the cloud. In disaster scenarios where cloud connection is usually not available, the fog network could performance SLAM as well. If any working FN is lost, then the remaining FNs need to take over the tasks and continue SLAM processes.
- **RAS**: reliability is essential to fog-enabled robot SLAM. All the network, software, and hardware need to be reliable. Rescue robots are usually deployed in harsh environments like earthquakes, fire, storms, etc., thus very robust communication is required. Availability, measured by uptime, is required for continuous support to robot SLAM and serviceability, meaning correct operation of the fog-enabled robot SLAM, which imposes the requirement of self-healing, autonomous configuration, etc.
- **Agility**: the fog network needs to response quickly when robots are moving in the environment, or the environment is dynamic and changes fast. In the case of rescue robots saving lives, it is critical for the FNs to map and locate as fast as they can.
- **Hierarchy**: when multiple robots are used to building map jointly, then a master FN merges the maps from different robots. A master FN is a higher-layer node comparing to other FNs and may coordinate other FN behavior. Moreover, if there are other applications associated with SLAM, like rescue, then the SLAM function may need to interact with higher-layer services like path planning, monitoring, business operation, etc.
- **Programmability**: SLAM may dynamically change with environment, network topology, and so on, so the fog network and nodes may be a highly adaptive program at the hardware and software layers. The computing FNs or the group FNs may be re-tasked automatically to accommodate the changes.

2.4.2 A Fog-enabled Solution

2.4.2.1 System Architecture

Figure 2.13 illustrates the fog-enabled robot SLAM system architecture(Yang et al., 2020b), which has a three-tier computing network. As illustrated, the bottom tier is the device level, which consists of multiple robots with limited computing, communication, storage, and energy resources. The middle tier is the local level, which consists of distributed FNs. The FNs exist near the robots. They can connect with each other via wired or wireless networks and collectively construct the fog network. The FNs can collect data from them, support real-time data processing, transmit data to the upper tiers, etc. The upper tier is the regional/global level, which consists of the cloud. As the robots are mobile, they connect to the fog network via wireless access, possibly wi-fi, or 4G/5G. The fog network may connect to a cloud, but it is not necessary, particularly in disaster scenarios where cloud is not available.

The function interface between robots and FNs is called a southbound interface, which defines information of the data plane and control plane. The data plane carries key frames

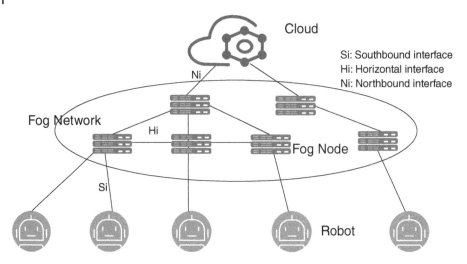

Figure 2.13 The illustration of fog-enabled robot SLAM system architecture (Yang et al., 2020b).

and poses information that is processed by robots or FNs. The control plane carries commons that manage the whole workflow such as initializing, stopping, getting key-frames, and setting poses. FNs can communicate with each other via a horizontal interface. Multiple FNs can collaborate and realize coordinated multiple points (CoMPs), a distributed storage system, or a distributed computing system to get more powerful capabilities of communication, storage, and computing. The horizontal interface carries the control and data information supporting coordinating functions. Based on FNs or a fog network that is much closer to robots, the latency of robot processing can be shortened to a 10 ms level. FNs can communicate with the cloud via a northbound interface.

In robot SLAM, except for collecting sensing data, all other steps can be offloaded to the fog network. In the case that cameras are used, the video streams require high data rate wireless transmission, e.g., several to tens of Mbps. Meanwhile, the front-end odometry needs (relatively) much lower computing capacity than other steps, and the output data rate of the front-end processing may be as low as tens kbps. Therefore, it is supposed to offload the steps, which are after the odometry step, to the fog network, as shown in Figure 2.14. First, an FN initializes the odometry process on a robot and then the robot starts front-end processing to abstract key frames and position itself. The robot reports key frame streams, which are the main visual captures representing the environment, to the fog network to optimize the map. The fog network sends the position information back to each robot, and the robot will correct the position drift accumulated. When the cloud is available then it can merge maps from different robots. If the cloud is unavailable, then one master FN may merge maps from different robots through the collaboration between FNs.

Figure 2.15 further shows the function view of fog-enabled robot SLAM. The function modules include task process, communication, storage, deployment, security, and interfaces. Considering the low-latency requirement and stream-oriented processing, it may apply real-time streaming processing framework such as Storm or Spark. Taking Storm as an example, the task process can be composed of application, task scheduler, cluster coordination, supervisor, and worker. To be compatible with the legacy mechanism, the

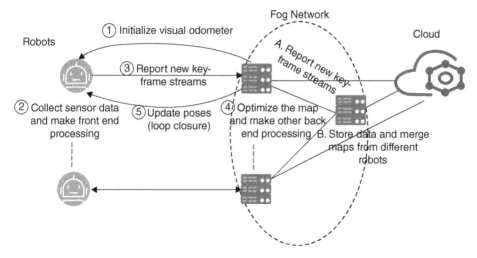

Figure 2.14 The workflow of fog-enabled robot SLAM (Yang et al., 2020b).

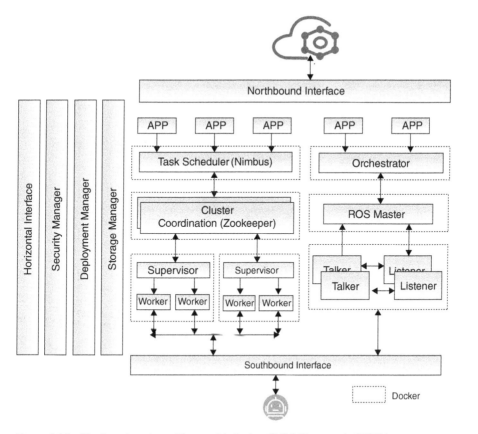

Figure 2.15 The function view of fog-enabled robot SLAM (Yang et al., 2020b).

deployment manager may adopt a robot operation system (ROS) as well. The storage manager can support distributed storage system and centralized databases. Distributed storage systems are suitable for the delay-insensitive applications while demanding large storage. Centralized databases are suitable for delay-sensitive systems. One can select different storage schemes through the deployment manager. The security manager can manage different security functions such as authentication, encryption, privacy protection, and so on. The southbound interface provides information exchange of control plane, data plane, and manage plane between FNs and robots. The horizontal interface supplies information exchange between FNs to realize fog-net-based collaborative functions. The northbound interface provides information exchange between FNs and cloud-end to support delay-insensitive, computing-intensive, and storage-heavy tasks.

2.4.2.2 Practical Implementation

Based on the architecture proposed above, we have developed a demo of fog-enabled robot SLAM with OpenLTE, as shown in Figure 2.16. It is mainly composed of the mobile robot [Figure 2.16(a)] and an FN [Figure 2.16(b)].

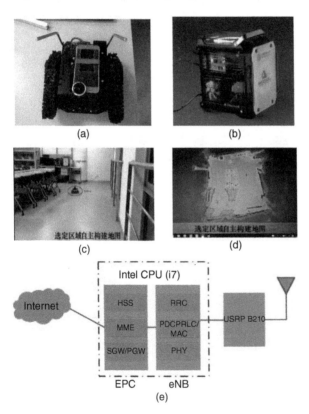

Figure 2.16 The developed demo of fog-enabled robot SLAM with OpenLTE. (a) mobile robot; (b) an FN; (c) experiment area; (d) SLAM map built on FN; (e) illustration of the OpenLTE architecture implemented (Yang et al., 2020b).

The mobile robots are based on four-wheel differential driving platforms, and all of them are equipped with sensors including an inertial measurement unit (IMU), wheel encoder sensor, and so on. Raspberry Pi 3b and STM32 MUC are used for calculation and control, respectively. The IMU and wheel encoder sensor first input desired linear velocity and angular velocity to the motion controller, then Raspberry Pi 3b and STM32 MUC will convert them into the motion of robot and control the wheel rotation. In addition, a camera is mounted on the central axis of each robot for mapping and navigation. To reduce the cost, the camera can be replaced by a mobile phone, as shown in Figure 2.16(a). The camera continuously takes pictures or videos of the environment around the robot, which are then transmitted to the FNs for the collaborative SLAM.

The FNs are customized servers installed software-defined OpenLTE components. OpenLTE is a software defined LTE system that runs on the general purpose processor (GPP) platform. It includes software-defined eNodeB (SDN-eNB) and software-defined EPC (SDN-EPC). Both the eNB and EPC could be a miniPC, a notebook, or a server. USRP B210 is used to convert the LTE RF signal to the baseband. The FNs process the baseband signal and implement eNB and EPC functions. In addition, as the FNs are computer in nature, they can easily carry out information processing like SLAM computation offloading. We installed a customized SIM card on the robot mobile phone so it could communicate with the FNs by LTE signal. Compared to wi-fi, LTE has the advantages of supporting handover, better QoS, etc. A smooth handover is required when a robot moves to anther cell. FNs can connect to the cloud, but it is not necessary.

The function and control are based on ROS, which is a widely used open-source software platform for robots. ROS is a robotics middleware providing services for heterogeneous computing clusters such as hardware abstraction, low-level device control, implementation of commonly used functionality, message-passing between processes, and package management. It has a modular design, and each module can be deployed on different FNs in a distributed manner. Thanks to its distributed feature, components of a service can be run on any desired nodes. For example, visual processing requires a lot of computation resources, so it runs on FN. While front-end odometry requires much lower computation resources, and thus it is better run on a Raspberry Pi carried on the robot.

Figure 2.17 illustrates the abstract system architecture of fog-enabled robot SLAM with OpenLTE, which is an instance of Figure 2.13.

In conclusion, the system architecture of fog-enabled robot SLAM with OpenLTE consists of a hardware layer, a software backplane layer, a fog application support layer, and an applications layer, as shown in Figure 2.18.

- **Hardware layer**: it mainly includes eNB/EPC hardware, the robot hardware, cameras, laser radar, other sensors and actuators, etc.
- **Software backplane layer**: it manages the FNs hardware and FNs themselves, so they include operation system like Linux and Android, communication software, FN and services discovery, etc. With OpenLTE, communication software includes eNB and EPC functions. FN discovery is the basic requirement of a fog network, by which neighbor FNs can be discovered and then it is possible to perform resource sharing and distributed

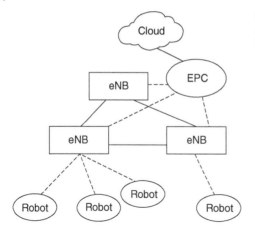

Figure 2.17 The function view of fog-enabled robot SLAM (Yang et al., 2020b).

Application	Robot SLAM	Robot AI	......
Support	Fog Application Support		
Software Backplane	ROS		
	eNB & EPC OS, and software	Robot OS	Node & service discovery, HW virtualization, container
Hardware Layer	eNB & EPC HW	Robot HW	Camera, Laser Radar, Sensors, Actuators...

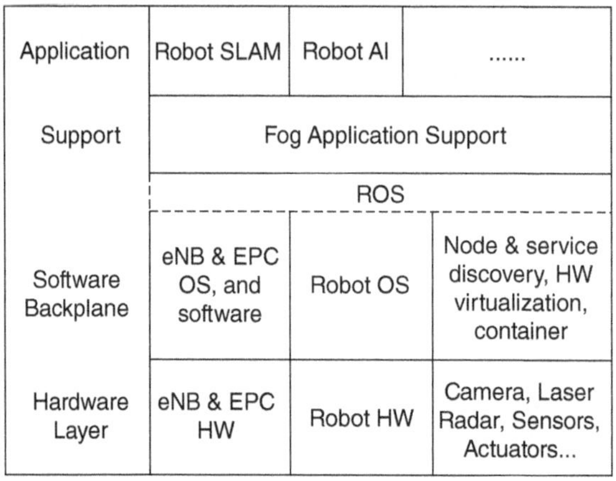

Figure 2.18 The function view of fog-enabled robot SLAM (Yang et al., 2020b).

computing. This layer also includes hardware virtualization and container. With virtualization technologies, computing and storage resources on an FN can be sliced for different applications with good isolations. Another important component of backbone software layer is ROS. A function module may publish message that others may subscribe and then data flows among different nodes. ROS enables data exchange between robots and FNs easily.

- **Fog application support layer**: this layer supports fog applications and has a broad spectrum. It supports security, storage, message, runtime engine, etc.
- **Application layer**: a bound of applications that may be supported by fog networks include robot SLAM, AI, etc.

Finally, we summarize how the proposed fog-enabled robot SLAM can solve and satisfy the challenges and requirements of traditional robot SLAM, as shown in Figure 2.19 (Yang et al., 2020b).

Challenges & Requirements	Functions
Security	1. Robots ID authentication should be supported on SI.
	2. Data information encryption should be supported on SI, NI, and HI.
	3. The sensitive information included in key frames and map data should be encrypted or erased before stored into databases.
Scalability	1. Task scheduler can orchestrate computation, storage, communication resources based on the task QoS requirements, status of robots, FNs, and cloud-cnd.
	2. Task scheduler can orchestrate computation, storage, communication resources based on the task QoS requirements, status of robots, FNs, and cloud-cnd.
	3. Task scheduler should consider computing-communication tradeoff when making decision on task scheduling.
	4. The FNs can form a collaborative network to support niore powerful communication, computing, and storage systems.
Open	1. The information model, message primitives, and protocol stack of application layer on SI, NI. and HI should be standardized based on the cominon robot applications.
	2. The policies of task scheduling, resources orchestration, and inodules deployment can be open and defined by users.
	3. All kinds of resources should be visualized by GUI.
Autonomy	1. The network topology should be flexible to support different application scenarios, such as stand-alone robots, network of robot and FNs, network of robot, FNs, and cloud-end network.
	2. Both robots and FNs can finish task independently.
RAS	1. Robots can independently work without the connection with FNs and cloud-end.
	2. Robots can learn more experience and knowledge from big history data in FNs or cloud-end.
Hierarchy	1. The solution can be one (robots only), two (robots and fog nodes), or three (robots, FNs, and cloud-end) layers.
	2. Each level can work independently.
	3. Fog nodes can provide distributed computing, storage solution for higher QoS applications. Each level can work independently.
Programmability	1. The work flow of applications can be programmed to achieve different QoS requirements.
	2. The policies of task scheduling, resources allocation, information process mechanism can be programmed by users.

Figure 2.19 Challenges and requirements solved and satisfied by the fog-enabled robot SLAM (Yang et al., 2020b).

2.5 Conclusion

This chapter presents a novel infrastructure and two exemplary fundamental service architecture of a multi-tier computing network, aiming at achieving a cloud-to-things continuum of services to enable intelligent IoT. With the task scheduling framework of CAST, efficient cost aware task scheduling, as well as a simple but effective incentive mechanism for motivating resource sharing, can be realized. With the service architecture of FA2ST, cross-layer resource discovery, fine-grained resource management and dynamic service provision can be supported under the dynamic networks. Two major intelligent

IoT applications of industrial edge intelligence and collaborative SLAM by robot swarm are extensively investigated to demonstrate the superior performance achieved by the proposed multi-tier computing networks and service architectures.

References

Igor Bisio, Chiara Garibotto, Aldo Grattarola, Fabio Lavagetto, and Andrea Sciarrone. Exploiting context-aware capabilities over the internet of things for industry 4.0 applications. *IEEE Network*, 32(3):101–107, 2018.

Thomas D Burd and Robert W Brodersen. Processor design for portable systems. *Journal of VLSI signal processing systems for signal, image and video technology*, 13(2-3):203–221, 1996.

Nanxi Chen and Siobhán Clarke. A dynamic service composition model for adaptive systems in mobile computing environments. In *International Conference on Service-Oriented Computing*, pages 93–107. Springer, 2014.

Nanxi Chen, Nicolás Cardozo, and Siobhán Clarke. Goal-driven service composition in mobile and pervasive computing. *IEEE Transactions on Services Computing*, 11(1):49–62, 2016.

Nanxi Chen, Yang Yang, Tao Zhang, Ming-Tuo Zhou, Xiliang Luo, and John K Zao. Fog as a service technology. *IEEE Communications Magazine*, 56(11): 95–101, 2018a.

Xu Chen, Qian Shi, Lei Yang, and Jie Xu. Thriftyedge: Resource-efficient edge computing for intelligent iot applications. *IEEE network*, 32(1):61–65, 2018b.

Mung Chiang and Tao Zhang. Fog and iot: An overview of research opportunities. *IEEE Internet of Things Journal*, 3(6):854–864, 2016.

Karel De Vogeleer, Gerard Memmi, Pierre Jouvelot, and Fabien Coelho. The energy/frequency convexity rule: Modeling and experimental validation on mobile devices. In *International Conference on Parallel Processing and Applied Mathematics*, pages 793–803. Springer, 2013.

Yun Chao Hu, Milan Patel, Dario Sabella, Nurit Sprecher, and Valerie Young. Mobile edge computing-a key technology towards 5G. *ETSI White Paper*, 11 (11):1–16, 2015.

Mianheng Jiang. Urbanization and informatization: Historical opportunity for China's development. *Informatization Construction*, 2:4–5, 2010. (in Chinese).

Slađana Jošilo and György Dán. Selfish decentralized computation offloading for mobile cloud computing in dense wireless networks. *IEEE Transactions on Mobile Computing*, 18(1):207–220, 2018.

Slacđana Jošilo and György Dán. A game theoretic analysis of selfish mobile computation offloading. In *IEEE INFOCOM 2017-IEEE Conference on Computer Communications*, pages 1–9. IEEE, 2017.

Y. Kang, J. Hauswald, C. Gao, A. Rovinski, T. Mudge, J. Mars, and L. Tang. Neurosurgeon: Collaborative intelligence between the cloud and mobile edge. In *2017 ASPLOS*, 2017.

A Krizhevsky, I. Sutskever, and G. Hinton. Imagenet classification with deep convolutional neural networks. Number 6. 2017.

Yang Liu, Changqiao Xu, Yufeng Zhan, Zhixin Liu, Jianfeng Guan, and Hongke Zhang. Incentive mechanism for computation offloading using edge computing: A stackelberg game approach. *Computer Networks*, 129:399–409, 2017.

Zening Liu, Kai Li, Liantao Wu, Zhi Wang, and Yang Yang. Cats: Cost aware task scheduling in multi-tier computing networks. *Journal of Computer Research and Development*, 57(9), 2020. (in Chinese).

Yuyi Mao, Changsheng You, Jun Zhang, Kaibin Huang, and Khaled B Letaief. A survey on mobile edge computing: The communication perspective. *IEEE Communications Surveys & Tutorials*, 19(4):2322–2358, 2017.

Carla Mouradian, Diala Naboulsi, Sami Yangui, Roch H Glitho, Monique J Morrow, and Paul A Polakos. A comprehensive survey on fog computing: State-of-the-art and research challenges. *IEEE Communications Surveys & Tutorials*, 20(1):416–464, 2017.

Duong Tung Nguyen, Long Bao Le, and Vijay Bhargava. Price-based resource allocation for edge computing: A market equilibrium approach. *IEEE Transactions on Cloud Computing*, 2018.

OpenFog Consortium. Openfog reference architecture technical paper. https://www .iiconsortium.org/pdf/OpenFog_Reference:Architecture_2_09_17.pdf, 2017.

Lingjun Pu, Xu Chen, Jingdong Xu, and Xiaoming Fu. D2d fogging: An energy-efficient and incentive-aware task offloading framework via network-assisted d2d collaboration. *IEEE Journal on Selected Areas in Communications*, 34(12):3887–3901, 2016.

Vivienne Sze, Yu-Hsin Chen, Tien-Ju Yang, and Joel S Emer. Efficient processing of deep neural networks: A tutorial and survey. *Proceedings of the IEEE*, 105(12):2295–2329, 2017.

S. Teerapittayanon, B. McDanel, and H. T. Kung. Branchynet: Fast inference via early exiting from deep neural networks. In *2016 23rd ICPR*, 2016.

Seiya Tokui, Kenta Oono, Shohei Hido, and Justin Clayton. Chainer: a next-generation open source framework for deep learning. In *Proceedings of workshop on machine learning systems (LearningSys) in the twenty-ninth annual conference on neural information processing systems (NIPS)*, volume 5, pages 1–6, 2015.

Jinjiang Wang, Yulin Ma, Laibin Zhang, Robert X Gao, and Dazhong Wu. Deep learning for smart manufacturing: Methods and applications. *Journal of Manufacturing Systems*, 2018.

Kun Wang, Yihui Wang, Yanfei Sun, Song Guo, and Jinsong Wu. Green industrial internet of things architecture: An energy-efficient perspective. *IEEE Communications Magazine*, 54(12):48–54, 2016.

Yixuan Wang, Kun Wang, Huawei Huang, Toshiaki Miyazaki, and Song Guo. Traffic and computation co-offloading with reinforcement learning in fog computing for industrial applications. *IEEE Transactions on Industrial Informatics*, 15(2):976–986, 2019.

Yang Yang. Multi-tier computing networks for intelligent iot. *Nature Electronics*, 2(1):4–5, 2019.

Yang Yang, Shuang Zhao, Wuxiong Zhang, Yu Chen, Xiliang Luo, and Jun Wang. DEBTS: Delay energy balanced task scheduling in homogeneous fog networks. *IEEE Internet of Things Journal*, 5(3):2094–2106, 2018.

Yang Yang, Xiliang Luo, Xiaoli Chu, and Ming-Tuo Zhou. Fog computing architecture and technologies. In *Fog-Enabled Intelligent IoT Systems*, pages 39–60. Springer, 2020a.

Yang Yang, Xiliang Luo, Xiaoli Chu, and Ming-Tuo Zhou. Fog-enabled multi-robot system. In *Fog-Enabled Intelligent IoT Systems*, pages 99–131. Springer, 2020b.

Wanghong Yuan and Klara Nahrstedt. Energy-efficient cpu scheduling for multimedia applications. *ACM Transactions on Computer Systems (TOCS)*, 24(3):292–331, 2006.

Liekang Zeng, En Li, Zhi Zhou, and Xu Chen. Boomerang: On-demand cooperative deep neural network inference for edge intelligence on the industrial internet of things. *IEEE Network*, 33(5):96–103, 2019.

Huaqing Zhang, Yanru Zhang, Yunan Gu, Dusit Niyato, and Zhu Han. A hierarchical game framework for resource management in fog computing. *IEEE Communications Magazine*, 55(8):52–57, 2017.

Weiwen Zhang, Yonggang Wen, Kyle Guan, Dan Kilper, Haiyun Luo, and Dapeng Oliver Wu. Energy-optimal mobile cloud computing under stochastic wireless channel. *IEEE Transactions on Wireless Communications*, 12(9): 4569–4581, 2013.

Yanru Zhang and Zhu Han. *Contract Theory for Wireless Networks*. Springer, 2017.

3

Cross-Domain Resource Management Frameworks

3.1 Introduction

Due to the explosive increase in devices and data, and unprecedentedly powerful computing power and algorithms, IoT applications are becoming increasingly intelligent, and are shifting from simple data sensing, collection, and representation tasks towards complex information extraction and analysis. To support future diverse intelligent IoT applications, multi-tier computing resources are required, as well as efficient cross-domain resource management, as discussed in the previous chapter. Under this novel computing and service architecture for intelligeng IoT, fog computing plays a crucial role in managing cross-domain resources, as it is the bridge connecting centralized clouds and distributed network edges or things (Yang, 2019).

Just as centralized data centers build the cloud infrastructure, geo-distributed fog nodes (FNs) are the building blocks of a fog network. FNs embody a variety of devices between end-users and cloud data centers, including routers, smart gateways, access points (APs), base stations (BSs), as well as portable devices such as drones, robots, and vehicles with computing and storage capabilities. Therefore, fog computing will not only be at the network perimeter but also span along the cloud-to-things continuum, pooling these distributed resources to support applications. Through efficient resource sharing and collaborative task processing among FNs and between FNs and cloud/end, fog computing can support delay-sensitive, computation-intensive, and privacy-protective IoT applications (Yang et al., 2020).

To realize the collaboration among FNs and tap the potential of fog computing, efficient cross-domain resource management is necessary and crucial. Obviously, a FN can offload its task to neighboring FNs to reduce workload and accelerate task processing. Once the task has been implemented, the FN can collect results from the neighboring FNs. Such a collaboration procedure involves data transmission, task execution/computation, and data storage, thus demanding communication, computation, and caching resources, which are called 3C resources (Luo et al., 2019).

However, such cross-domain resource management is not easy and faces the following challenges. First, with the introduction of computation and caching resources, the resource management problem transforms from single (single-domain) resource management to

Intelligent IoT for the Digital World: Incorporating 5G Communications and Fog/Edge Computing Technologies,
First Edition. Yang Yang, Xu Chen, Rui Tan, and Yong Xiao.
© 2021 John Wiley & Sons Ltd. Published 2021 by John Wiley & Sons Ltd.

joint (cross-domain) resource management (Muñoz et al., 2015, Ndikumana et al., 2018, Xing et al., 2019). Second, due to the diversity of applications and services, different users or tasks may have different performance requirements in terms of delay, energy, cost, etc. Third, due to the heterogeneity of FNs, they typically have different resources and capabilities in type and size. Last but not least, since massive FNs are generally distributed in geographical location and logistic network level, it may be impossible to manage these nodes/resources in a centralized manner (Ali et al., 2018).

To solve these challenges, in this chapter, we focus on cross-domain resource management among FNs to support intelligent IoT applications. To be specific:

In Section 3.2, we study the joint computing and communication resource management for delay-sensitive applications. We focus on the delay-minimization task scheduling problem and distributed algorithm design in Multiple Tasks and Multiple Helpers (MTMH) fog computing networks under two different scenarios of non-splittable tasks and splittable tasks. In this book, we don't distinguish between non-splittable tasks and indivisible tasks, and splittable tasks are alternative to divisible tasks.

For non-splittable tasks, we propose a potential game based analytical framework and develop a distributed task scheduling algorithm called POMT (paired offloading of multiple tasks). For splittable tasks, we propose a generalized Nash equilibrium problem based analytical framework and develop a distributed task scheduling algorithm called POST (parallel offloading of splittable tasks). Analytical and simulation results show that the POMT algorithm and POST algorithm can offer the near-optimal performance in system average delay and achieve a higher number of beneficial task nodes.

In Section 3.3, we study the joint computing, communication, and caching resource management for energy-efficient applications. We propose F3C, a fog-enabled 3C resource sharing framework for energy-efficient IoT data stream processing by solving an energy cost minimization problem under 3C constraints. Nevertheless, the minimization problem is proven to be NP-hard via reduction to a Generalized Assignment Problem (GAP). To cope with such challenge, we propose an efficient F3C algorithm based on an iterative task team formation mechanism which regards each task's 3C resource sharing as a sub-problem solved by the elaborated minimum cost flow transformation. Via utility improving iterations, the proposed F3C algorithm is shown to converge to a stable system point. We conduct extensive performance evaluations, which demonstrate that our F3C algorithm can achieve superior performance in energy saving compared to various benchmarks.

In Section 3.4, we make a case study of the energy-efficient resource management in tactile internet (TI). We focus on an energy-efficient design of fog computing networks that support low-latency TI applications. We investigate two performance metrics: service delay of end-users and power usage efficiency of FNs. In this book, we don't distinguish between power usage and energy consumption, and power usage is an alternative to energy consumption.

We quantify the fundamental trade-off between these two metrics and then extend our analysis to fog computing networks involving cooperation between FNs. We introduce a novel cooperative fog computing mechanism, referred to as offload forwarding, in which a set of FNs with different computing and energy resources can cooperate with each other. The objectives of this cooperation are to balance the workload processed by different FNs, further reduce the service delay, and improve the efficiency of power usage. We develop a distributed optimization framework based on dual decomposition to achieve the optimal trade-off. Our framework does not require FNs to disclose their private information or conduct back-and-forth negotiations with each other. Finally, to evaluate the performance of our proposed mechanism, we simulate a possible implementation of a city-wide self-driving bus system supported by fog computing in the city of Dublin. The fog computing network topology is set based on a real cellular network infrastructure involving 200 base stations deployed by a major cellular operator in Ireland. Numerical results show that our proposed framework can balance the power usage efficiency among FNs and reduce the service delay for users by around 50% in urban scenarios.

3.2   Joint Computation and Communication Resource Management for Delay-Sensitive Applications

Figure 3.1 illustrates a general heterogeneous fog network, which consists of many randomly distributed FNs with diverse computation resources and capabilities. At any specific time slot, the FNs can be further classified as (i) task nodes (TNs), which have a task to process, (ii) helper nodes (HNs), which have spare resources to help their neighboring TNs to process tasks, or (iii) busy nodes (BNs), which are busy with processing previous tasks and thus have no spare resources to help their neighboring TNs to process tasks. Such a general fog network is referred to as MTMH fog network. It is worth noting that TNs, HNs, and BNs are not fixed. To be specific, TNs and BNs can be HNs when they are idle and available in the following time slots, and vice versa. It is also worth noting that such a MTMH network is quite general and many different application scenarios can be abstracted. To give some examples: for fog-enabled robot systems, TNs can be the robots, and HNs can be the idle robots, edge nodes in the room, factory. For fog-enabled wireless communication networks, TNs can be the mobile devices, and HNs can be the AP, BS in the vicinity. For fog-enabled intelligent transportation systems, TNs can be the vehicles, and HNs can be the roadside units. For fog-enabled smart home, TNs can be the heterogeneous sensors, devices, and HNs can be the routers, local edge nodes in the house (Yang et al., 2020).

As already mentioned, tasks with strict delay requirements may not be accomplished in time on the local devices, and thus they need to be offloaded to neighboring available HNs. Take the cases in Figure 3.1 as examples. For non-splittable tasks, a task can be either executed by the local TN/device, or entirely offloaded to one neighboring HN. A HN, if its spare resources and capabilities permit, can accommodate multiple tasks from different TNs. For example, as shown in Figure 3.1(a), TN-1 executes its task on a local device, which may be due to (1) HN-2 does not have sufficient computation/storage resources for accommodating the task from TN-1, or (2) the communication channel between TN-1 and TN-2 is not good at this slot. TN-2 offloads its task to HN-1, because HN-1 has more computation resources and better communication channel than HN-2. HN-4 accommodates multiple tasks from TN-3 and TN-4 simultaneously as it has sufficient resources and capabilities. For splittable tasks, a task can be divided into multiple independent subtasks and offloaded to multiple HNs for parallel processing. For example, as shown in Figure 3.1(b), TN-1 processes its task on a local device, which may be due to the fact that the neighboring BNs are busy with processing previous tasks and TN-1 is also out of the coverage of HN-1. TN-2 is in the coverage

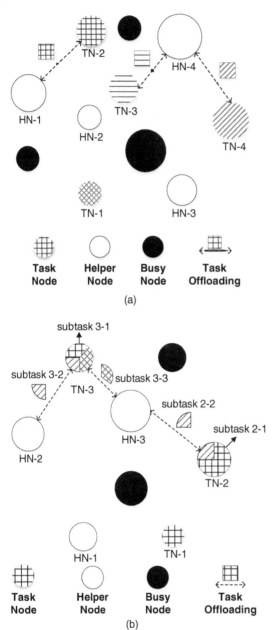

Figure 3.1 A general heterogeneous MTMH fog network. The FNs have different computation resources and capabilities, which are distinguished by the size of cycles. (a) Non-splittable tasks (b) Splittable tasks.

of HN-3, and thus it divides the task into two subtasks, i.e. subtask 2-1 and subtask 2-2. It processes subtask 2-1 by itself, and meanwhile offloads subtask 2-2 to HN-3 for parallel processing. TN-3 can access both HN-2 and HN-3. It divides its task into three subtasks, i.e. subtask 3-1, subtask 3-2 and subtask 3-3, and thus task 3 can be processed on a local device, HN-2, and HN-3 meanwhile. HN-3 accommodates subtask 2-2, subtask 3-3 from TN-2, TN-3 simultaneously, as it has sufficient resources and capabilities.

3.2.1 2C Resource Management Framework

3.2.1.1 System Model

For the sake of analysis, the quasi-static scenario is considered, where the time is divided into identical small time slots, such that the network condition and user distribution can be treated as static in each time slot (Chen et al., 2015b). At any specific time slot, denote the set of TNs (i.e. tasks) and the set of HNs by $\mathcal{N} = \{1, 2, ..., N\}$ and $\mathcal{K} = \{1, 2, ..., K\}$, respectively. For convenience, we use the task n interchangeably for the task of TN n in the following context. Then, the task scheduling problem is indeed the task allocation problem between N TNs and K HNs. To be specific, the task allocation strategy can be mathematically described by the matrix $\mathbf{A} \in [0, 1]^{N \times (K+1)}$, whose (n, k)th entry is denoted by $a_n^k \in [0, 1]$, $n \in \mathcal{N}$, $k \in \{0\} \cup \mathcal{K}$. a_n^k is the portion of task n processed on HN k. It is worth noting that for non-splittable tasks, $a_n^k \in \{0, 1\}$, $n \in \mathcal{N}$, $k \in \{0\} \cup \mathcal{K}$. Obviously, $a_n^0 + \sum_{k \in \mathcal{K}} a_n^k = 1$ should be satisfied. Notably, index 0 represents the local TN. Specifically, taking TN n as an example, a_n^0 of the task n is processed on local device and a_n^k of the task n is offloaded to HN k. Rewrite $\mathbf{A} = (\mathbf{a}_1^T, \mathbf{a}_2^T, ..., \mathbf{a}_N^T)^T$, where $\mathbf{a}_n = (a_n^0, a_n^1, \cdots, a_n^K)$ is the task allocation strategy of TN n. In addition, we introduce a matrix $\mathbf{B} = (\mathbf{b}_1^T, \mathbf{b}_2^T, ..., \mathbf{b}_N^T)^T$ to represent the connectivity between TNs and HNs, where $\mathbf{b}_n = (b_n^1, \cdots, b_n^K)$. Specifically, if $b_n^k = 1$, then TN n can access HN k; otherwise, TN n cannot access HN k. The connectivity matrix can be determined by the distance or the signal to interference plus noise ratio (SINR).

In this section, we focus on the delay-minimization task scheduling problem within the computation offloading. When it comes to the computation offloading, a three-phase protocol usually needs to be considered, among which the joint communication and computation resource management is generally involved. The TNs first transmit tasks to HNs, and then after the computation of HNs, the results are transmitted back to TNs from HNs. Thus, to characterize the delay experienced by tasks, both communication delay and computation delay need to be involved. Here, it is worth noting that similar to most previous works (Chen et al., 2015b, Nowak et al., 2018, Shah-Mansouri and Wong, 2018), the phase in which HNs transmit results back to TNs can be ignored because the size of results is negligible compared to that of tasks.

With regard to the communication aspect, we assume that all HNs occupy orthogonal wireless channels, i.e. no interference between different HNs (Dinh et al., 2017, Yang et al., 2018b). In this work, the channel is pre-allocated, which is beyond the scope of discussion. We also assume that the HNs serve the TNs via time division multiple access (TDMA) or frequency division multiple access (FDMA) scheme, which is consistent with most current wireless standards (Zhao et al., 2018). Thus, the TNs will share or compete for the communication resources, e.g., time frames (Yang et al., 2018a) or resource blocks (Pang et al., 2017), when they transmit tasks to the same HN. Here, taking the TDMA scheme as an example, the TNs are assumed to equally share the time slots. Further, we assume that a HN begins computing tasks until all subtasks offloaded to it have finished transmitting. The benefits of these assumptions are two-folds. First, it circumvents the complex transmission scheduling or resource allocation problem, and thus the problem can be simplified (Dinh et al., 2017, Meng et al., 2017, Xing et al., 2019). Second, the resulted transmission time can be taken as an estimate or upper bound of the actual transmission time (Nowak et al., 2018).

Thus, the time of transmitting subtask from TN n to HN k can be represented as

$$T_n^{k,\text{trans}} = \sum_{n \in \mathcal{N}} a_n^k \frac{z_n}{R_n^k},$$

where z_n is the size of task n in bits, R_n^k is the data rate of transmitting task n to HN k.

With regard to the computation aspect, the computation time can be characterized by dividing the computation workload by the computation capability of devices, which are both in CPU cycles. Typically, computation load of each subtask is proportional to the size of subtask (Muñoz et al., 2015). Thus, the time of computing TN n's subtask a_n^0 on a local device, i.e. local computing, is given by

$$T_n^{0,\text{comp}} = a_n^0 \frac{z_n \gamma_n}{f_n},$$

where γ_n is the processing density of task n, i.e. the CPU cycles required to process a unit bit of data, and f_n is the computation capability of TN n, i.e., the CPU clock frequency (in CPU cycles per second). Different from local computing, for subtasks offloaded to HNs, i.e. HN computing, there may be competition or sharing among tasks for the computation capability of HNs. That's because the computation capability of single HN is usually limited and can not be exclusively used by a single task. Here, it is assumed that every HN assigns its spare computation resources to all the existing tasks in proportion to their computation workloads (Dinh et al., 2017, Nowak et al., 2018, You et al., 2016). Again, this assumption avoids the complex computation scheduling problem and the resulted computation time can be taken as an estimate or upper bound of the actual computation time (Nowak et al., 2018). Therefore, the time of computing task n's subtask on HN k can be written as

$$T_n^{k,\text{comp}} = \sum_{n \in \mathcal{N}} a_n^k \frac{z_n \gamma_n}{f_k},$$

where f_k is the spare computation capability of HN k.

In conclusion, for local computing, since no communication is necessary, the time required to process TN n on a local device can be represented as

$$T_n^0 = a_n^0 \frac{z_n \gamma_n}{f_n}.$$

For HN computing, it involves both communication delay and computation delay. According to the communication model and the computation model introduced above, the time required to process TN n on HN k can be represented as

$$T_n^k = \sum_{n \in \mathcal{N}} a_n^k z_n \left(\frac{1}{R_n^k} + \frac{\gamma_n}{f_k} \right).$$

For ease of exposition, we define $O_n^k \triangleq z_n(1/R_n^k + \gamma_n/f_k)$, which represents the amount of time that task n contributes to the total time for offloading. Thus, $T_n^k = \sum_{n \in \mathcal{N}} a_n^k O_n^k$.

3.2.1.2 Problem Formulation

In our problem setting, every TN needs to make decisions of task scheduling or allocation by themselves and aim to minimize the time required to process its own task. The problem formulation for non-splittable tasks and splittable tasks are as follows, respectively.

Non-splittable tasks and paired offloading

$$\min_{a_n} T_n \qquad (3.1a)$$

$$\text{s.t. } a_n^0, a_n^k \in \{0,1\}, \forall k \in \mathcal{K}, \qquad (3.1b)$$

$$a_n^0 + \sum_{k \in \mathcal{K}} a_n^k = 1, \forall n \in \mathcal{N}, \qquad (3.1c)$$

$$a_n^k \leq b_n^k, \forall k \in \mathcal{K}, \qquad (3.1d)$$

where,

$$T_n = a_n^0 T_n^0 + \sum_{k \in \mathcal{K}} a_n^k T_n^k. \qquad (3.2)$$

Constraints (3.1b) and (3.1c) imply that a task must be either processed on its own TN or entirely offloaded to one neighboring HN, i.e. non-splittable tasks. Constraint (3.1d) guarantees that a task can be only offloaded to the HNs, to which it can connect.

Splittable tasks and parallel offloading

$$\min_{a_n, T_n} T_n \qquad (3.3a)$$

$$\text{s.t. } a_n^0 \frac{z_n \gamma_n}{f_n} \leq T_n, \forall n \in \mathcal{N} \qquad (3.3b)$$

$$a_n^k \left[\sum_{n \in \mathcal{N}} a_n^k O_n^k \right] \leq a_n^k T_n, \forall k \in \mathcal{K} \qquad (3.3c)$$

$$a_n^0, a_n^k \geq 0, \forall k \in \mathcal{K} \qquad (3.3d)$$

$$a_n^0 + \sum_{k \in \mathcal{K}} a_n^k = 1, \forall n \in \mathcal{N} \qquad (3.3e)$$

$$a_n^k \leq b_n^k, \forall k \in \mathcal{K}. \qquad (3.3f)$$

Constraint (3.3b) shows that the local computing time should be less than the total time, while constraint (3.3c) indicates that the HN computing time should also be less than the total time. Constraints (3.3d) and (3.3e) ensure that all subtasks are completed.

Problem (3.1) and problem (3.3) are difficult to solve, because the objectives among TNs are conflicting and thus the task scheduling strategies are coupled. Intuitively, the time required to process some task depends on not only its own task scheduling strategy, but also other TNs' task scheduling strategies, for the reason that there exists competition for the communication resources and computation capabilities of HNs among TNs. For the ease of presentation and analysis, we denote the task scheduling strategies of all tasks except task n by $\mathbf{A}_{-n}$, $\mathbf{A}_{-n} = (\boldsymbol{a}_1^T, \cdots, \boldsymbol{a}_{n-1}^T, \boldsymbol{a}_{n+1}^T, \cdots, \boldsymbol{a}_N^T)^T$. Further, compared to the problem (3.1), the problem (3.3) is much more difficult. That's because the problem (3.3) not only needs to tackle the node selection problem, but also needs to solve the task division problem, which will result in a more complicated solution space. To solve these problems, we will resort to game theory and develop distributed algorithm.

3.2.2 Distributed Resource Management Algorithm

3.2.2.1 Paired Offloading of Non-splittable Tasks

For non-splittable tasks, a task can be either processed on its own TN or entirely offloaded to one neighboring HN. For the ease of presentation and analysis, we redefine the task scheduling strategy of TN n as a scalar a_n. If task n is processed on local device, then $a_n = 0$; if task n is offloaded to HN k for processing, then $a_n = k$. Thus, the task scheduling strategies of all tasks can be represented as $\boldsymbol{a} = (a_1, a_2, \cdots, a_N)$, the task scheduling strategies of all tasks except task n can be written as $\boldsymbol{a}_{-n} = (a_1, \cdots, a_{n-1}, a_{n+1}, \cdots, a_N)$, and the time required to process task n can be also simplified as

$$
T_n(a_n, \mathbf{a}_{-n}) = \begin{cases} \dfrac{z_n \gamma_n}{f_n}, & a_n = 0 \\ \displaystyle\sum_{m \neq n} O_m^{a_m} I_{\{a_m = a_n\}} + O_n^{a_n}, & a_n > 0 \end{cases}, \tag{3.4}
$$

where $I_{\{x\}}$ is an indicator function. Specifically, if x is true, $I_{\{x\}} = 1$; otherwise, $I_{\{x\}} = 0$.

To proceed, we reformulate the problem (3.1) as a non-cooperative game called POMT. We define the POMT game as $G_1 = (\mathcal{N}, (S_n)_{n \in \mathcal{N}}, (c_n)_{n \in \mathcal{N}})$, where $S_n : \{0\} \cup \{k \in \mathcal{K} | b_n^k = 1\}$ is the task scheduling strategy space of TN n, and c_n is the cost function which can be expressed as

$$
c_n(a_n, \mathbf{a}_{-n}) = \begin{cases} \dfrac{z_n \gamma_n}{f_n}, & a_n = 0 \\ \displaystyle\sum_{m \neq n} (O_m^{a_m} + O_n^{a_n}) I_{\{a_m = a_n\}} + O_n^{a_n}, & a_n > 0 \end{cases}. \tag{3.5}
$$

Notably, different from (3.4), we include the influence on others incurred by TN n (the first $O_n^{a_n}$ term in the second line of (3.5)) when constructing the cost function, i.e. local altruistic behavior (Zheng et al., 2014, Zhang et al., 2017). This is because the way to define the individual process time of tasks as the cost of TNs usually greatly damages the system performance. Therefore, to balance the TNs' selfishness and the system performance, local altruistic behavior among TNs is introduced.

The construction above guarantees that the POMT game is an exact (cardinal) potential game (Monderer and Shapley, 1996), and thus it possesses at least one pure-strategy Nash equilibrium (NE) and the finite improvement property (Yang et al., 2019), which is summarized in the following theorem.

Theorem 3.1 *The POMT game G_1 possesses at least one pure-strategy NE and guarantees the finite improvement property.*

Utilizing the finite improvement property, we can develop an algorithm to find an NE of any potential game, which can work in a distributed way. Therefore, we can develop a corresponding distributed task scheduling algorithm called POMT for the POMT game, as shown in Algorithm 1. For more details, please refer to (Yang et al., 2019).

3.2.2.2 Parallel Offloading of Splittable Tasks

Different from non-splittable tasks, the potential game based analytical framework is no longer applicable to splittable tasks, because the task scheduling strategy space of TNs is infinite. We will resort to the GNEP theory.

Algorithm 1 POMT Algorithm

1: **initialization**:
2: every TN n chooses the pairing strategy $a_n(0) = 0$
3: **end initialization**
4: **repeat** for every TN n and every decision slot in parallel:
5: broadcast the parameters message to the achievable HNs
6: receive the measurements of $\sum\limits_{n \in \mathcal{N}} O_n^k I_{\{a_n=k\}}$ from the achievable HNs
7: compute the best response function $b_n(\mathbf{a}_{-n}(t))$
8: **if** $a_n(t) \notin b_n(\mathbf{a}_{-n}(t))$ **then**
9:     send RTU message to the corresponding HN for contending for the pairing strategy update opportunity
10:     **if** not permitted to update the pairing strategy **then**
11:         update the pairing strategy $a_n(t+1) \in b_n(\mathbf{a}_{-n}(t))$ for the next decision slot
12:     **else**
13:         maintain the current pairing strategy $a_n(t+1) = a_n(t)$ for the next decision slot
14:     **end if**
15: **else**
16:     maintain the current pairing strategy $a_n(t+1) = a_n(t)$ for next slot
17: **end if**
18: **until** END message is received from the HNs

To proceed, we first reformulate the problem (3.3) into a GNEP called POST. As already mentioned, every TN needs to solve its own optimization problem (3.3) with the task scheduling strategy $\mathbf{a}_n$ and the time T_n (which is to be optimized) being the variables. The feasible strategy set of (3.3) are all $\mathbf{s}_n \triangleq (\mathbf{a}_n, T_n)$ satisfying the given constraints. Further, denote the strategies of all TNs except n by $\mathbf{s}_{-n} = (\mathbf{s}_m)_{m \neq n}$. Since every TN's strategy depends on the rival TNs' strategies, the problem (3.3) is a GNEP (Facchinei and Kanzow, 2007), which is denoted by G_2.

The existence of generalized Nash equilibrium (GNE) for GNEP can be guaranteed if the GNEP is convex (Facchinei and Kanzow, 2007). But for the POST problem, the existence of GNE is not straightforward for the reason that the optimization problem (3.3) is non-convex. To understand this point, let us see the constraint (3.3c). For $a_n^k = 0$, this constraint is automatically fulfilled, and thus vanishes. Such a constraint is called vanishing constraint, which is non-convex (Hoheisel, 2012).

To overcome this challenge, we resort to the Brouwer's fixed point theorem. We first study the structural properties of solutions to problem (3.3), which is shown as follows.

Lemma 3.1 *For the optimal solution $(\mathbf{a}_n^*, T_n^*)$ to the problem (3.3), it satisfies*

$$a_n^{0*} \frac{z_n \gamma_n}{f_n} = T_n^*, \tag{3.6}$$

and

$$\sum_{m \neq n} a_m^k O_m^k + a_n^{k*} O_n^k = T_n^*, \forall k: a_n^{k*} > 0. \tag{3.7}$$

Lemma 3.2 *For TN n, the solution $\boldsymbol{a}_n$ to problem (3.3) for a given $\boldsymbol{A}_{-n}$ is given by (3.8), which is a continuous function.*

$$
a_n^k(\boldsymbol{a}_{-n}) = \begin{cases}
\dfrac{1 + \sum\limits_{l \in \mathcal{A}_n} \frac{\sum\limits_{m \neq n} a_m^l O_m^l}{O_n^l}}{1 + \frac{z_n \gamma_n}{f_n} \sum\limits_{l \in \mathcal{A}_n} \frac{1}{O_n^l}} & , k = 0 \\[4ex]
0 & , \sum\limits_{m \neq n} a_m^k O_m^k \geq \dfrac{1 + \sum\limits_{l \in \mathcal{A}_n} \frac{\sum\limits_{m \neq n} a_m^l O_m^l}{O_n^l}}{1 + \frac{z_n \gamma_n}{f_n} \sum\limits_{l \in \mathcal{A}_n} \frac{1}{O_n^l}} \frac{z_n \gamma_n}{f_n} \\[4ex]
\dfrac{\frac{1}{O_n^k}\left(\frac{z_n \gamma_n}{f_n} - \sum\limits_{m \neq n} a_m^k O_m^k\right)}{} & \\
\quad - \dfrac{\frac{1}{O_n^k}\frac{z_n \gamma_n}{f_n} \sum\limits_{l \in \mathcal{A}_n} \frac{1}{O_n^l}\left(\frac{z_n \gamma_n}{f_n} - \sum\limits_{m \neq n} a_m^l O_m^l\right)}{1 + \frac{z_n \gamma_n}{f_n} \sum\limits_{l \in \mathcal{A}_n} \frac{1}{O_n^l}} , & \sum\limits_{m \neq n} a_m^k O_m^k < \dfrac{1 + \sum\limits_{l \in \mathcal{A}_n} \frac{\sum\limits_{m \neq n} a_m^l O_m^l}{O_n^l}}{1 + \frac{z_n \gamma_n}{f_n} \sum\limits_{l \in \mathcal{A}_n} \frac{1}{O_n^l}} \frac{z_n \gamma_n}{f_n}
\end{cases}
\tag{3.8}
$$

$$
T_n(\boldsymbol{a}_{-n}) = \frac{1 + \sum\limits_{l \in \mathcal{A}_n} \frac{\sum\limits_{m \neq n} a_m^l O_m^l}{O_n^l}}{1 + \frac{z_n \gamma_n}{f_n} \sum\limits_{l \in \mathcal{A}_n} \frac{1}{O_n^l}} \frac{z_n \gamma_n}{f_n}.
$$

Here, $\mathcal{A}_n$ is the set of active HNs of TN n, i.e, the HNs to which TN n offloads subtasks, and can be represented as

$$
\mathcal{A}_n = \left\{ k \in \mathcal{K} \,\Big|\, \sum_{m \neq n} a_m^k O_m^k < \frac{1 + \sum\limits_{l \in \mathcal{A}_n} \frac{\sum\limits_{m \neq n} a_m^l O_m^l}{O_n^l}}{1 + \frac{z_n \gamma_n}{f_n} \sum\limits_{l \in \mathcal{A}_n} \frac{1}{O_n^l}} \frac{z_n \gamma_n}{f_n} \right\}.
\tag{3.9}
$$

Lemma 3.3 *There is exactly one set $\mathcal{A}_n$, and it is of the form*

$$
\mathcal{A}_n = \left\{ k \in \mathcal{K} \,\Big|\, \sum_{m \neq n} a_m^k O_m^k < \frac{1 + \sum\limits_{l=1}^{k} \frac{\sum\limits_{m \neq n} a_m^l O_m^l}{O_n^l}}{1 + \frac{z_n \gamma_n}{f_n} \sum\limits_{l=1}^{k} \frac{1}{O_n^l}} \frac{z_n \gamma_n}{f_n} \right\},
\tag{3.10}
$$

in the case of $\sum\limits_{m \neq n} a_m^1 O_m^1 \leq \sum\limits_{m \neq n} a_m^2 O_m^2 \leq \cdots \leq \sum\limits_{m \neq n} a_m^K O_m^K$.[1]

Based on the lemmas above, we can adopt the Brouwer fixed point theorem to prove that the POST game has at least a GNE, as summarized in the following theorem.

Theorem 3.2 *The POST game G_2 possesses at least one GNE.*

Adopting the Gauss–Seidel-type method, which is the most popular method of finding GNE points in practice, we can develop a corresponding distributed task scheduling

---

1 For TN n and HN k such that $b_n^k = 0$, we have $\sum\limits_{m \neq n} a_m^K O_m^k = \infty$.

Algorithm 2 POST Algorithm

1: **initialization**:
2: every TN n chooses local computing, i.e. $\mathbf{a}_n(0) = [1, 0, \cdots, 0]$
3: **end initialization**
4: **repeat** for every TN n and every decision slot in parallel:
5: broadcast the parameters message to the achievable HNs
6: receive the measurements of $\sum_{m \neq n} a_m^k O_m^k, O_n^k$ from the achievable HNs
7: calculate the optimal solution to problem (3.3) according to (3.6)–(3.10)
8: **if** $T_n(t) - T_n^*(t) > \epsilon$ **then**
9:    send RTU message to HNs for contending for the strategy update opportunity
10:    **if** not permitted to update the task offloading strategy **then**
11:       update the strategy $\mathbf{a}_n(t+1) = \mathbf{a}_n^*(\mathbf{A}_{-n}(t))$ for next slot
12:    **else**
13:       maintain the current strategy $\mathbf{a}_n(t+1) = \mathbf{a}_n(t)$ for next slot
14:    **end if**
15: **else**
16:    maintain the current strategy $\mathbf{a}_n(t+1) = \mathbf{a}_n(t)$ for next slot
17: **end if**
18: **until** END message is received from the HNs

---

algorithm called POST for the POST game, as shown in Algorithm 2. For more details, please refer to (Liu et al., 2020).

3.2.3 Delay Reduction Performance

3.2.3.1 Price of Anarchy

We first make some theoretical analysis of the proposed task scheduling algorithms. In game theory, the price of anarchy (PoA) is usually used to evaluate the efficiency of an NE solution. It answers the question that how far is the overall performance of an NE from the socially optimal solution. The PoA is defined as the ratio of the maximum social welfare, i.e. the total utility, achieved by a centralized optimal solution, to the social welfare, achieved by the worst case equilibrium (Han et al., 2012). Here, since we consider the cost function, we define the PoA as the ratio of the system-level (average) delay, achieved by the worst case equilibrium, over the system-level (average) delay, achieved by the centralized optimal solution. In this case, the smaller the PoA, the better performance is.

To be specific, let Γ be the set of NEs of the POMT game G_1 and $\mathbf{a}^* = \{a_1^*, a_2^*, \cdots, a_N^*\}$ be the centralized optimal solution that minimizes the system-level average delay. Then, the PoA is defined as

$$\text{PoA} = \frac{\max\limits_{\mathbf{a} \in \Gamma} \sum\limits_{n \in \mathcal{N}} T_n(\mathbf{a})}{\sum\limits_{n \in \mathcal{N}} T_n(\mathbf{a}^*)}. \tag{3.11}$$

For the POMT game G_1, we have the following lemma.

Lemma 3.4 *For the POMT game G_1, the PoA of the system-level average delay satisfies that*

$$1 \leq PoA \leq \frac{\sum_{n=1}^{N} \min\left\{O_{\max}, \frac{z_n \gamma_n}{f_n}\right\}}{\sum_{n=1}^{N} \min\left\{O_{n,\min}, \frac{z_n \gamma_n}{f_n}\right\}}, \tag{3.12}$$

where $O_{\max} \triangleq \max_{k \in \mathcal{K}_n} \sum_{n \in \mathcal{N}} O_n^k$, $O_{n,\min} \triangleq \min_{k \in \mathcal{K}_n} O_n^k$, $\mathcal{K}_n = \{k \in \mathcal{K}|b_n^k = 1\}$.

Similarly, for the POST game G_2, we have the following lemma.

Lemma 3.5 *For the POST game G_2, the PoA in terms of the system average delay satisfies that*

$$1 \leq PoA \leq \frac{\sum_{n=1}^{N} \min\left\{O_{\max}, \frac{z_n \gamma_n}{f_n}\right\}}{\sum_{n=1}^{N} \min\left\{O_{n,\min}, \frac{z_n \gamma_n}{f_n}\right\}} \times (1 + |\mathcal{K}_n|), \tag{3.13}$$

where $|\mathcal{K}_n|$ is the cardinality of set $\mathcal{K}_n$.

For the detailed proof, please refer to (Yang et al., 2019, Liu et al., 2020).

3.2.3.2 System Average Delay

We finally evaluate the performance of the proposed POMT algorithm and POST algorithm via simulations. Consider an 80×80 m square area, and divide it into 4×4, 16 grids in total. Deploy an HN in the center of every grid, which not only possesses rich communication resource, but also provides strong computation capability. Every HN has a coverage range of 20 m, a bandwidth of 5 MHz, and a computation capability of 5 GHz (Chen et al., 2015b). The TNs are randomly scattered in the area. To account for the heterogeneity of TNs and tasks: (1) the size and processing density of tasks are randomly chosen from [500, 5000] KB and [500, 3000] cycle/bit, which are in accordance with the real measurements in practice[2] (Kwak et al., 2015); (2) the computation capability of TNs is randomly selected from the set [0.8, 0.9, 1.0, 1.1, 1.2] GHz[3] (Nowak et al., 2018). In addition, the transmission power of all TNs is set as 100 mW and the channel noise power is set to be −100 dBm. The channel gain g_n^k is modeled by $(d_n^k)^{-\alpha}$, where d_n^k is the distance between TN n and HN k and $\alpha = 4$ is the path loss factor (Chen et al., 2015b). Unless stated otherwise, ϵ is chosen as 10^{-3}. All simulation results are averaged over 500 simulation rounds, if not specified.

Figure 3.2 compares the proposed POMT algorithm with the following baseline solutions in terms of the system average delay:

• Local computing (local): every TN chooses to process its task on local device.

---

2 For example, the data size and processing density of face recognition application are about 5000 KB and 3000 cycle/bit respectively (Kwak et al., 2015).
3 Here, we model the TNs as various mobile devices with limited computation capability. For example, smartphones with ARM Cortex-A8 processors are with about 1 GHz clock speed (Shah-Mansouri and Wong, 2018).

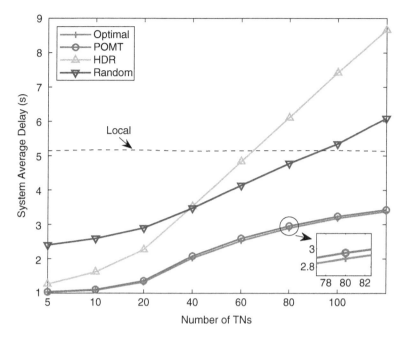

Figure 3.2 System average delay with different number of TNs (non-splittable tasks).

- HDR scheduling (HDR): every TN chooses to offload its task to the HN with the highest data rate (HDR). It imitates the myopic behavior of individuals.
- Random scheduling (random): every TN randomly chooses to process its task on local device or offload it to a randomly selected HN.
- optimal scheduling (optimal): the centralized optimal solution in terms of system average delay is obtained, utilizing the cross-entropy method (Liu et al., 2018). The cross-entropy method is an efficient method for finding near-optimal solutions to complex combinatorial optimization problems (Rubinstein and Kroese, 2013).

As illustrated in Figure 3.2, the system average delay increases as the number of TNs increases, except for the local computing. This is because that the communication resources and computation capabilities of HNs become insufficient when the number of TNs increases. As a result, fewer TNs choose to offload tasks to HNs rather than to process tasks on a local device, and this results in the rise of the system average delay. For local computing, it does not rely on the resources of HNs, and thus it remains unchanged. In addition, it can be observed that the proposed POMT algorithm can always achieve the near-optimal performance in terms of system average delay. Specifically, it can reduce 27–74%, 40–67% and 49–53% of system average delay, compared to the local computing, HDR scheduling and random scheduling, respectively.

Figure 3.3 compares the POST algorithm with the following baselines in terms of the system average delay:

- Local computing (local): every TN processes its task on local device.

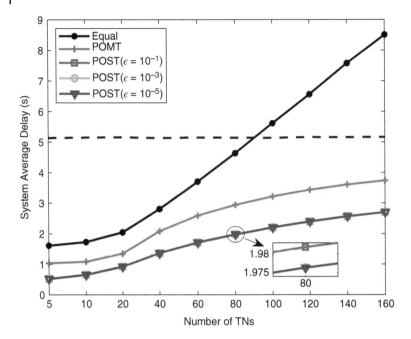

Figure 3.3 System average delay with different number of TNs, under different schemes and ϵ (splittable tasks).

- Equal scheduling (equal): every TN divides its task into multiple subtasks with equal size, and all subtasks are processed on local TN and all available HNs in parallel. This is a heuristic solution that utilizes all available computation resources.
- POMT: this solution can be taken as a worst case or upper bound for our problem.

As shown in Figure 3.3, the system average delay increases with the number of TNs increasing, and the POST algorithm can always offer the best performance and reduce over 50% and 25% delay than the equal scheme and the POMT scheme, under the ideal condition, i.e. no extra time cost for dividing tasks and assembling results. It is also interesting to note that the value of ϵ almost has no impact on the system average delay. In particular, as ϵ is smaller than the magnitude of 10^{-3}, the system average delay achieved by the POST algorithm maintains the same. This indicates that the POST algorithm possesses good robustness.

To further study the impact of extra time cost for dividing tasks and assembling results, Figure 3.4 plots the POST algorithm with different extra time costs. Typically, the time cost for dividing tasks and assembling results is proportional to the complexity of tasks, and inversely proportional to the computation capability of devices. Thus, the time cost for dividing tasks and assembling results is set to be proportional to that of local computing, such as 10% (POST+10% local), 20% (POST+20% local), and 30% (POST+30% local).

It is interesting to notice that the parallel processing (POST) does not always perform better than the non-parallel processing (POMT). It depends on the extra time cost for dividing tasks and assembling results and the network size. Generally speaking, the smaller the extra time cost for dividing tasks and assembling results is, then the more advantageous the POST

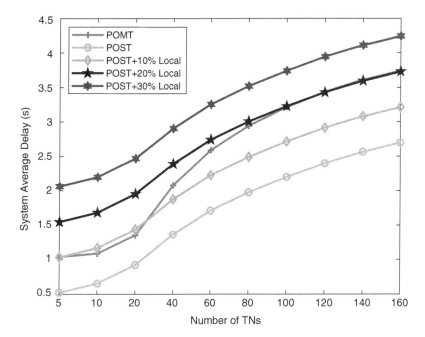

Figure 3.4 System average delay with different number of TNs, under different schemes and extra time costs (splittable tasks).

algorithm is. While, the larger the number of TNs is, then the more advantageous the POST algorithm is. For example, if the time cost of dividing tasks and assembling results is 10% of that of local computing, the POST algorithm can obtain lower system average delay than the POMT scheme, as long as the number of TNs is larger than 20. And, the performance gain becomes larger as the number of TNs rises. While, if dividing tasks and assembling results take up 20% of the time cost of local computing, the POST algorithm performs worse than the POMT scheme in terms of the system average delay, when the number of TNs is less than 100. While, the performance of POST algorithm and the performance of POMT scheme almost maintain the same, as TNs are more than 100.

It is also worth noting that it seems that whatever the number of TNs, the POMT scheme almost can not achieve better performance than the POST+10% local scheme. It may be claimed that dividing tasks into subtasks and offloading them to multiple HNs for parallel processing is a better choice, as long as the extra time cost for dividing tasks and assembling results takes up less than 10% of the time cost of local computing.

3.2.3.3 Number of Beneficial TNs

The beneficial TNs are those who can reduce its delay compared with local computing, via offloading tasks to HNs, i.e. $\{n \in \mathcal{N} | T_n < T_n^0\}$. This metric answers the question that how many TNs can benefit from computation offloading, i.e. reduce its delay via computation offloading. Therefore, it can not only reflect the individual-level delay performance, but also depict the satisfaction level of TNs to the task scheduling from individual respects.

As shown in Figure 3.5, the number of beneficial TNs achieved by the POMT algorithm increases with the number of TNs increasing and levels off. Moreover, the POMT

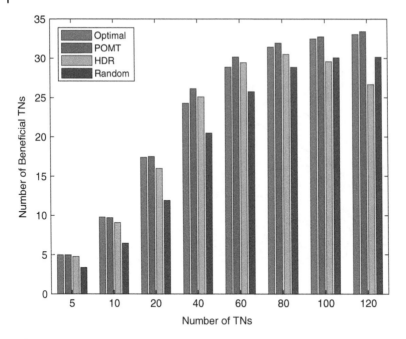

Figure 3.5 Number of beneficial TNs with different number of TNs (non-splittable tasks).

algorithm can always achieve a higher number of beneficial TNs than the other three schemes, especially when the number of TNs is large. This is because, in the case of a large number of TNs, there exists serious contention among TNs for the resources of HNs. While all TNs offloading tasks to HNs will further worsen such contention and finally damage each other's interests, and thus the number of beneficial TNs reduces. In contrast, the POMT algorithm can coordinate the contention among TNs and achieve more beneficial TNs until the network is saturated, i.e. the number of beneficial TNs levels off. As for the optimal solution, it may sacrifice individuals' interests for better system performance, and thus fewer TNs would get benefits from computation offloading.

As demonstrated in Figure 3.6, the number of beneficial TNs increases as the number of TNs increases. In addition, the number of beneficial TNs achieved by the POST algorithm is much larger than that achieved by the equal scheme and the POMT scheme, especially when the number of TNs is large, i.e. the communication resources and computation resources are insufficient. The reason is two-fold: (1) by dividing the whole task into multiple subtasks and offloading them to multiple HNs for parallel processing, the efficiency can be further improved, compared with the POMT scheme; (2) by formulating the task offloading problem as a game, a better negotiation can be achieved among TNs, compared with the equal scheme. Therefore, more TNs can benefit from computation offloading and the POST algorithm can also offer much better performance in terms of the system average delay, compared with the equal scheme and POMT scheme. These advantages are much more evident when there are a large number of TNs in the network, because there will be more serious contention among TNs. Again, as can be observed, ϵ almost has no influence on the algorithm performance in terms of the number of beneficial TNs, as long as ϵ is small enough.

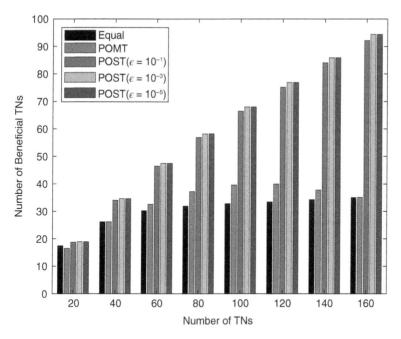

Figure 3.6 Number of beneficial TNs with different number of TNs (splittable tasks).

3.2.3.4 Convergence

Figure 3.7 demonstrates the number of decision slots required for the POMT algorithm and the POST algorithm to converge, with different numbers of TNs and under different ϵ. As shown, the number of decision slots increases sub-linearly with the number of TNs, regardless of ϵ. This indicates that both the POMT algorithm and the POST algorithm scale well with the size of network. In addition, the number of decision slots increases slightly as ϵ decreases. This is because that it takes more effort for the POST algorithm to get closer to the NE (when ϵ is smaller). However, as ϵ is small enough, i.e. below the magnitude of 10^{-3}, the number of decision slots no longer increases. Further, the number of decision slots required for the POST algorithm is usually more than the number of decision slots required for the POMT algorithm. The reason behind is intuitive and straightforward. As the competition among TNs gets more serious and complex for dividing original tasks into multiple subtasks, it is more difficult to reach an equilibrium among TNs.

3.3 Joint Computing, Communication, and Caching Resource Management for Energy-efficient Applications

In the previous Section 3.2, we studied the joint computation and communication resource management for delay-sensitive applications. We analyzed how much gain can be achieved in terms of the delay reduction by collaboration between the resource sharing/management among FNs. However, we just assumed the coarse-granularity 2C resource management, where computation and communication resources are equally

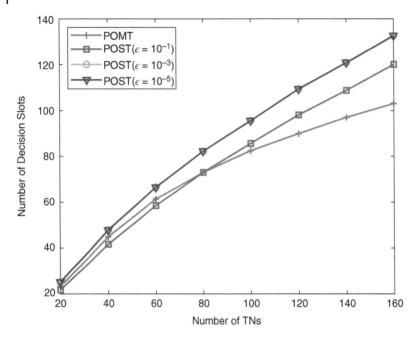

Figure 3.7 Number of decision slots with different number of TNs.

shared among FNs. A common question naturally arises: how do we perform efficient 3C resource sharing/management? How do we conduct the fine-granularity 3C resource management to further efficiently leverage 3C resources? And, how much gain can be achieved in terms of the energy saving? Owing to devices' heterogeneous 3C resources and various task service requirements, it is highly non-trivial to accomplish efficient resource scheduling for diverse tasks in a satisfactory way. Although most previous studies have involved the sharing of 1C or 2C resources (Iosifidis et al., 2016, Syrivelis et al., 2015, Chi et al., 2014, Chen et al., 2015a, Jiang et al., 2015, Chen et al., 2016, Stojmenovic and Sheng, 2014, Destounis et al., 2016, Qun and Ding, 2017, Huang et al., 2017), the design of an efficient holistic 3C resource sharing mechanism in fog computing is much less understood.

Motivated by the grand challenge above, in this section, we target the far-ranging IoT applications with different data stream processing tasks and propose F3C, an efficient fog-enabled 3C resource sharing framework for energy-efficient sensing data stream processing. In Figure 3.8, we present the illustration of collaborative task execution in fog computing. Suppose our fog architecture is comprised of two network layers: a cloud layer and a fog layer, in which the cloud layer with an IoT sensing platform runs a series of IoT applications, while the fog layer accommodates a variety of FNs, such as edge servers and mobile IoT devices embedded with various sensors. In Figure 3.8, an IoT platform in the cloud layer generates four IoT tasks to the fog layer, requiring adjacent FNs to process the sensed data stream. As is shown, task 1 is executed by device 1, device 2, device 3 and edge server 3, and task 2 is handled by device 3 and device 4 while task 3 is carried out by edge server 1 and edge server 2. In the end, all the task results need to be sent back to the IoT platform through cellular links for further data analytics or result aggregation.

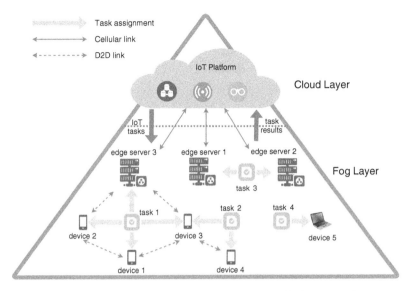

Figure 3.8 An illustration of collaborative task execution in fog computing.

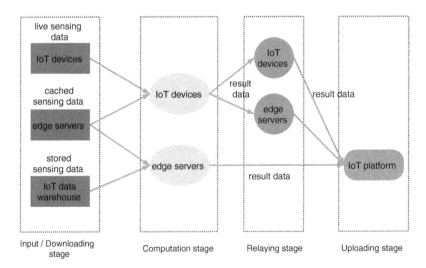

Figure 3.9 Cooperative F3C task execution.

As illustrated in Figure 3.9, the execution of an IoT data processing task at the fog layer could involve 3C resource sharing across multiple FNs.

To maximize the system-wide efficiency of the F3C collaborative task execution in FNs, we concentrate on an energy minimization problem in unit time. Unfortunately, such optimization problem is proven to be NP-hard. When we take further steps to dissect the task execution structure formed by several FNs, it occurs that a network flow model appropriately conforms to the 3C resource sharing for each task among cooperative FNs as a task team. Hence, for our F3C algorithm design, the proposed task team formation algorithm is tailored to accomplish such collaboration of multiple FNs in a task team and seek iterative task team formation for all the tasks. The iterative task team formation finally converges to a stable system point, where each task owns a stable set of processing FNs to achieve energy efficiency and no task will change its team formation.

3.3.1 Fog-enabled 3C Resource Management Framework

In our F3C framework, an IoT data stream processing task execution involves four stages: input/dowloading, computation, relaying, and uploading as presented in Figure 3.9. First at the input/downloading stage, required sensing data of a task can be partially or totally offered through several sources: FNs or some online data warehouse in the IoT platform, since the data can be sensed by sensors in FNs or has been sensed and stored in some online data warehouse before. Then at the computation stage, all sensing data can be allocated to one or multiple FNs for processing to leverage the resources in its proximity for fast data stream processing at the edge. Eventually at uploading stage, the result will be sent back to the IoT platform directly (e.g., for further analytic or result aggregation) through cellular connections or first relayed to adjacent FNs with high-quality cellular connections at the relaying stage to help to upload. To sum up, in the F3C framework there are three roles for FNs to play: caching devices, computing devices and relaying devices.

3.3.1.1 System Resources

We first describe the 3C resources in F3C framework as follows.

IoT data cache

We assume a set of IoT sensing data streams $\mathcal{K} = \{1, ..., K\}$ distributively generated or cached in FNs. These sensing data streams can be fetched from the corresponding sensing IoT device and the gateway edge server caches through local device-to-device (D2D) links or downloaded from the online IoT data warehouse (if some data has been sensed and stored before) through cellular links by computing devices.

IoT devices and edge servers

In our architecture, we homogenize IoT devices and gateway edge servers as FNs, which though vary in types and amount of resources. Let $\mathcal{N} = \{1, 2, ..., N\}$ characterize the set of all FNs. Then, for $\forall n \in \mathcal{N}$, it owns various sorts of resources represented by a tuple:

$$C_n = (C_n^{\text{cpu}}, C_n^{\text{com}}, C_n^{\text{down}}, C_n^{\text{up}}),$$

where C_n^{cpu}, C_n^{down} and C_n^{up} respectively denote device n's available computation capacity (CPU cycles per unit time), downloading and uploading data rate (bits per unit time). Vector $C_n^{\text{com}} = \{C_{n1}^{\text{com}}, ..., C_{nN}^{\text{com}}\}$ indicates the D2D transmission rates involving n and other FNs in $\mathcal{N}$ where C_{nm}^{com} represents the D2D transmission rate from device n to device m. Additionally for the expression of energy consumption of each device n, we adopt power parameters $P_n^X (X = \text{cpu, down, up, tr, re})$ which respectively represent CPU power, cellular receiving power, cellular transmission power, D2D transmission power and D2D receiving power per data bit.

3.3.1.2 Task Model

There is a set of data stream processing tasks launched by the IoT platform, which is denoted as $\mathcal{T} = \{1, 2, ..., T\}$ and demands execution of low-latency and energy-efficiency by FNs. For each task $t \in \mathcal{T}$, it requires different kinds of resources which can be given by the following tuple:

$$D_t = (D_t^{\text{ca}}, D_t^{\text{cpu}}),$$

where parameter $D_t^{\text{ca}} = k$ ($k \in \mathcal{K}$) indicates that task t demands IoT sensing data stream k for input[4] . D_t^{cpu} represents the computation density (CPU cycles per data bit per unit time) to process the input data stream of task t. In general, the data stream processing is conducted based on some atomic data stream block (e.g., an image/video frame of certain size for object recognition, a mini-batch of dataset for deep learning model training). Thus, for a data stream k, we also introduce B_k to denote the size of the atomic data stream block and L_k to indicate required processing rate, i.e. the number of blocks per unit time. Let $\mathcal{N}_t$ denote the subset of FNs that are feasible for handling task t, which is subject to the geo-distribution coverage requirement of task t. For a task t, we denote C_{nk}^{ca} as the maximum data stream block generation rate (i.e., blocks per unit time) to be supported by device $n \in \mathcal{N}_t$ for data $k \in \mathcal{K}$. When $C_{nk}^{\text{ca}} = 0$, it means that device n does not have sensing data k. Similarly we define the variable C_{0k}^{ca} by regarding the online IoT data warehouse as the data caching device 0.

3.3.1.3 Task Execution

As already mentioned, each task execution goes through multiple procedures including input/downloading, computation, relaying, and uploading, each of which incurs energy cost on FNs. Therefore, we formulate the energy consumption in the corresponding four stages as below.

D2D input/downloading stage

At this stage, required sensing data of the task can be offered by FNs or the IoT data warehouse. FNs in proximity can provide partial or total sensing data locally cached, while the data can also be downloaded from the IoT data warehouse where it has been stored before. Thus, we need to calculate the energy cost incurred by data downloading and data D2D transmission respectively.

Downloading cost: We denote the decision variable that describes the data stream rate (stream blocks per unit time) of task t downloaded by device n from IoT data warehouse as $l_{t \to n}^{\text{down}}$. Then the downloading cost per unit time of task t is given as follows.

$$E_t^{\text{down}} = \sum_{n \in \mathcal{N}_t} P_n^{\text{down}} l_{t \to n}^{\text{down}} B_{D_t^{\text{ca}}}, \tag{3.14}$$

subject to:

$$\sum_{t \in \mathcal{T}} l_{t \to n}^{\text{down}} B_{D_t^{\text{ca}}} \leq C_n^{\text{down}}, \forall n \in \mathcal{N}_t, l_{t \to n}^{\text{down}} \in \mathbb{N}, \tag{3.15}$$

$$l_{t \to n}^{\text{down}} \leq C_{0D_t^{\text{ca}}}^{\text{ca}}, \forall n \in \mathcal{N}_t, l_{t \to n}^{\text{down}} \in \mathbb{N}, \tag{3.16}$$

---

4 For simplicity, here we regard processing different data streams as different tasks.

where $\mathbb{N}$ represents the set of non-negative integers. Constraint (3.15) guarantees that the maximum affordable downloading data stream rate of n should be larger than the total generating rate of all the sensing data from IoT data warehouse. Constraint (3.16) ensures the downloaded stream block rate for processing should not exceed the maximum block generation rate.

D2D input cost: To describe D2D communication energy incurred by data being obtained from the nearby FNs (e.g., the sensing IoT devices with cached data), let $l^{ca}_{t:m\rightarrow n} \in \mathbb{N}$ be the decision variable to represent the stream blocks of task t transmitted from caching device m to computing device n per unit time. For each $n \in \mathcal{N}$, we define $\Phi(n)$ as a set of FNs having feasible connections to n. The energy cost of D2D input communication of task t per unit time can be given by:

$$E^{in}_t = \sum_{n \in \mathcal{N}_t, m \in \Phi(n)} \sum l^{ca}_{t:m\rightarrow n} B_{D^{ca}_t} (P^{tr}_m + P^{re}_n), \tag{3.17}$$

subject to:

$$\sum_{t \in \mathcal{T}} l^{ca}_{t:m\rightarrow n} B_{D^{ca}_t} \leq C^{com}_{mn}, \forall m \in \mathcal{N}_t, n \in \Phi(m), l^{ca}_{t:m\rightarrow n} \in \mathbb{N}, \tag{3.18}$$

$$l^{ca}_{t:m\rightarrow n} \leq C^{ca}_{mD^{ca}_t}, \forall m \in \mathcal{N}_t, n \in \Phi(m), l^{ca}_{t:m\rightarrow n} \in \mathbb{N}. \tag{3.19}$$

Constraint (3.18) ensures that the maximum D2D data rate between m and n is larger than the total input data rate of all the tasks from m to n. The physical meaning of constraint (3.19) is similar to constraint (3.16) above.

Computation stage
At this stage, let decision variable $c^{cpu}_{n\rightarrow t}$ represent the amount of CPU capacity allocated by device n to task t, and $l^{cpu}_{t\rightarrow n} \in \mathbb{N}$ denote the stream blocks allocated by task t to device n for computation per unit time. Thus, the cost incurred by processing task t per unit time is given by:

$$E^{cpu}_t = \sum_{n \in \mathcal{N}_t} P^{cpu}_n D^{cpu}_t l^{cpu}_{t\rightarrow n} B_{D^{ca}_t}, \tag{3.20}$$

subject to:

$$0 < \sum_{t=1}^{T} c^{cpu}_{n\rightarrow t} \leq C^{cpu}_n, \forall n \in \mathcal{N}_t, \tag{3.21}$$

$$\sum_{n \in \mathcal{N}_t} l^{cpu}_{t\rightarrow n} = L_{D^{ca}_t}, \forall t \in \mathcal{T}, \tag{3.22}$$

$$l^{down}_{t\rightarrow n} + \sum_{m \in \Phi(n)} l^{ca}_{t:m\rightarrow n} = l^{cpu}_{t\rightarrow n}, \forall n \in \mathcal{N}_t, \tag{3.23}$$

$$l^{cpu}_{t\rightarrow n} B_{D^{ca}_t} D^{cpu}_t \leq c^{cpu}_{n\rightarrow t}, \forall n \in \mathcal{N}_t. \tag{3.24}$$

Constraint (3.21) ensures that each device's computation capacity won't be exceeded, while constraint (3.22) guarantees all the input data of task t should be completely processed on several computation FNs. Constraint (3.23) ensures that the allocated data size of task t to device n from both the cached devices and the IoT data warehouse equals the amount of data allocated to n for computation, namely, keeping the balance between input size and computation size for each computing device n. Particularly, constraint (3.24) is responsible for that the maximum affordable computing rate in n for task t needs to be no smaller than the total input stream processing rate. Thus the input sensing data stream can be processed in a timely way.

Relaying and uploading stage

We denote that the output processed data size of task t is proportional to its input data size with a ratio ρ_t. Moreover, let decision variable $l^{out}_{t:n\to m}$ denote the output size of task t that computational device n will relay to device m for uploading per unit time. As a result, the energy consumption of output and uploading of task t can be accounted as follows.

$$E^{out}_t = \sum_{n\in\mathcal{N}_t}\sum_{m\in\Phi(n)} l^{out}_{t:n\to m} B_{D^{ca}_t}(P^{tr}_n + P^{re}_m + P^{up}_m),\tag{3.25}$$

subject to:

$$\sum_{t\in\mathcal{T}} l^{out}_{t:n\to m} B_{D^{ca}_t} \leq C^{com}_{nm}, \forall n\in\mathcal{N}_t, m\in\Phi(n),\tag{3.26}$$

$$\sum_{t\in\mathcal{T}}\sum_{n\in\mathcal{N}_t} l^{out}_{t:n\to m} B_{D^{ca}_t} \leq C^{up}_m, \forall m\in\mathcal{N}_t,\tag{3.27}$$

$$l^{cpu}_{t\to n} = \frac{1}{\rho_t}\sum_{m\in\Phi(n)} l^{out}_{t:n\to m}, \forall n\in\mathcal{N}_t.\tag{3.28}$$

Here, we need to ensure the affordable data stream output rate between computing device n and relaying device m is not greater than the D2D data transmission rate in constraint (3.26). Also the total cellular uploading rate of a device m at the final stage will not exceed its cellular transmission capacity in constraint (3.27). Likewise, constraint (3.28) is responsible for keeping the balance between input and output data stream rate for each computing device n.

Considering the energy consumption at various stages, we derive the total energy consumption of task t as $E_t = E^{down}_t + E^{in}_t + E^{cpu}_t + E^{out}_t$ in the F3C model which is integrated as a general resource sharing framework in fog collaboration. It can also be reduced to a 2C or 1C sharing model due to its generality as shown in the examples of Figure 3.10 and 3.11. For instance in Figure 3.10, which depicts a type of 1C resource sharing, i.e.

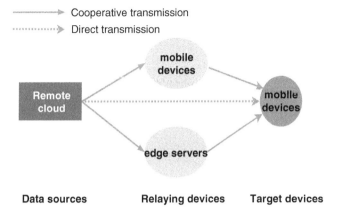

Figure 3.10   1C resource sharing example.

D2D connection   **Figure 3.11**   2C resource sharing example.

mobile devices

mobile devices

edge servers

Resource-rich devices   **Resource-poor devices**

communication resource only, in a cooperative downloading scenario, target devices are assisted by neighboring FNs with larger transmission bandwidth to download content or directly download content from remote cloud. As another illustration, Figure 3.11 displays computation offloading via wireless connections, i.e. computing and communication (2C) resource sharing, in which CPU-sufficient FNs allow tasks to be offloaded from CPU-poor target devices and offer task processing assistance.

In a nutshell, the proposed F3C framework enabling FNs to share three types of resources is able to generalize existing 1C/2C resource sharing models.

3.3.1.4 Problem Statement

We concentrate on an energy minimization problem to quantify the system-wide efficiency of the F3C framework per unit time as follows.

$$\min \ \sum_{t=1}^{T} E_t = E_t^{\text{down}} + E_t^{\text{in}} + E_t^{\text{cpu}} + E_t^{\text{out}}, \tag{3.29}$$

subject to: (3.15), (3.16), (3.18), (3.19), (3.21), (3.22), (3.23), (3.24), (3.26), (3.27), (3.28),

where detailed formulas of E_t^{down}, E_t^{in}, E_t^{cpu} and E_t^{out} are respectively referred to (3.14), (3.17), (3.20) and (3.25).

In order to analyze the solvability and complexity of problem (3.29), we unfold the hardness analysis in the following.

Theorem 3.3   *The energy minimization problem (3.29) for multiple task execution in F3C is NP-hard.*

The proof is available in (Luo et al., 2019). The NP-hardness of the energy minimization problem (3.29) for multiple task execution in F3C implies that for large inputs it is impractical to obtain the global optimal solution in a real-time manner. Thus, efficient approximate algorithm design with low-complexity is highly desirable and this motivates the F3C algorithm design in the following sections.

3.3.2 Fog-enabled 3C Resource Management Algorithm

3.3.2.1 F3C Algorithm Overview

Since the optimization problem in Section 3.3.1 is proven to be NP-hard, a common and intuitive solution is to design a feasible and computation efficient approach to approximately minimize the system-wide energy. Considering the combinatorial explosion phenomenon due to three dimensional resource collocation between devices and tasks, we adopt the divide-and-conquer principle and decompose the overall F3C algorithm design issue into two key sub-problems.

First of all, we observe that the IoT data stream processing of a single task as shown in Figure 3.9 between FNs through various connections resembles the network flow structure. For a task t, the sensing data stream from several source nodes in $\mathcal{N}_t$ are transmitted to proximal devices for computation and then the processed results from computing nodes will be relayed back to the IoT platform, which is destination node.

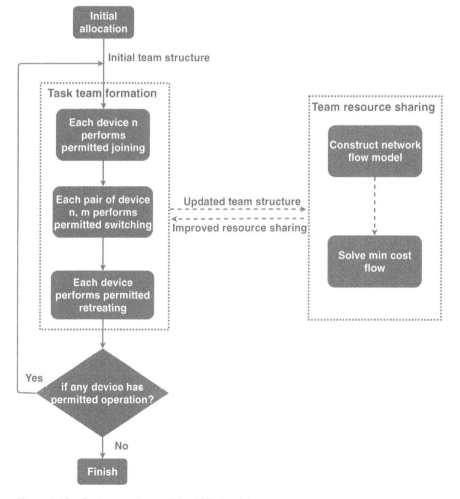

Figure 3.12 Basic procedures of the F3C algorithm.

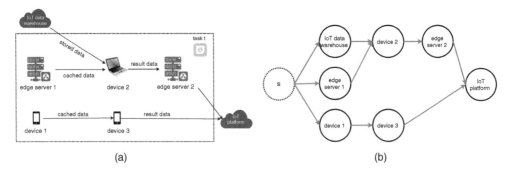

Figure 3.13 Minimum cost flow network transformation in single task. (a) Cooperative execution of task *t* by practical FNs (b) Transformed flow network from *t*'s cooperative execution.

In consequence we can set up a network flow model denoted by an auxiliary graph G to represent cooperative execution for a single task t and figure out the sub-problem, namely, the optimal resource allocation of each task t given available FNs in $\mathcal{N}_t$. In particular, as in Figure 3.13(a), the F3C resource sharing solution to each sub-problem for task t will obtain a set of caching devices like edge server 1 and device 1, a set of computing devices including device 2 and device 3, and a set of relaying devices for result data transmission back to IoT platform such as edge server 2. While computing device 3 chooses to transmit result data directly by itself. In particular, the IoT data warehouse can be regarded as a source of sensing data in case there exists no cached data in the available FNs of task t. We transform the cooperative FNs in Figure 3.13(a) to Figure 3.13(b), which is the minimum cost flow solved in the network flow model of task t. A virtual node s in Figure 3.13(b) represents the source of total processing data stream.

Then, for the multi-task scenario, we can divide the set of FNs into multiple task teams, with each task team executing a single task. Since within a task team the optimal resource sharing can be derived based on the min cost flow solution, accordingly we devise an efficient cooperative team formation mechanism to gradually improve the system-wide performance.

As shown in Figure 3.12, the basic procedures of our scheme are elaborated as follows:

- We first carry out an initial F3C resource allocation for all the tasks (e.g., using a random task assignment scheme), obtaining an initial cooperative FN assignment for the tasks, i.e. task team formation.
- Then we define that, for each FN, it has three possible operations to perform to improve the task team formation: joining, switching, or retreating (which will be formally defined in Section 3.3.2.3 later on). These operations are permitted to be carried out if they can improve the system-wide performance without damaging any individual task's utility.
- When a device performs a permitted operation, it incurs a change of the structure of some task teams. Thus we will construct a new network flow model to optimize the resource sharing for each updated task team.
- All the devices iteratively perform possible operations until there exists no permitted operation, i.e. no change of all the task team formations.

Since the minimum cost flow problem can be efficiently solved in practice and the task team formation can converge within small number of iterations, the overall F3C algorithm can converge in a fast manner and hence is amendable for practical implementation.

3.3.2.2 F3C Algorithm for a Single Task

In this section, we focus on the optimal energy minimization of the single task sub-problem, i.e. considering the resource pooling for a single task t with its available FNs $\mathcal{N}_t$. By constructing the network flow model to solve min cost flow, we can obtain optimal F3C resource sharing for single task t.

Intuitively, in the optimal resource sharing strategy for a task, the task members (i.e. involved FNs) cooperate to complete the task by sharing 3C resources and minimize their processing energy overhead. Let $\Phi(n, t) = \Phi(n) \cap N_t$ represent the set of D2D connected devices of n in task t's coverage. We can solve such problem by calculating the min cost denoted as E_t^G of the network flow model characterized by an auxiliary graph G as illustrated in Figure 3.14.

Minimum cost flow network construction: Corresponding to the input/downloading, computing, relaying, and uploading stages in each task processing, we design graph G of the network flow model of each task t by the following steps:

- We first set up an artificial source node s and a sink node d (i.e. IoT platform).
- Given a set of caching devices $\mathcal{N}_t^{ca} \subseteq \mathcal{N}_t$, including an IoT data warehouse, which has cached input data of task t, for $\forall m \in \mathcal{N}_t^{ca}$, we create a caching node m and an edge from s to m with the maximum capacity $C_{mD_t^{ca}}^{ca}$ and zero weight.
- As for the set of computation devices $\{n : n \in \mathcal{N}_t^{cpu} = \mathcal{N}_t\}$, we create a corresponding computing node n and edges $\{e_{mn} : m \in \mathcal{N}_t^{ca} \cap \Phi(n, t)\}$ in which the input of data stream processing rate of n (i.e. the capacity) is not allowed to exceed the transmission bandwidth between m and n if m denotes IoT data warehouse. The weight equals to the sum of computation energy in n and transmission energy between m and n per data stream block. Particularly, if $m = n$, it means caching device m directly computes the data on its own such that the weight of e_{mn} is zero.

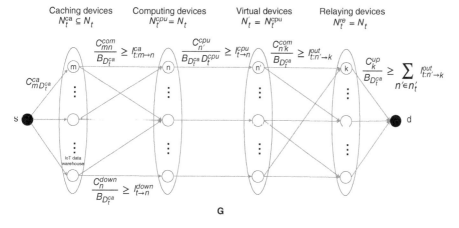

Figure 3.14 An auxiliary graph G of minimum cost flow for task t.

- To account for the total the CPU capacity constraint of a computing device, we add a virtual duplicated node n' for each node $n \in \mathcal{N}_t^{\text{cpu}}$ where n and n' have the same attributes and set up an edge $e_{nn'}$ for each pair of n and n'. Thus we set the capacity constraint of $e_{nn'}$ that the sensing data which n receives for computing per unit time is no larger than the CPU capacity of device n'. Intuitively the weight of $e_{nn'}$ is zero.
- For each relaying device $k \in \mathcal{N}_t^{\text{re}} = \mathcal{N}_t$, we create a relaying node k and corresponding edges $\{e_{n'k} : n' \in \Phi(k, t) \cap \mathcal{N}_t\}$, while meeting the constraint that the D2D data rate between n' and k (i.e. the capacity of $e_{n'k}$) should be faster than the result data output rate. Since the output data size will scale down by a factor of ρ_t after computation, we hence scale up the edge capacity here by a factor of $1/\rho_t$ from the perspective of input data stream. On the other hand, the weight of $e_{n'k}$ equals to the D2D transmission energy per data stream block between n' and k multiplied by ρ_t if $n' \neq k$ is to reflect the actual output overhead, otherwise it becomes zero.
- Finally, we create edges $\{e_{kd} : k \in \mathcal{N}_t^{\text{re}}\}$ where d denotes the sink node, i.e. IoT platform while meeting the capacity constraint by setting the edge capacity as the scaled maximum uploading data rate of relaying device k. And the corresponding weight equals to be the uploading energy consumption per data stream block between k and IoT platform, likewise multiplied by ρ_t to balance the network flow.

For a task t, we set up its network flow model, i.e. the auxiliary graph G as elaborated above and figure out the minimum cost flow in G of sending a flow of volume $L_{D_t^{\text{ca}}}$ from the source node s so as to obtain the optimal solution of 3C resource sharing with the set of available FNs $\mathcal{N}_t^{\text{ca}} \cup \mathcal{N}_t^{\text{cpu}} \cup \mathcal{N}_t^{\text{re}}$. Specifically, we adopt a network simplex algorithm to solve the minimum cost flow in G which is known to have a polynomial running time (Orlin, 1997). Hence the complexity of solving minimum cost flow in F3C is derived to be $\mathcal{O}((|\mathcal{N}_t^{\text{ca}}| + |\mathcal{N}_t^{\text{cpu}}| + |\mathcal{N}_t'| + |\mathcal{N}_t^{\text{re}}|)|E|^2)$ as (Orlin, 1997) where E denotes the set of edges in G. Moreover, it is known that when the input data flows are integer variables, the solved network flow solution is also integral (Orlin, 1997), which is highly desirable in our case such that the divided data streams across different processing devices are integral (i.e. stream block numbers per unit time).

3.3.2.3 F3C Algorithm For Multiple Tasks
We then consider the general case of multiple tasks. Given the single task solution above, the key idea of solving the multi-task problem is to efficiently form different task teams such that a task team is responsible for a single task. In the following, we will design efficient team formation operations to adjust the team association of the FNs, in order to gradually improve the overall system performance iteratively.

First, we introduce some critical mechanisms and definitions about forming task teams in the following.

Definition 3.1 In our system, a **task team** C_t is termed as a subset of $\mathcal{N}_t$ assigned to collectively process task t, where we can define corresponding utility function as $v(C_t) = -E_t^G$ which takes a minus sign over the outcome of solving minimum-cost flow problem characterized by auxiliary graph G for the task team C_t.

Definition 3.2 **Team structure** $\text{TS} = \{C_t : t \in \mathcal{T}\}$ is defined as the set of teams of all the tasks, where $C_t = \{n : n \in \mathcal{N}_t\}$, such that the total utility of the generated team structure can be denoted as $v(\text{TS}) = \sum_{t=1}^{T} v(C_t)$.

For the whole system, which kind of team structure it prefers to form depends on $v(\text{TS})$ and each $v(C_t), \forall t \in \mathcal{T}$. Given that there might exist devices finally idle, i.e. not joining any team, we take no consideration of single device's utility. Hence, the preference on team structure TS is defined as follows.

Definition 3.3 Given two different team structures TS^1 and TS^2, we define a preference order as $\text{TS}^1 \triangleright \text{TS}^2$ if and only if $v(\text{TS}^1) > v(\text{TS}^2)$ and for $C_t^1 \in \text{TS}^1$ and $C_t^2 \in \text{TS}^2$ with $C_t^1 \neq C_t^2$ of each task t, we have $v(C_t^1) \geq v(C_t^2)$, indicating team structure TS^1 is preferred over TS^2 by all the tasks.

Next, we can solve the energy minimization problem by constantly improving total utility $v(\text{TS})$ in accordance with preference order $\triangleright$, resulting in termination with a stable team structure TS^* such that no device in the system will deviate from the current team structure when it gets to TS^*.

Permitted operations in task team formation
Obviously the change of team structure TS basically results from each FN's participation between different tasks. For each device it is allowed to perform some utility improving operations to join different teams based on $\triangleright$. Thus we define three possible permitted operations for FNs to perform in the task team formation procedure.

Definition 3.4 A **switching operation** between device n in team C_t and device m in team $C_{t'}$ means that devices n and m switch to each other's teams. Causing a change of team structure from TS^1 to TS^2, the switching operation is permitted if and only if $\text{TS}^2 \triangleright \text{TS}^1$.

Definition 3.5 A **joining operation** by device n which isn't included in the current team structure to join team C_t, which causes a change of team structure from TS^1 to TS^2, is permitted if and only if $\text{TS}^2 \triangleright \text{TS}^1$.

Definition 3.6 A **retreating operation** by device n to quit the current team structure, which causes a transformation of team structure from TS^1 to TS^2 is permitted if and only if $\text{TS}^2 \triangleright \text{TS}^1$.

After each improvement, the utility of the system increases by $\Delta = v(\text{TS}^2) - v(\text{TS}^1)$. With the iteration of every permitted operation by every device $n \in \mathcal{N}$, the team formation process will terminate to be stable where no device will deviate from the current team structure to form any other team structure.

Definition 3.7 A resource sharing strategy is at a **stable system point** if no device will change its strategy to form any other team structure at that point with the strategies of the other devices unchanged.

Algorithm 3 Initial Allocation Algorithm

Input: Set of devices $\mathcal{N}$, tasks $\mathcal{T}$ and sensing data $\mathcal{K}$
Output: Initial TS $= \{C_1, ..., C_T\}$
  1: **initialize** the set of devices having joined teams as $\mathcal{V} = \emptyset$
  2: **for** $t = 1$ to $t = T$ in $\mathcal{T}$ **do**
  3:   **sort** the set of devices $\mathcal{N}_t$ randomly and label them as 1 to N_t in turn
  4:   **initialize** the set of members of team C_t as $C_t = \emptyset$ and $v(C_t) = 0$
  5:   **for** $n = 1$ to $n = N_t$ **do**
  6:     **if** $n \notin \mathcal{V}$ **then**
  7:       **set** $C'_t = C_t \cup \{n\}$
  8:       **set up** auxiliary graph G'_t based on C'_t
  9:       **solve** min-cost flow in G'_t to obtain $v(C'_t)$
 10:       **if** $v(C'_t) > v(C_t)$ **then**
 11:         **set** $C_t = C'_t$ and $\mathcal{V} = \mathcal{V} \cup \{n\}$
 12:       **end if**
 13:     **end if**
 14:   **end for**
 15: **end for**

Task team formation algorithm

Next, we design a task team formation algorithm to achieve energy efficient collaboration of FNs and seek feasible 3C resource sharing for all the tasks. In our scenario, the task team formation process consists of two steps: *initialized allocation* and *task team formation*, as respectively in Algorithm 3 and Algorithm 4.

In the first stage, the *initialization allocation* procedure is as follows.

- For each task $t \in \mathcal{T}$, we first sort devices $\mathcal{N}_t$ randomly.
- Then we iteratively update task team C_t by adding an unorganized device in $\mathcal{N}_t$ sequentially if the derived utility $v(C_t)$ keeps increasing.
- After the initial team formation of all the tasks completes, we achieve an initial team structure TS $= \{C_1, ..., C_T\}$.

In the second step, each device n performs all possible permitted switching, joining or retreating operations until no device has permitted operation as described in Algorithm 4. In particular, each device n maintains a history set h_n to record the team composition it has joined before and the corresponding team utility value so that repeated calculations can be avoided.

Theorem 3.4 *The proposed task team formation process in Algorithm 4 will converge to a stable system point.*

The proof is available in (Luo et al., 2019). As we have proved that the task team formation (i.e. Algorithm 4) eventually terminates through finite utility improving iterations, let Z denote the number of utility improving iterations. Thus, the computational complexity of Algorithm 4 is $\mathcal{O}(Z|\mathcal{N}||E|^2)$ since we have derived the computational complexity of solving min cost flow in each iteration as $\mathcal{O}((|\mathcal{N}_t^{ca}| + |\mathcal{N}_t^{cpu}| + |\mathcal{N}_t'| + |\mathcal{N}_t^{re}|)|E|^2)$. As a result the

---

Algorithm 4 Task Team Formation Algorithm

Input: Set of devices $\mathcal{N}$, tasks $\mathcal{T}$ and sensing data $\mathcal{K}$
Output: Stable system point TS*
1: **Perform** initial allocation algorithm
2: **while** until no operation of any device is permitted: **do**
3:    **for** each device $n \in \mathcal{N}$ **do**
4:       **if** n in current structure TS **then**
5:          n performs **retreating operation** if permitted, and updates h_n and TS
6:       **else**
7:          n performs **joining operation** to each C_t which satisfies $n \in \mathcal{N}_t$ if permitted, then records the corresponding increment and updates h_n
8:          n joins the task team C_t which brings the highest utility increment
9:       **end if**
10:   **end for**
11:   **repeat**
12:     randomly pick device n in C_t and m in $C_{t'}$ where $t \neq t'$, perform **switching operation** if permitted and update h_n, h_m and TS
13:   **until** no switching operation is permitted by any pair of FNs n and m
14: **end while**

---

task team formation algorithm has a polynomial complexity and hence is computational efficient. Moreover, the performance evaluation in Section 3.3.3 shows that the proposed F3C algorithm can achieve superior performance over various benchmarks.

3.3.3 Energy Saving Performance

In this section, we carry out simulations to evaluate the energy saving performance of the proposed task cooperative team formation algorithm. From the perspective of device and task availability, all the devices and tasks are distributed randomly within an entire 1×1 KM area and typical parameters of the tasks (e.g., data compression/encoding (Chen et al., 2017)) are listed in Table 3.1, with a task's data uniformly distributed among its fog devices and the platform.

We compare our algorithm to several schemes in the following:

- *Non-cooperation scheme*: let each task be executed only by the best computing FN (IoT device or edge server).
- *Random cooperation scheme*: in this scheme, we randomly assign cooperative devices to each task if feasible.
- *Brute greedy scheme*: first, we randomly generate massive task execution ordering samples (1000 samples). According to each execution order, each task can select the most beneficial resources among the residual devices by solving the minimum cost flow. Then, we choose the lowest sum of task execution energy among all the ordering samples as the final overhead in this scheme.

In the following we will use the random cooperation scheme as the benchmark of 100% energy consumption ratio.

Table 3.1 Simulation setting.

Parameter	Value
Cellular data rate	[50, 100] Mbps
Cellular power	600 mW
Maximum D2D bandwidth	20 Mhz
D2D power	200 mW
CPU frequency	[1, 10] Ghz
CPU load	[2, 20]%
CPU power	900 mW
Processing density of tasks	[2000, 4000] cycle/bit
Stream block size	10 KB
Stream block number	[500, 2000]
Output size ratio	(0, 0.3]

3.3.3.1 Energy Saving Performance with Different Task Numbers

In this subsection, we fix the FN number as 300.

As described in Figure 3.15, based on the random cooperation scheme as a benchmark, our F3C algorithm fulfills the significant average energy consumption ratio. Under the simulation setting, with the task number T varying from 10 to 35, our algorithm accomplishes a significantly lower average energy consumption ratio as 31.6%, 44.7%, 48.0%, 33.9%, and 45.7% compared to the random cooperation scheme. Compared with the non-cooperation scheme and the brute greedy scheme, our algorithm is more efficient and achieves up to 30% and 10% performance gain in terms of energy consumption and reduction, respectively.

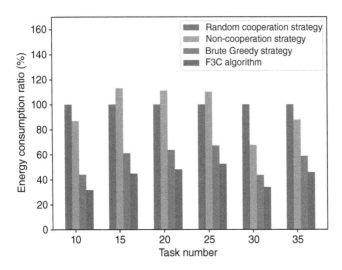

Figure 3.15 Performance gain in energy with different task numbers.

3.3.3.2 Energy Saving Performance with Different Device Numbers

In this subsection we explore the impact of different FN numbers on the performance gain by fixing the task number as 25.

In Figure 3.16, where the results of random cooperation scheme are regarded as baseline, our F3C algorithm still outperforms the other comparing schemes. For example, our algorithm can achieve up to 68%, 53%, and 10% performance gain over random cooperation, non-cooperation and the brute greedy scheme for energy cost reduction, respectively.

In Figure 3.17, we present the average participated device number comparison under the number of FNs = {100, 200, 300, 400, 500, 600} between the F3C algorithm and the brute

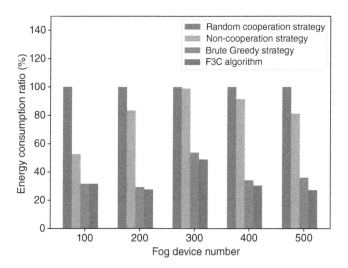

Figure 3.16 Performance gain in energy with different device numbers.

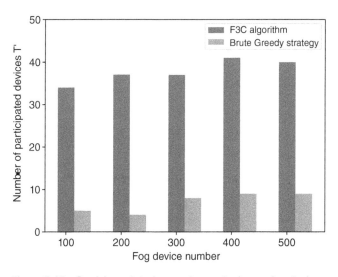

Figure 3.17 Participated device number under increasing device numbers.

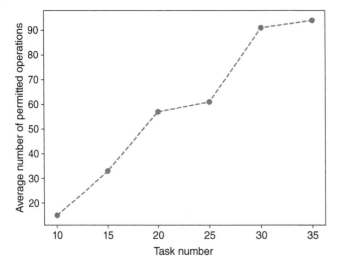

Figure 3.18 Average permitted operation number under different task numbers.

greedy scheme. The participated device number T' of the two schemes increases as the FN number grows; however, our algorithm is more efficient and achieves a much higher participation level for 3C resource sharing, which leads to more significant performance improvement.

We finally show the average iteration numbers of our algorithm in Figure 3.18 with the number of tasks varying from 10 to 35, and the average iteration numbers of our algorithm in Figure 3.19 with the number of FNs ranging from 100 to 600. The results show that the convergence speed of the proposed task team formation algorithm is fast and grows (almost) linearly as the task and device size increase.

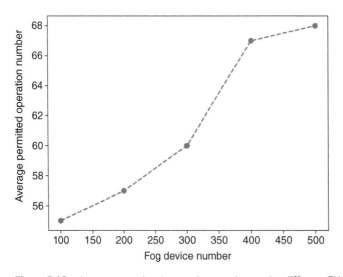

Figure 3.19 Average permitted operation number under different FN numbers.

3.4   Case Study: Energy-efficient Resource Management in Tactile Internet

In the previous two sections, we have studied joint 2C/3C resource management for delay-sensitive/energy-efficient applications. In this section, we will make a case study of energy-efficient resource management in the TI. We will investigate the response time and power efficiency trade-off and further implement a fog computing-supported self-driving bus system in the city of Dublin.

With the widespread deployment of high-performance computing infrastructure and advancement of networking and communication technology, it is believed that the vision of the TI will soon become a reality, transforming the existing content-delivery-based internet into skill-set delivery-based networks (Simsek et al., 2016). According to the Next-Generation Mobile Network (NGMN) Alliance (NGMN Alliance, 2015), TI is defined as the capability of remotely delivering real-time control over both real and virtual objects as well as physical haptic experiences through the internet. It will be able to contribute to the solution of many complex challenges faced by our society, enabling novel services and applications that cannot fit well in the current state-of-the-art networking and cloud computing architectures. Examples of these applications include long-distance education with immersive learning experience, high-precision remote medical diagnosis and treatment, high-sensitive industry control and automation, collision-avoidance for high-speed autonomous vehicles, high-fidelity Virtual/Augment Reality (VR/AR), etc. Recent analysis shows that the TI has the potential to generate up to $20 trillions to the global market, accounting for around 20% of the global GDP (Maier et al., 2016). Because the TI will provide critical services, it needs to be extremely reliable, ultra-responsive, and widely available. More specifically, according to the International Telecommunication Union (ITU), the TI must support latencies as low as 1 ms, reliability of around one second of outage per year, enhanced security, as well as sufficient computational resources within the range of a few kilometers from each end-user (ITU-T, Aug. 2014).

Fog computing has recently been introduced as a promising solution to accommodate the stringent requirements of the TI. However, the theoretical foundations for optimizing distributed fog computing systems to meet the demands of the TI are still lacking. In particular, computationally intensive services requiring low latencies generally demand more energy consumption from FNs. At the same time, many TI applications involve portable devices such as robots, drones and vehicles with limited power supplies. In addition, the fast growing power consumption of information and communication technologies and its impact on climate change have recently raised significant concerns in both industry and academia (Patel et al., 2014, Vaquero and Rodero-Merino, 2014). Existing cloud data centers in the US have already constituted more than 2% of the country's total electricity usage. Power consumption is expected to be significantly increased with the deployment of a large number of fog computing servers throughout the world. How to improve the efficiency of the power usage for fog computing networks while taking into consideration the stringent requirements of TI services is still an open problem.

Another issue is that, in contrast to other applications, workload generated by the TI can exhibit much higher temporal and geographical variations due to the bursty nature of human-generated traffic. For example, an autonomous vehicle that is trying to pass another will create a much larger computational workload and require much lower latency service compared to other vehicles that stick to pre-planned routes. How to efficiently distribute and orchestrate the workload of different FNs for parallel execution under real-time constraints is still an open issue.

In this section, we take steps towards addressing the above issues. In particular, we study energy-efficient workload offloading for fog computing systems that support the TI applications and services. We focus on optimizing two important performance metrics: (1) service delay, including the round-trip transmission latency between users and FNs, queuing delays, and workload transmission and forwarding latency among FNs as well as that between FNs and CDCs; and (2) FN power efficiency, measured by the amount of power consumed by FNs to process a unit of workload. We perform detailed analysis under different scenarios and derive the optimal amount of workload to be processed by FNs so as to minimize the response time under a given power efficiency. We quantify the fundamental trade-off between these two metrics.

To address the issue of skewed workload distribution among FNs, we study a cooperative setting in which the workload of an FN can be partially processed by other nodes in proximity. We observe that the response time and power-efficiency trade-off is closely related to the cooperation strategy among FNs. Accordingly, we propose a novel cooperation strategy called offload forwarding, in which each FN can forward a part or all of its unprocessed workload to other nearby FNs, instead of always forwarding workload that exceeds its processing capability to a remote CDC. We study the offload allocation problem in which all FNs jointly determine the optimal amount of workload to be forwarded and processed by each other to further reduce the response time.

Based on our analysis, we observe that, for most TI applications, it is generally impossible to optimize the workload distribution among FNs in a centralized fashion due to the following reasons: (1) deploying a central controller to calculate the amounts of workload processed by every FN may result in intolerably high information collection and coordination delay as well as high computation complexity at the controller; (2) the workload received by each FN can be highly dynamic, and constantly exchanging information about workload as well as computational resource availability among FNs can result in network congestion; and (3) FNs may not want to reveal their private information. Motivated by these observations, we propose a novel distributed optimization framework for cooperative fog computing based on dual decomposition. Our proposed framework does not require FNs to have back-and-forth negotiation or disclose their private information.

Motivated by the fact that the self-driving vehicle has been considered as one of the key use cases for the TI (ITU-T, Aug. 2014, Simsek et al., 2016), as a case study, we evaluate the performance of our framework by simulating a city-wide implementation of a self-driving bus system supported by a fog computing network. We analyz over 2500 traffic images of 8 existing bus routes operated in the city of Dublin and consider the scenario that these traffic data can be submitted and processed by a fog computing network deployed in a real wireless network infrastructure consisting of over 200 base stations of a major cellular operator in Ireland to ensure safe and efficient decision making and driving guidance for all the buses. We evaluate the service delay and power efficiency of fog computing networks in different areas of the city with different densities of FN deployment. Numerical results show that our algorithms can almost double the workload processing capacity of FNs in urban areas with high density of FN deployment. To the best of our knowledge, this is the first work that studies distributed workload allocation among cooperative FNs with energy efficiency awareness.

3.4.1 Fog-enabled Tactile Internet Architecture

A generic fog computing-supported TI architecture consisting of four major components, as illustrated in Figure 3.20 (Simsek et al., 2016, ITU-T, 2014, Maier et al., 2016):

(1) Operator: A human operator and/or Human-to-machine (H2M) interface that can manipulate virtual and/or real objectives using various input signals such as human gestures, touch, and voice. In some applications such as self-driving vehicles and autonomous robots, operators can also correspond to pre-calculated policies that can mimic the human behaviors/decision making processes. Operators can also expect feedback within a given time duration depending on the particular applications. For example, a real-time AR/VR game may require as low as 1 ms response time. Other applications, such as remotely controlled robots, can tolerate up to one second response time.

(2) Responder: One or multiple teleoperators (remotely controlled robots, machines, drones, etc.) that can be directly controlled by the operators. Responders interact with the environment and send feedback signals to the operators.

(3) Fog: The composition of a number of low-cost FNs that are characterized by limited computing capabilities and power supplies. FNs are deployed closer to end-users. According to the types of users served by FNs, the fog can be further divided into control fog and tactile fog. Control fog consists of FNs that can support computation-intensive services, such as analog-to-digital conversion, coding, signal processing, data compression, etc. Tactile fog consists of the FNs that are responsible for processing, compressing and sending feedback data generated through the interactions between responders and the environment. FNs may

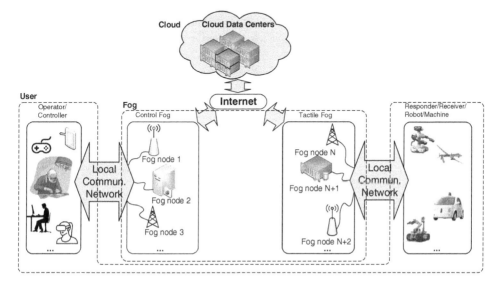

Figure 3.20 Fog computing-supported tactile internet architecture.

take the form of mini-servers within the wireless edge network infrastructure, reside in base stations (BSs), roadside units (RSUs), etc. Each FN serves a distinct set of users in its coverage area. Users first submit their workloads to the closest FN. Each FN will then need to carefully decide the amount of workload it can process locally. If multiple closely located FNs can communicate with each other (e.g., using high-speed backhaul connection links in a cellular network), some FNs may forward part of their workload to other nearby nodes to further improve the processing capability and balance the workload among nodes.

(4) Cloud: Large-scale CDCs equipped with powerful processing units. These data centers are often built in remote areas, far from end users.

FNs and users (operators or responders) may correspond to the same type of devices. For example, in some distributed mobile computing grid systems (e.g., IBM's world community grid project), the idle computing resources of some computers (i.e. FNs) can be used by other computers (i.e. users) to perform computationally intensive tasks. We consider a fog computing system that contains a set of N FNs $\mathcal{F} = \{1, 2, \ldots, N\}$. Any user can be associated with one or more of these FNs. The association between users and FNs can be made based on physical proximity, channel conditions, or prior agreements between users and network service provider. For example, if the fog is deployed by the cloud provider, users can send their service requests to CDCs following the same procedure of a traditional cloud computing system. The cloud provider can then delegate one or more nearby FNs to process the workload submitted by these users. Each FN j can process a non-negative portion α_j of its received workload using its local resources. Remaining workload, if any, is forwarded to the cloud. Note that $\alpha_j = 1$ means that FN j will process all its received workload. The workload arrival rate at each FN j, denoted by λ_j, is assumed to be fixed. We focus on two performance metrics:

(1) (Service) response time of end-users: The response time includes the round-trip time for transmitting the workload between a user and the associated FN as well as the queuing delay at the fog. Given their proximity to to users, FNs are likely to exhibit smaller transmission times than remote CDCs. However, due to their limited resources, FNs that process a large amount of workload will likely have a long queuing delay. Therefore, it is important to balance the workload offloaded by FNs. Note that the response time associated with FN i, denoted as $R_i(\alpha_i)$, depends on the portion of workload locally processed by FN i.

(2) Power efficiency of FNs: We consider the power efficiency by the amount of power spent on processing a unit of received workload. Maximizing the power efficiency amounts to minimizing the power consumption for processing a given workload. It is known that the total amount of power consumed by any electronic device (e.g., an FN) depends on the power usage effectiveness (PUE) as well as the static and dynamic power consumption. The PUE is the input power from the power grid divided by the power consumption of the given device. Static power consumption, also called leakage power, is mainly caused by the leakage currents, and is unrelated to the usage of the computing resources at an FN. Dynamic power consumption is the result of the circuit activity and is determined by the activity of computing resources. Let e_i and w_i^S be the PUE and static power consumption of FN i, respectively. Let w_i^D be the dynamic power consumed by FN i to offload each unit of workload. We can write the total power consumption of FN i per time unit as $w_i = e_i(w_i^S + w_i^D \alpha_i \lambda_i)$.

The power efficiency of FN i can then be written as

$$\eta_i(\alpha_i) = \frac{w_i}{\alpha_i \lambda_i} = e_i \left(\frac{w_i^S}{\alpha_i \lambda_i} + w_i^D \right). \tag{3.30}$$

One of the main objective of this section is to develop workload allocation strategies to determine the appropriate portion of workload to be processed locally so as to minimize the response time under given power-efficiency constraints. Formally, each FN i tries to find the optimal value α_i^* by solving the following optimization problem:

$$\alpha_i^* = \arg \min_{\alpha_i \in [0,1]} R_i(\alpha_i) \tag{3.31}$$
$$\text{s.t.} \quad \eta_i(\alpha_i) \leq \bar{\eta}_i,$$

where $\bar{\eta}_i$ is the maximum power efficiency that can be supported by the hardware of FN i. In Section 3.4.2, we will give a more detailed discussion of the response time of FN i under different scenarios.

As mentioned earlier, different FNs can have different workload arrival rates. Therefore, allowing FNs to cooperate with each other and jointly process their received workload can further improve the overall workload processing capability. Specifically, FNs that receive more workload than their processing capabilities can seek help from nearby FNs with surplus computing resources. The main objective in this case is to minimize the average response time of users associated with all cooperative FNs. The total amount of workload processed by each FN in this case will not only depend on its own received workload, but also on the workload forwarded from other FNs. We can write the response time of FN i under cooperation as $R_i^C(\boldsymbol{\alpha})$ where $\boldsymbol{\alpha} = \langle \alpha_1, \alpha_2 \ldots, \alpha_N \rangle$. The optimal workload distribution under cooperative fog computing can then be written as

$$\boldsymbol{\alpha}^* = \arg \min_{\boldsymbol{\alpha}} \sum_{i \in \mathcal{F}} R_i^C(\boldsymbol{\alpha}) \tag{3.32}$$
$$\text{s.t.} \quad \eta_i(\alpha_i) \leq \bar{\eta}_i, 0 \leq \alpha_i \leq 1, \forall i \in \mathcal{F}.$$

Later on, we provide a more detailed discussion of the strategies for cooperative fog computing.

3.4.2 Response Time and Power Efficiency Trade-off

3.4.2.1 Response Time Analysis and Minimization

Let τ_j^u be the average round trip time (RTT) between FN j and its users. Typically, FNs and CDCs have fixed locations. Thus, we assume that the average workload transmission time between each FN j and the cloud can be regarded as a constant denoted as τ^f. Node j can directly forward its received workload to CDCs through the backbone IP network (Chiang, 2016). In this case, the fog computing network becomes equivalent to the traditional cloud computing network with all the workload being processed by the cloud. As mentioned before, CDCs are generally installed with high-performance workload processing units, and therefore their processing times are much smaller than the workload transmission time(Keller and Karl, 2016). For simplicity, we ignore the processing time of CDCs. In this case, the response time of FN j can be written as $R_j^{W1} = \tau_j^u + \tau^f$.

Since in this case FN j does not activate any computing resources to process its received workload, the power efficiency will not depend on the response time. In another extreme case, node j may process all its received workload using its local computing resources, i.e. $\alpha_j = 1$. If we follow a commonly adopted setting and consider an M/M/1 queuing system for each FN to process the received request, we can write the response time of FN j as $R_j^{W2}(\lambda_j) = \tau_j^u + \frac{1}{\mu_j - \lambda_j}$ where μ_j os the maximum amount of workload that can be processed by the on-board computing resources of FN j. We have $\lambda_j \leq \mu_j$.

Compared to CDCs, each FN can only have limited computing resources. It is generally impossible to always allow each FN to process all the received workload. We now consider the cases that FN j processes only a portion α_j, $0 \leq \alpha_j < 1$, of its received workload and forwards the remaining $1 - \alpha_j$ of its workload to CDCs, i.e. we still require $\alpha_j \lambda_j < \mu_j$. We can write the expected response time for FN j as:

$$R_j^{W3}(\alpha_j) = \tau_j^u + \alpha_j \left(\frac{1}{\mu_j - \alpha_j \lambda_j} \right) + (1 - \alpha_j)\tau^f. \tag{3.33}$$

Consider the solution of problem (3.31) by substituting the response time equation in (3.33). We can observe that problem (3.31) is a convex optimization problem, and hence can be solved using standard approaches. We omit the detailed derivation and directly present the solutions of these problems as follows. The minimum response time for users associated with FN j is $R_j^{W3}(\alpha_j^*)$, where α_j^* has the following closed-form solution:

$$\alpha_j^* = \begin{cases} 1, & \text{if } \mu_j < \frac{\lambda_j}{\tau^f + 1}, \\ \frac{1}{\lambda_j}\left(\frac{w_j^S e_j}{\overline{\eta}_j - e_j w_j^D} \right), & \text{if } \mu_j \geq \frac{\chi_j}{2\chi_j - \lambda_j(1 - \tau^f)}, \\ \frac{\mu_j}{\lambda_j} - \frac{\mu_j}{\lambda_j}\sqrt{1 - \frac{\lambda_j}{\mu_j}(1 - \tau^f)}, & \text{otherwise}, \end{cases} \tag{3.34}$$

where $\chi_j \triangleq \frac{w_j^S e_j}{\overline{\eta}_j - e_j w_j^D}$ is the maximum amount of workload that can be processed by FN j under power efficiency constraint $\eta_j(\alpha_j) \leq \overline{\eta}_j$.

3.4.2.2 Trade-off between Response Time and Power Efficiency

In Figure 3.21(a), we consider a single FN serving five users, and we compare the response time under different amounts of workload (number of requests) processed by the FN. There exists an optimal amount of workload to be processed by FN j that minimizes the response time. As observed in (3.30), the power consumption for the FN to process one unit of workload decreases with the total amount of processed workload. In many practical applications, there is a maximum tolerable response time for users. We can therefore observe that the power efficiency maximization solution for the FN in this case will be achieved when the response time approaches the maximum tolerable service delay θ. In Figure 3.21(a), we use a solid line to highlight the segment between the response time minimization solution and the power efficiency maximization solution at the maximum tolerable response time θ. We can observe a fundamental trade-off between the response time and the power efficiency of the FN. This trade-off can be characterized by substituting (3.33) into (3.31), as shown in Figure 3.21(b). We can observe that starting from the power consumption minimization

Figure 3.21 (a) Response time under different amounts of workload processed by FN *j*, (b) response time under different power efficiency values.

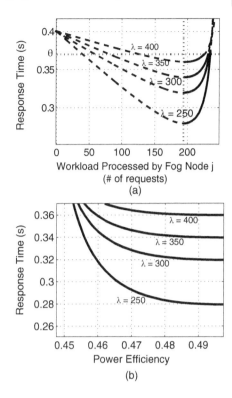

point, the response time decreases with the power consumption for the FN to process each unit (request) of workload. As the power consumption of the FN continues to grow, the rate of reduction in the response time decreases. This means that, for non-delay-sensitive applications such as voice/image processing services (e.g., voice/image recognition), the FN can choose a low power consumption solution as long as the resulting response time is tolerable for the users. On the other hand, for delay-sensitive applications such as online gaming and virtual reality (VR), it is ideal for the FN to choose a high power consumption solution to satisfy users' low latency requirement. In Figure 3.21 (b), we also present the trade-off solutions with different workload arrival rates at FN *j*. We can observe that the response time increases with the workload arrival rate under a given power efficiency. The higher the workload arrival rate, the smaller the changes in the response time. As the amount of workload processed by an FN approaches its maximum processing capability, the response time approaches infinite. In other words, allowing the FN to handle all its arriving workload cannot always reduce the response time for end-users, especially when the amount of workload to be processed by the FN cannot be carefully chosen.

3.4.3 Cooperative Fog Computing

3.4.3.1 Response Time Analysis for Cooperative Fog Computing with *N* FNs

In cooperative fog computing network, we introduce an FN cooperation strategy, referred to as offload forwarding. In this strategy, each FN can forward part or all of its offloaded workload to multiple neighboring FNs in the fog and/or help multiple other FNs process their workloads. Each node j divides its received workload into $N+1$ partitions: $\varphi_{j1}, \varphi_{j2}, \ldots, \varphi_{jN}, \varphi_{jc}$ where φ_{jc} is the workload forwarded to the remote CDC and φ_{ji}, $i \in \mathcal{F} \setminus \{j\}$, is the workload forwarded to FN i (this includes φ_{jj}, the workload processed by node j itself). We denote $\boldsymbol{\varphi}_{j\bullet} \triangleq \langle \varphi_{jk} \rangle_{k \in \mathcal{F}}$. Note that it is not necessary for each FN to always forward a non-zero workload to other FNs, i.e. $\varphi_{ji} = 0$ means that FN i does not process any workload for FN j. We refer to $\boldsymbol{\varphi}_{j\bullet}$ as the request vector of FN j. We also refer to $\boldsymbol{\varphi}_{\bullet i} = \langle \varphi_{ji} \rangle_{j \in \mathcal{F}}$ as the service vector of FN i. Let $\boldsymbol{\varphi} = \langle \varphi_{ji} \rangle_{i,j \in \mathcal{F}}$ be the workload processing matrix for the entire fog. We have $0 \le \varphi_{jk} \le 1$ and $\sum_{k \in \mathcal{F}} \varphi_{jk} \le 1, \forall j \in \mathcal{F}$. The response time of FN $j \in \mathcal{F}$ can then be written as

$$R_j^C(\xi_j, \boldsymbol{\phi}_{j\bullet}) = \tau_j^u + \frac{1}{\sum_{i \in \mathcal{F}} \lambda_i} \sum_{i \in \mathcal{F}} \phi_{ji} \left(\tau_{ji} + \frac{1}{\mu_i - \sum_{k \in \mathcal{F}} \phi_{ki}} \right) + \varphi_{jc} \tau^c, \tag{3.35}$$

where $\varphi_{jc} = 1 - \sum_{i \in \mathcal{F}} \varphi_{ji}$, $\phi_{jk} = \lambda_j \varphi_{jk}$ is the amount of workload processed by FN k for FN j. Note that if FN j cannot help other FNs to process their workload, but forward its own workload to other FNs to process, we have $\phi_{kj} = 0$ and $\phi_{ji} \ne 0 \, \forall k, i \in \mathcal{F} \setminus \{j\}$. We can rewrite the optimization problem in (3.32) as follows:

$$\min_{\boldsymbol{\phi}_{1\bullet}, \ldots, \boldsymbol{\phi}_{N\bullet}} \sum_{j=1}^{N} R_j^C(\xi_j, \boldsymbol{\phi}_{j\bullet}) \tag{3.36}$$

$$\text{s.t.} \sum_{k \in \mathcal{F}} \phi_{jk} + \phi_{jc} = \lambda_j, \tag{3.37}$$

$$\sum_{k \in \mathcal{F}} \phi_{kj} \le \chi_j \text{ and } 0 \le \phi_{jk} \le \lambda_j, \forall k, j \in \mathcal{F}. \tag{3.38}$$

It can be observed that, in order for each FN j to calculate the portions of workload to be forwarded to other FNs, FN j needs to know the workload processing capabilities and the workload arrival rates of all the other FNs, which can be private information and impossible to be known by FN j. In the next section, we will propose a distributed optimization framework to allow all the FNs to jointly optimize the average response time of the fog without disclosing their private information.

3.4.3.2 Response Time and Power Efficiency Trade-off for Cooperative Fog Computing Networks

In Figure 3.22(a), we present the minimum response time of the fog in a cooperative fog computing network derived from solving problem (3.36). Note that the workload processed by each FN can consist of both its own received workload and the workload sent from other FNs. We can observe that the response time of the fog is closely related to the amount of workload processed by each FN. We also use a black grid to highlight the area between the response time minimization solution and the power efficiency maximization solution with a given maximum tolerable response time in Figure 3.22(a). By substituting the power efficiency defined in (3.30) into (3.36), we can also present the relationship between the fog's response time and each FN's power efficiency for a two-node cooperative fog computing network with offload forwarding in Figure 3.22(b). Similar to the single-node fog computing, we can observe a fundamental trade-off between the response time of all the users served by the fog and the power efficiency of each FN. In addition, we can observe that

Figure 3.22 Response time under different amounts of processed workload and power consumption (PC) for each FN to offload one unit of workload.

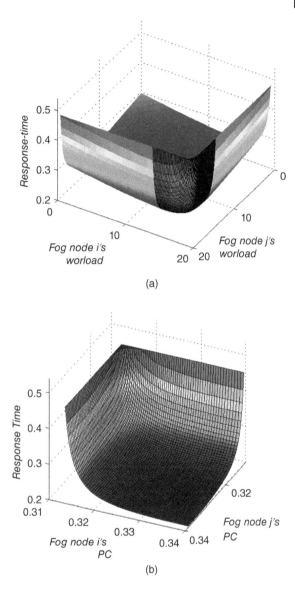

(a)

(b)

by allowing offloading forwarding, even if the power consumption of each FN to process each unit of workload has been limited to a very small value, it is still possible to achieve the response time constraint if there exist other nearby FNs with users that are more delay tolerant.

3.4.4 Distributed Optimization for Cooperative Fog Computing

As mentioned previously, deciding the proper amount of workload to be processed by each FN is essential to achieve the optimal response time and power efficiency trade-off for fog computing networks. Unfortunately, solving problem (3.36) involves carefully deciding the amount of workload processed and forwarded by every individual FN according to global information such as the computational capacities of all the FNs and the round-trip work-load transmission latency between any two FNs as well as that between FNs and the cloud. Deploying a centralized controller to collect all this global information and calculate the optimal service and request vectors for all the FNs may result in a huge communication overhead and intolerably high information collection and processing delay. In addition, it can also be observed that (3.36) is non-smooth and therefore cannot be solved by traditional optimization approaches that can only handle smooth objective functions.

To address the above challenges, we need to develop a distributed framework that can solve problem (3.36) with the following two main design objectives:

(O1) Distributed and scalable: We would like to develop a framework that can separate the optimization problem in (3.36) into N sub-problems, each of which can be solved by each FN using its local information. The framework should also be scalable in the sense that the computation complexity for each FN to solve its sub-problem should not increase significantly with the number of FNs that have the potential to cooperate with each other.

(O2) Privacy preserving: Each FN may not be willing to reveal its private proprietary information such as the maximum computational capacity and the round-trip workload transmission latency to others.

We propose a novel distributed optimization framework based on dual decomposition in which problem (3.36) will be first converted into its Lagrangian form and then the con-verted problem will be decomposed into N sub-problems, each of which can be solved by an individual FN using its local information. The optimization of all the sub-problems will be coordinated through dual variables sent to a workload forwarding coordinator (WFC) which can be established by the cloud data centers or deployed as one of the virtualized components in the cloud.

3.4.5 A City-wide Deployment of Fog Computing-supported Self-driving Bus System

3.4.5.1 Simulation Setup for Traffic Generated by a Self-driving Bus

In this section, we consider a possible implementation of a fog computing-supported self-driving bus system in the city of Dublin as a case study. A self-driving vehicle relies on a combination of sensors including cameras, sonars, radars, light detection, ranging systems (LiDARs), etc., to sense the surrounds and decide driving behaviors. More specif-ically, image and sensing data collected by the sensors will be processed by a computer or processor to extract useful information such as the types of objects as well as their specific semantics that may affect the driving decisions in various scenarios. For example, an autonomous vehicle must be able to detect and recognize various types of unintelligent objects such as traffic/road work signs and traffic lights as well as intelligent objects including surrounding vehicles, animals and pedestrians. It is known that accurate and low-latency object recognition require significant computing power and energy supply (Teichman and Thrun, 2011). How to develop effective object recognition methods for autonomous driving vehicles is out of the scope of this section.

In this section, we focus on the scenario that each self-driving bus relies on an FN in proximity to process the traffic image and feedback the driving decision. We focus on the workload transmission and forwarding between vehicles and the FNs. It is known that, for each self-driving vehicle, the amount of data that needs to be collected and processed is different when it drives into different areas. For example, the traffic conditions in the city center will be much more complex than that in the countryside. To take into consideration of the geographical diversity of the traffic data generated by each bus, we analyze the statistic feature of the traffic generated by buses operated at eight existing routes in the city of Dublin. We generate driving videos from over 2500 stitched high-resolution street view images in the considered bus routes extracted from Google street view (Google Street View). We then apply H.265 coding to compress the generated driving videos and keep track of the recorded frame rates of the compressed video to simulate the possible traffic data streaming of each self-driving bus. H.265 is a high efficiency video coding technique that can remove both temporal and spatial redundancies between adjacent frames using the enhanced hybrid spatial-temporal prediction model (Sullivan et al., 2012). In other words, the frame rates generated by H.265 can reflect different levels of traffic complexity as well as the impact of the driving speed and surrounding traffics at different locations. We fit the recorded frame rates of each bus route using a Kernel-based Probability Density Function. The bus driving routes, considered areas, and FN distribution are shown in Figure 3.23(a). The deployment densities of FNs are listed in Figure 3.23(b). The recorded frame sizes and fitted probability distributions of traffic generated by the self-driving bus when driving in different considered areas are presented in Figure 3.23(c).

3.4.5.2 Simulation Setup for a Fog Computing Network

We simulate a possible implementation of FNs, e.g., mini-computing servers, over 200 BSs (including GSM and UMTS BSs) deployed by a primary telecom operator in the city of Dublin. The actual distribution and the deployment density of FNs are shown in Figure 3.23. Each bus always submits the traffic images (frame-by-frame) taken by its on-board camera to its closest FN. In this case, the FN installed at each BS will be responsible for receiving and processing the images sent by each bus and feeding back the driving decision to each bus when the processing (e.g., objective recognition, tracking, and prediction) is finished. We assume each FN can process at most 400 frames at the same time and the maximum tolerable response time of each bus is 500 ms. We consider two scenarios of offloading forwarding. In the first one, each FN can only forward its workload to its closest FN. In the second scenario, each FN can forward part of its received workload to other FNs within a 500 m range. We assume there exist local communication links among FNs and the round trip workload forwarding time between any two FNs within forwarding distance is the same given by $\tau_{ij} = 20$ ms.

3.4.5.3 Numerical Results

To evaluate the performance improvement that can be achieved by allowing offload forwarding among FNs, we first compare the number of frames that can be processed by each FN in the five areas highlighted in Figure 3.23(a). We can observe in Figure 3.24 that allowing each FN to cooperate with all the other FNs within a 500 m range can significantly improve the number of frames processed by FNs. We can also observe that even when each FN can only cooperate with its closest FN, the average number of frames processed by each FN can be almost doubled compared to the case without cooperation among FNs. Note that

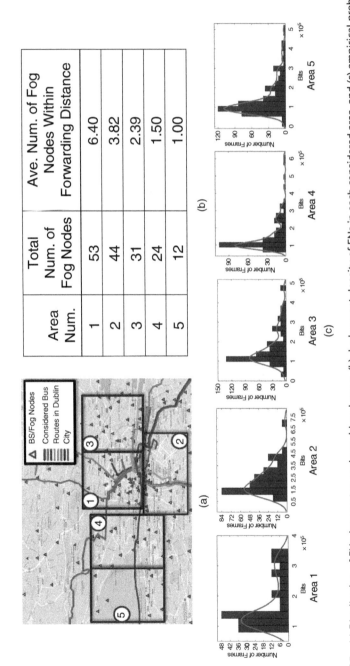

Area Num.	Total Num. of Fog Nodes	Ave. Num. of Fog Nodes Within Forwarding Distance
1	53	6.40
2	44	3.82
3	31	2.39
4	24	1.50
5	12	1.00

(b)

(c)

Figure 3.23 (a) Distribution of FNs, bus routes, and considered areas, (b) deployment density of FNs in each considered area, and (c) empirical probability distribution of traffics generated by self-driving buses in each considered area.

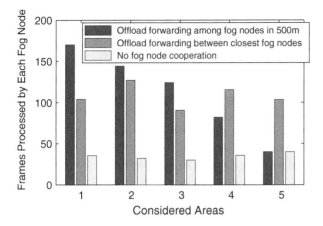

Figure 3.24 Average workload (number of requests) processed by each FN at different areas of consideration.

in Figure 3.24 we can also observe that in areas 4 and 5, allowing FNs within a 500 m range cannot achieve higher workload offloading performance than only allowing each FN to cooperate with its closest neighboring FN. This is because in both of these two considered rural areas, some FNs cannot have any other FN located within the 500 m range.

In Figure 3.25, we consider the average workload processing capability of all five considered areas. We investigate the impact of FN workload arrival rates on the total amount of workload to be offloaded by the fog computing network. We can observe that the average number of frames that can be offloaded by each FN increases almost linearly when the workload arrival rate is small. However, with the workload arrival rate continuing to grow, the total amount of offloaded workload that can be offloaded by the fog computing network approaches a fixed value limited by the maximum response time that can be tolerated by end-users.

In Figure 3.26, we present the average response time and power efficiency trade-off curves with and without offload forwarding. We observe that our proposed offload forwarding significantly reduces the response time of end-users especially when the power efficiency

Figure 3.25 Average workload (number of requests) processed by each FN with different workload arrival rates.

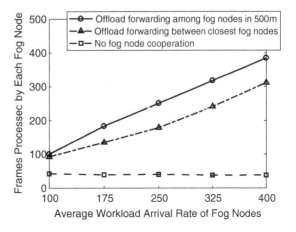

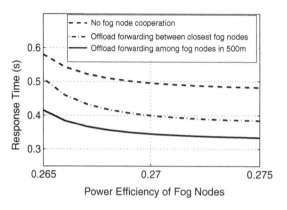

constraint of FN j is low. With an increase in the power efficiency of FNs, the response time that can be provided by the FNs approaches a fixed value limited by the maximum workload processing capability of the fog.

3.5 Conclusion

To solve these challenges, in this chapter, we focus on cross-domain resource management among FNs to support intelligent IoT applications. To be specific:

In Section 3.2, we study the joint computing and communication resource management for delay-sensitive applications. We focus on the delay-minimization task scheduling problem and distributed algorithm design in the MTMH fog computing networks under two different scenarios of non-splittable tasks and splittable tasks. For non-splittable tasks, we propose a potential game based analytical framework and develop a distributed task scheduling algorithm called POMT. For splittable tasks, we propose a generalized Nash equilibrium problem based analytical framework and develop a distributed task scheduling algorithm called POST. Analytical and simulation results show that the POMT algorithm and POST algorithm can offer the near-optimal performance in system average delay and achieve more number of beneficial task nodes.

In Section 3.3, we study the joint computing, communication, and caching resource management for energy-efficient applications. We propose F3C, a fog-enabled 3C resource sharing framework for energy-efficient IoT data stream processing by solving an energy cost minimization problem under 3C constraints. Nevertheless, the minimization problem is proven to be NP-hard via reduction to a GAP. To cope with such challenge, we propose an efficient F3C algorithm based on an iterative task team formation mechanism which regards each task's 3C resource sharing as a sub-problem solved by the elaborated min cost flow transformation. Via utility improving iterations, the proposed F3C algorithm is shown to converge to a stable system point. We conduct extensive performance evaluations, which demonstrate that our F3C algorithm can achieve superior performance in energy saving compared to various benchmarks.

In Section 3.4, we make a case study of the energy-efficient resource management in TI. We focus on an energy-efficient design of fog computing networks that support low-latency

TI applications. We investigate two performance metrics: service delay of end-users and power usage efficiency of FNs. We quantify the fundamental trade-off between these two metrics and then extend our analysis to fog computing networks involving cooperation between FNs. We introduce a novel cooperative fog computing mechanism, referred to as offload forwarding, in which a set of FNs with different computing and energy resources can cooperate with each other. The objective of this cooperation is to balance the workload processed by different FNs, further reduce the service delay and improve the efficiency of power usage. We develop a distributed optimization framework based on dual decomposition to achieve the optimal trade-off. Our framework does not require FNs to disclose their private information nor conduct back-and-forth negotiations with each other. Finally, to evaluate the performance of our proposed mechanism, we simulate a possible implementation of a city-wide self-driving bus system supported by fog computing in the city of Dublin. The fog computing network topology is set based on a real cellular network infrastructure involving 200 base stations deployed by a major cellular operator in Ireland. Numerical results show that our proposed framework can balance the power usage efficiency among FNs and reduce the service delay for users by around 50% in urban scenarios.

References

Mudassar Ali, Nida Riaz, Muhammad Ikram Ashraf, Saad Qaisar, and Muhammad Naeem. Joint cloudlet selection and latency minimization in fog networks. *IEEE Transactions on Industrial Informatics*, 14(9):4055–4063, 2018.

Min Chen, Yixue Hao, Yong Li, Chin Feng Lai, and Di Wu. On the computation offloading at ad hoc cloudlet: Architecture and service models. *IEEE Communications Magazine*, 53(6):18–24, 2015a.

Xu Chen, Lei Jiao, Wenzhong Li, and Xiaoming Fu. Efficient multi-user computation offloading for mobile-edge cloud computing. *IEEE/ACM Transactions on Networking*, 24(5):2795–2808, 2015b.

Xu Chen, Lingjun Pu, Lin Gao, Weigang Wu, and Di Wu. Exploiting massive d2d collaboration for energy-efficient mobile edge computing. *IEEE Wireless Communications*, 24(4):64–71, 2017.

Zhuoqun Chen, Yangyang Liu, Bo Zhou, and Meixia Tao. Caching incentive design in wireless d2d networks: A stackelberg game approach. In *IEEE International Conference on Communications*, 2016.

F. Chi, X. Wang, W. Cai, and V. C. M. Leung. Ad hoc cloudlet based cooperative cloud gaming. In *2014 IEEE 6th International Conference on Cloud Computing Technology and Science*, pages 190–197, Dec 2014. doi: .

Mung Chiang. Fog networking: An overview on research opportunities. *arXiv preprint arXiv:1601.00835*, 2016. URL http://arxiv.org/pdf/1601.00835.

Apostolos Destounis, Georgios S Paschos, and Iordanis Koutsopoulos. Streaming big data meets backpressure in distributed network computation. 2016.

Thinh Quang Dinh, Jianhua Tang, Quang Duy La, and Tony QS Quek. Offloading in mobile edge computing: Task allocation and computational frequency scaling. *IEEE Transactions on Communications*, 65(8):3571–3584, 2017.

Francisco Facchinei and Christian Kanzow. Generalized Nash equilibrium problems. *4OR*, 5(3):173–210, 2007.

Google Street View. URL https://www.google.com/streetview/.

Zhu Han, Dusit Niyato, Walid Saad, Tamer Başar, and Are Hjørungnes. *Game theory in wireless and communication networks: theory, models, and applications.* Cambridge University Press, 2012.

Tim Hoheisel. Mathematical programs with vanishing constraints: a new regularization approach with strong convergence properties. *Optimization*, 61 (6):619–636, 2012.

Yaodong Huang, Xintong Song, Fan Ye, Yuanyuan Yang, and Xiaoming Li. Fair caching algorithms for peer data sharing in pervasive edge computing environments. In *IEEE International Conference on Distributed Computing Systems*, pages 605–614, 2017.

George Iosifidis, Gao Lin, Jianwei Huang, and Leandros Tassiulas. Efficient and fair collaborative mobile internet access. *IEEE/ACM Transactions on Networking*, 25(3):1386–1400, 2016.

ITU-T. The Tactile Internet. ITU-T technology watch report, Aug. 2014. URL https://www.itu.int/dms_pub/itu-t/opb/gen/T-GEN-TWATCH-2014-1-PDF-E.pdf.

Jingjie Jiang, Yifei Zhu, Bo Li, and Baochun Li. Rally: Device-to-device content sharing in lte networks as a game. In *IEEE International Conference on Mobile Ad Hoc and Sensor Systems*, pages 10–18, 2015.

Matthias Keller and Holger Karl. Response-time-optimized distributed cloud resource allocation. *arXiv preprint arXiv:1601.06262*, 2016. URL http://arxiv.org/abs/1601.06262.

Jeongho Kwak, Yeongjin Kim, Joohyun Lee, and Song Chong. DREAM: Dynamic resource and task allocation for energy minimization in mobile cloud systems. *IEEE Journal on Selected Areas in Communications*, 33(12): 2510–2523, 2015.

Zening Liu, Yang Yang, Ming-Tuo Zhou, and Ziqin Li. A unified cross-entropy based task scheduling algorithm for heterogeneous fog networks. In *Proceedings of the 1st ACM International Workshop on Smart Cities and Fog Computing*, pages 1–6, 2018.

Zening Liu, Yang Yang, Kunlun Wang, Ziyu Shao, and Junshan Zhang. POST: Parallel offloading of splittable tasks in heterogeneous fog networks. *IEEE Internet of Things Journal*, 2020.

Siqi Luo, Xu Chen, and Zhi Zhou. F3c: Fog-enabled joint computation, communication and caching resource sharing for energy-efficient iot data stream processing. In *2019 IEEE 39th International Conference on Distributed Computing Systems (ICDCS)*, pages 1019–1028. IEEE, 2019.

Martin Maier, Mahfuzulhoq Chowdhury, Bhaskar Prasad Rimal, and Dung Pham Van. The tactile internet: vision, recent progress, and open challenges. *IEEE Communications Magazine*, 54(5):138–145, May 2016.

Xianling Meng, Wei Wang, and Zhaoyang Zhang. Delay-constrained hybrid computation offloading with cloud and fog computing. *IEEE Access*, 5: 21355–21367, 2017.

Dov Monderer and Lloyd S Shapley. Potential games. *Games and economic behavior*, 14(1):124–143, 1996.

Olga Muñoz, Antonio Pascual-Iserte, and Josep Vidal. Optimization of radio and computational resources for energy efficiency in latency-constrained application offloading. *IEEE Transactions on Vehicular Technology*, 64(10): 4738–4755, 2015.

Anselme Ndikumana, Nguyen H Tran, Manh Ho Tai, Zhu Han, Walid Saad, Dusit Niyato, and Choong Seon. Joint communication, computation, caching, and control in big data multi-access edge computing. 2018. URL `https://arxiv.org/abs/1803.11512v1`.

NGMN Alliance. 5G white paper, Feb. 2015. URL `https://www.ngmn.org/uploads/media/NGMN_5G_White_Paper_V1_0.pdf`.

Daniel Nowak, Tobias Mahn, Hussein Al-Shatri, Alexandra Schwartz, and Anja Klein. A generalized nash game for mobile edge computation offloading. In *2018 6th IEEE International Conference on Mobile Cloud Computing, Services, and Engineering (MobileCloud)*, pages 95–102. IEEE, 2018.

James B. Orlin. A polynomial time primal network simplex algorithm for minimum cost flows. *Mathematical Programming*, 78(2):109–129, 1997.

Ai-Chun Pang, Wei-Ho Chung, Te-Chuan Chiu, and Junshan Zhang. Latency-driven cooperative task computing in multi-user fog-radio access networks. In *2017 IEEE 37th International Conference on Distributed Computing Systems (ICDCS)*, pages 615–624. IEEE, 2017.

Milan Patel, B Naughton, C Chan, N Sprecher, S Abeta, A Neal, et al. Mobile-edge computing introductory technical white paper. *White Paper, Mobile-edge Computing (MEC) industry initiative*, 2014.

L. I. Qun and X. U. Ding. An efficient resource allocation algorithm for underlay cognitive radio multichannel multicast networks. *Ieice Transactions on Fundamentals of Electronics Communications & Computer Sciences*, 100 (9):2065–2068, 2017.

Reuven Y Rubinstein and Dirk P Kroese. *The Cross-Entropy Method: A Unified Approach to Combinatorial Optimization, Monte-Carlo Simulation and Machine Learning*. Springer Science & Business Media, 2013.

Hamed Shah-Mansouri and Vincent WS Wong. Hierarchical fog-cloud computing for iot systems: A computation offloading game. *IEEE Internet of Things Journal*, 5(4):3246–3257, 2018.

M. Simsek, A. Aijaz, M. Dohler, J. Sachs, and G. Fettweis. 5g-enabled Tactile Internet. *IEEE J. Sel. Areas Commun.*, 34(3):460–473, Mar. 2016. ISSN 0733-8716. doi: .

Ivan Stojmenovic and Wen Sheng. The fog computing paradigm: Scenarios and security issues. In *Federated Conference on Computer Science & Information Systems*, 2014.

G. J. Sullivan, J. R. Ohm, W. J. Han, and T. Wiegand. Overview of the high efficiency video coding (HEVC) standard. *IEEE Trans. Circuits and Systems for Video Technology*, 22(12):1649–1668, Dec 2012. ISSN 1051-8215. doi: .

D Syrivelis, G Iosifidis, D Delimpasis, and K Chounos. Bits and coins: Supporting collaborative consumption of mobile internet. In *IEEE Conference on Computer Communications (INFOCOM), Kowloon, 2015*, pages 2146–2154, 2015.

Alex Teichman and Sebastian Thrun. Practical object recognition in autonomous driving and beyond. In *IEEE Workshop on Advanced Robotics and its Social Impacts*, pages 35–38, San Francisco, CA, Oct. 2011.

Luis M. Vaquero and Luis Rodero-Merino. Finding your way in the fog: Towards a comprehensive definition of fog computing. *ACM SIGCOMM Comput. Commun. Rev.*, 44(5), October 2014. doi: .

Hong Xing, Liang Liu, Jie Xu, and Arumugam Nallanathan. Joint task assignment and resource allocation for D2D-enabled mobile-edge computing. *IEEE Transactions On Communications*, 67(6):4193–4207, 2019.

Yang Yang. Multi-tier computing networks for intelligent iot. *Nature Electronics*, 2(1):4–5, 2019.

Yang Yang, Kunlun Wang, Guowei Zhang, Xu Chen, Xiliang Luo, and Ming-Tuo Zhou. MEETS: Maximal energy efficient task scheduling in homogeneous fog networks. *IEEE Internet of Things Journal*, 5(5):4076–4087, 2018a.

Yang Yang, Shuang Zhao, Wuxiong Zhang, Yu Chen, Xiliang Luo, and Jun Wang. DEBTS: Delay energy balanced task scheduling in homogeneous fog networks. *IEEE Internet of Things Journal*, 5(3):2094–2106, 2018b.

Yang Yang, Zening Liu, Xiumei Yang, Kunlun Wang, Xuemin Hong, and Xiaohu Ge. POMT: Paired offloading of multiple tasks in heterogeneous fog networks. *IEEE Internet of Things Journal*, 6(5):8658–8669, 2019.

Yang Yang, Xiliang Luo, Xiaoli Chu, and Ming-Tuo Zhou. *Fog-Enabled Intelligent IoT Systems*. Springer, 2020.

Changsheng You, Kaibin Huang, Hyukjin Chae, and Byoung-Hoon Kim. Energy-efficient resource allocation for mobile-edge computation offloading. *IEEE Transactions on Wireless Communications*, 16(3):1397–1411, 2016.

Ning Zhang, Shan Zhang, Jianchao Zheng, Xiaojie Fang, Jon W Mark, and Xuemin Shen. QoE driven decentralized spectrum sharing in 5G networks: Potential game approach. *IEEE Transactions on Vehicular Technology*, 66(9): 7797–7808, 2017.

Shuang Zhao, Yang Yang, Ziyu Shao, Xiumei Yang, Hua Qian, and Cheng-Xiang Wang. FEMOS: Fog-enabled multitier operations scheduling in dynamic wireless networks. *IEEE Internet of Things Journal*, 5(2):1169–1183, 2018.

Jianchao Zheng, Yueming Cai, Yongkang Liu, Yuhua Xu, Bowen Duan, and Xuemin Sherman Shen. Optimal power allocation and user scheduling in multicell networks: Base station cooperation using a game-theoretic approach. *IEEE Transactions on Wireless Communications*, 13(12):6928–6942, 2014.

4

Dynamic Service Provisioning Frameworks

4.1   Online Orchestration of Cross-edge Service Function Chaining

4.1.1   Introduction

With the advancements in 5G communications and internet-of-things (IoT), billions of devices (e.g., mobile devices, wearable devices and sensors) are expected to be connected to the internet, which is indispensable for a wide variety of IoT applications, ranging from intelligent video surveillance and internet of vehicles (IoV) to augmented reality Taleb et al. (2017), Ananthanarayanan et al. (2017), Chen et al. (2018a), Zhou et al. (2019a), Chen et al. (2018c). As a result of the proliferation of these diverse applications, large volumes of multi-modal data (e.g., audio and video) of physical surroundings are continuously sensed at the device side. To process such a tremendous amount of data streams, which typically requires vast computing resources and very low latency, the concept of edge computing (EC) Taleb et al. (2017) has recently been proposed. As an extension of cloud computing, EC pushes cloud resources and services from the network core to the network edges that are in closer proximity to IoT devices and data sources, leading to significant reduction of end-to-end latency.

While recognizing the superiority of edge computing in reducing the user-perceived latency of IoT applications, it is important to note that low cost-efficiency may become the bottleneck towards sustainable EC ecosystems Jiao et al. (2018). Specifically, it is widely acknowledged that the power consumption typically dominates the operational cost of a data center, while the power efficiency of an edge node can be hundreds of times less efficient than that of a cloud server mle, as reported by Microsoft. Therefore, efficient cost management is of strategic importance for provisioning sustainable EC service. To this end, an emerging technology called a service function chain (SFC), which provisions services in an agile, flexible and cost-efficient manner, has been advocated Medhat et al. (2017). With an SFC, an EC service is decomposed into a chain of service functions (SFs) or microservices Chen et al. (2018a) with precedence order: each SF executes a certain function and all SFs are executed by following the specific precedence order. As an illustrative example, Figure 4.1 depicts three different SFCs for live video analytics which

Intelligent IoT for the Digital World: Incorporating 5G Communications and Fog/Edge Computing Technologies, First Edition. Yang Yang, Xu Chen, Rui Tan, and Yong Xiao.
© 2021 John Wiley & Sons Ltd. Published 2021 by John Wiley & Sons Ltd.

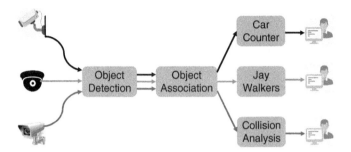

Figure 4.1 An example of three service function chains (SFCs) for live video analytics. The black, blue, and red arrows represent the video streams that require the SFC for car counter, jay walkers, and collision analysis, respectively.

is envisioned as the killer EC application by Microsoft Ananthanarayanan et al. (2017). In this example, since the two SF object detection and association are the common SFs of the three SFCs, they can be merged by running a single instance of the common SFs, allowing for resource sharing among SFCs and leading to significant cost saving.

However, when deploying various SFCs across multiple geographically dispersed edge clouds, which have recently gathered great attention Jiao et al. (2018, 2017), fully materializing the benefits of service function chaining is far from trivial, due to the following reasons. First, to fully unleash the potential of resource sharing among multiple SFCs, the traffic may need to traverse various edge clouds that host the corresponding SFs shared by those SFCs. Clearly, due to the geographical distance, cross-edge traffic routing increases the end-to-end latency, and thus *incurs a cost-performance trade-off that should be judiciously navigated*. Second, as a result of the geo-distribution and time-varying electricity price, the cost of provisioning a SF instance exhibits strong *spatial and temporal variabilities*. Intuitively, such diversities ought to be fully exploited to dynamically provisioning the SF instances, if we want to minimize the cost of running SF instances. Unfortunately, aggressively re-provisioning instances would greatly increase the switching cost caused by launching new SF instances, meaning that we should *strike a nice balance between the instance running and switching costs* when conducting dynamical instance provisioning. Finally, unlike traditional mega-scale cloud data centers that have abundant resources, the resource volume of an edge cloud is highly limited. As a result, the provisioned instances for each SF can be dispersed at multiple edge clouds, requiring us to carefully *split and route the traffic to those edge clouds, with an awareness to the spatial diversities on both resource cost and performance (in terms of network latency)*.

Keeping the above factors in mind, in this chapter, we advocate an online framework to orchestrate service function chaining across multiple edge clouds. Towards the goal of optimizing the holistic cost-efficiency of the cross-edge SFC system, the proposed framework unifies (1) the instances running costs incurred by the energy consumption and the amortized capital expenditure, (2) the instances switching cost due to launching new SF instances, (3) the cloud outsourcing cost due to the usage of the central cloud resources, in case of traffic flash crowd that outweighs the capacity of the edge clouds, and (4) the traffic routing cost that consists of wide-area network (WAN) bandwidth costs and penalties for cross-edge network latency. To minimize the unified cost, the proposed orchestration framework jointly applies two control knobs: (1) dynamical instance provisioning that periodically adapts the number of running SF instances at each edge cloud; and (2) cross-edge traffic routing that tunes the amount of traffic traversed among edge clouds. With the above setup, the cost minimization for cross-edge SFC deployment over the long-term is formulated as mixed integer linear programming (MILP).

However, for a practical cross-edge system, solving the above MILP is rather difficult, due to the following dual challenges. First, for the instance switching cost incurred every time when launching new SF instances, it couples the instance provisioning decisions over consecutive time periods. As a result, the long-term cost minimization problem is a time-coupling problem that involves future system information. However, in practice, parameters such as traffic arrivals typically fluctuate over time and thus cannot be readily predicted. Then, it is highly desirable to minimize the long-term cost in an online manner, without utilizing the future information as a priori knowledge. Second, even with an offline setting where all the future information is given as a priori knowledge, the corresponding cost minimization problem is proven to be NP-hard. This further complicates the design of an online algorithm that dynamically optimizes the long-term cost based on the historical and current system information.

Fortunately, by looking deep into the structure of the problem, we can address the above dual challenges by blending the advantages of a regularization method for online algorithm design and a dependent rounding technique for approximation algorithm design. In particular, by applying the regularization technique from the online learning literature, we temporally decompose the relaxed time-coupling problem into a series of one-shot fractional sub-problems, which can be exactly solved without requiring any future information. To round the regularized and fractional solution to feasible integer solutions of the original problem, a randomized dependent rounded scheme is carefully designed. The key idea of the rounding scheme is to compensate each rounded-down instance by another rounded-up instance Rajiv et al. (2006). This ensures that the solution feasibility is maintained without provisioning excessive SF instances, allowing significant improvement on the cost-efficiency. The proposed approximated online algorithm achieves a good performance guarantee, as verified by both rigorous theoretical analysis and extensive simulations based on realistic electricity prices and workload traces.

4.1.2 Related Work

As a promising approach to provision computing and network services, service function chaining has gathered great attention from both industry and academia. A large body of recent research was devoted to exploiting the opportunities or addressing the challenges in deploying SFCs across geo-distributed infrastructures Hantouti et al. (2018).

The majority of the existing literature focuses on deploying SFCs across geo-distributed mega-scale data centers. For example, Fei et al. Fei et al. (2017) explored how to achieve the goal of load balancing by dynamically splitting workload and assigning service functions to geo-distributed data centers. Zhou et al. Zhou (2018) proposed an online algorithm to place SFCs which arrive in an online manner, aiming at maximizing their total value. Abu-Lebdeh et al. Abu-Lebdeh et al. (2017) investigated the problem of minimizing the operational cost without violating the performance requirements, and proposed a tabu search based a heuristic search to solve the problem. Towards a fair trade-off between the cost-efficiency and quality of experience (QoE) of SFCs over multi-clouds, Benkacem et al. Benkacem et al. (2018) applied the bargaining game theory to jointly optimize the cost and QoE in a balanced manner. A more closely related work is Jia et al. (2018), in which Jia et al. studied how to jointly optimize instance provisioning and traffic routing to optimize the long-term system-wide cost, and proposed an online algorithm with a provable performance guarantee. However, our work is different from and complementary to Jia et al. (2018). Problem-wise, we extend the model in Jia et al. (2018) by considering: (1) the resource capacity constraints of edge clouds, and (2) a hybrid environment with both resource-limited edge clouds and resource-rich central cloud. Such extensions make the algorithm in Jia et al. (2018) not directly applicable to our problem. Algorithm-wise, we incorporate the technique of knapsack cover (KC) constraint to design a more intuitive and simplified online algorithm, yet still preserves the provable performance guarantee.

While significant progress has been made in deploying SFCs across geo-distributed data centers, the problem in the edge computing scenario is only beginning to receive attention. Xuan et al. Dinh-Xuan et al. (2018) proposed heuristic-based placement algorithms that aim to efficiently place the SFC in servers with regard to optimizing service delay and resource utilization. Gouareb et al. Gouareb et al. (2018) studied the problem of service function placement and routing across the edge clouds to minimize overall latency, defined as the queuing delay within the edge clouds and in network links. Laghrissi et al. Laghrissi et al. (2017) developed a spatial-temporal model for service function placement across edge clouds, and compared the performance of several placement strategies in terms of delay and cost. For the emerging paradigm of 5G service-customized network slices, an adaptive interference-aware service function chaining framework Zhang et al. (2019) is proposed to improve the throughput against performance interference among colocated SFs. In addition, recent efforts Gupta et al. (2018), Qu et al. (2018), Mouradian et al. (2019), Taleb et al. (2020) have all adopted an integer linear programming (ILP) formulation to derive computational-efficient heuristics for service function chaining. However, all these works assume a static environment, rather than the stochastic setup considered in this work. Furthermore, all those heuristics except Zhang et al. (2019) do not provide any performance guarantee.

4.1.3 System Model for Cross-edge SFC Deployment

In this section, we present the system model for cross-edge SFC deployment, and the problem formulation for the long-term cost minimization.

4.1.3.1 Overview of the Cross-edge System

As illustrated in Figure 4.2, we consider an edge service provider running an edge computing service on a set of I geographically dispersed edge clouds in proximity to the users (e.g., IoT devices), denoted as $\mathcal{I} = \{1, 2, \ldots, I\}$. Following the recent proposal of architecting edge computing with service function chaining, we assume that each edge computing service performs a series of consecutive service functions (a.k.a. microservice Chen et al. (2018a)) on the input traffic. For example, for the live video analytics Ananthanarayanan

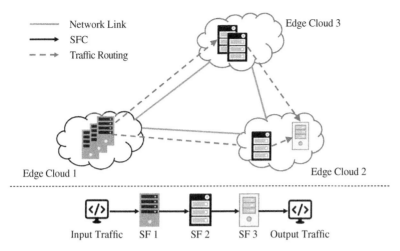

Figure 4.2   An illustration of cross-edge service function chain deployment.

et al. (2017), which is envisioned as the killer application of edge computing, an object recognition invokes three core vision primitives (i.e. object detection → object association → object classification) to be executed in sequence with the raw input image streams. To characterize such inherent sequential order of service functions, we adopt a service function chain (SFC) with precedence to model the consecutive service functions for each edge computing service Chen et al. (2018b). Specifically, we use $\mathcal{M} = \{1, 2, \ldots, M\}$ to denote the set of different SFs that can be selected to form diverse SFCs.

Each SF $m \in \mathcal{M}$ is instantiated in a virtual machine (VM) or container in the edge cloud, and the VM or container instances running different SFs are referred to as SF instances. Determined by the resource capacities of the underlying physical servers (e.g., CPU, GPU, memory and network I/O), the service capacity of each instance of SF m in edge cloud i is denoted as b_i^m, meaning the maximal data rate can be supported by an instance of SF m in edge cloud i. Considering the limited amount of resources at each cloud, we further use C_i^m to denote the resource capacity, i.e. maximal number of available instances of SF m in edge cloud i.

The input traffic arrival at each source edge cloud specifies an SFC, and traverses the corresponding SFs that may be deployed across various edge clouds to generate the output result, which is finally returned to the source edge cloud. Specifically, for the input traffic arrival at each source edge cloud $s \in \mathcal{I}$, we denote the requested SFC as SFC s. We further let $h_{mn}^s = 1$ if $m \to n$ is a direct hop of the SFC s (i.e. SF m is the predecessor of SF n in SFC s); and $h_{mn}^s = 0$ otherwise. Note that after the processing of each SF, the traffic rate of each SFC may change at different hops since the SF may increase or decrease the traffic amount. We use α_m^s to denote the change ratio of traffic rate of SFC s on SF m, meaning that the outgoing traffic rate of SFC s after passing an instance of SF m is on average α_m^s times the incoming traffic rate. For ease of presentation, we further use β_m^s to denote the cumulative traffic rate change ratio of SFC s before it goes through SF m, which is the ratio of the overall incoming traffic rate of SF m to the initial total input traffic rate. If we use m_F^s to denote the first SF of SFC s, then we have $\beta_m^s = 1$ if $m = m_F^s$, and $\beta_m^s = \sum_{n \in \mathcal{M}} h_{nm}^s \beta_n^s \alpha_m^s$ if $m \neq m_F^s$.

Without loss of generality, the system works in a time slotted fashion within a large time span of $\mathcal{T} = \{1, 2, ..., T\}$. Each time slot $t \in \mathcal{T}$ represents a decision interval, which is much longer than a typical end-to-end delay for the input traffic. At each time slot t, the input traffic rate (in terms of number of data packets per time slot) of each SFC s is denoted as $A_s(t)$. Note that $A_s(t)$ typically fluctuates over time. In the presence of a traffic flash crowd at service peak times, the resource required by the input traffic may exceed the overall resource capacity of the cross-edge clouds. To cope with this issue, we assume that a central cloud data center with sufficient resource capacity can be leveraged to absorb the extra input traffic.

4.1.3.2 Optimization Space

When running on top of geographically distributed edge clouds, the cross-edge system exhibits strong spatial and temporal variabilities on performance (in terms of the end-to-end latency of the input traffic) and cost. Specifically, at the spatial dimension, the operational cost of running an SF instance fluctuates over time, due to the time-varying nature of the electricity price. While at the spatial temporal dimension, both the operational cost and cross-edge network latency show geographical diversities. Clearly, such diversities ought to be fully exploited if we want to jointly optimize the cost-efficiency of the cross-edge system. Towards this goal, an effective approach is joint dynamical instance provisioning (i.e. dynamically adapting the number of running SF instances at each edge cloud) and cross-edge traffic routing (i.e. dynamically adapting the amount of traffic routed among edge clouds).

We now elaborate the control decisions that we tune to optimize the cost-efficiency of the cross-edge system. First, dynamical instance provisioning, we use $x_i^m(t)$ to denote the number of running instances of SF m provisioned in edge cloud i at time slot t. Since the resource capacity in an edge cloud is highly limited, it is impractical to relax the non-negative integer $x_i^m(t)$ into a real number. Instead, we enforce that $x_i^m(t) \in \{0, 1, 2, ..., C_i^m\}$, where C_i^m is the aforementioned maximal number of available instances of SF m in edge cloud i. To efficiently utilize the limited edge cloud resources, an instance of a SF is shared by multiple SFCs whose service chain includes this SF. Second, cross-edge traffic routing, here we use $y_{sij}^{mn}(t)$ to denote the amount of traffic of SFC s, routed from SF m in edge cloud i to SF n in edge cloud j. To denote the routing decisions of the input (output) traffic arrived at (back to) each edge cloud, we further introduce $z_{si}^m(t)$, which represents the total incoming traffic of SFC s to the instances of SF m in edge cloud i. Finally, for ease of problem formulation, we also use $a_s(t)$ and $u_s(t)$ to denote the amount of traffic arrived at each source edge cloud s routed to the edge clouds and the central cloud, respectively.

4.1.3.3 Cost Structure

Given the above control decisions, we are now ready to formulate the overall cost incurred by the cross-edge deployment of the SFCs, which include the instance operation cost, instance switching cost, traffic routing cost and the cloud outsourcing cost.

Instance running cost. Let $p_i^m(t)$ denote the cost of running an instance of SF m in edge cloud i and at time slot t, mainly attributed to the power consumption[1] as well as the

---

1 Here we focus on the static server power Zhou et al. (2018), since it contributes the majority of the whole server power consumption and overwhelms the dynamical server power. Nevertheless, since the static

amortized capital expenditure of the hosting edge server. Then the total instance running cost at time slot t is given by:

$$C_R(t) = \sum_{i \in I} \sum_{m \in M} p_i^m(t) x_i^m(t).$$

Instance switching cost. Launching a new instance of SF m requires transferring a VM image containing the service function to the hosting server, booting it and attaching it to devices on the host server. We use q_i^m to denote the cost of deploying a newly added instance of SF m at edge cloud i. Then the total switching cost at time slot t is given by:

$$C_S(t) = \sum_{i \in I} \sum_{m \in M} q_i^m [x_i^m(t) - x_i^m(t - 1)]^+,$$

where $[x_i^m(t) - x_i^m(t - 1)]^+ = \max\{x_i^m(t) - x_i^m(t - 1), 0\}$, denoting the number of newly launched instances of SF m at edge cloud i in time slot t. Note that here we assume that the cost of SF destruction is 0. If this is not the case, we can simply fold the corresponding cost into q_j^m incurred in the next SF launching operation. Without loss of generality, we let $x_i^m(0) = 0$.

Cloud outsourcing cost. In the case of a traffic flash crowd whose resource requirement exceeds the capacity of the cross-edge clouds, the extra demand would be outsourced to a remote central cloud with sufficient resource for processing. Here we use r_s to denote the aggregated cost of outsourcing one unit input traffic from the source edge cloud $s \in M$ to the remote central cloud. The cost r_s includes the resource usage cost of the remote cloud, the bandwidth usage cost and the performance cost incurred by the delay of the WAN connecting to the central cloud. Due to the high bandwidth usage cost and the performance cost of the WAN, processing the traffic in the remote central cloud is typically far more expensive than that across edge clouds. Given the amount $u_s(t)$ of input traffic outsourced to the central cloud from each source edge cloud s, the total cloud outsourcing cost at time slot t can be computed by:

$$C_O(t) = \sum_{s \in M} r_s u_s(t).$$

Traffic routing cost. When routing traffic of various SFCs across multiple edge clouds, two kinds of different cost would be incurred by the cross-edge WAN links. The first is the usage of the scarce and expensive WAN bandwidth. The second is the performance penalty incurred by the network latency of the cross-edge WAN links. Since both of these two kinds of cost are determined by the source edge cloud and destination edge cloud of the traffic routing process, we use a unified cost parameter d_{ij} to denote the overall cost (i.e. WAN bandwidth cost plus performance penalty) of routing one unit traffic from edge cloud i to edge cloud j.

For the traffic of SFC s processed by the cross-edge clouds, the total routing cost can be computed by summing up the followings: (1) the cost of routing the input traffic from the source edge cloud s to the first SF of the SFC s^2, $\sum_{m \in M} \sum_{i \in I} h_{0m}^s z_{si}^m(t) d_{si}$. Here we perform summation over all the edge clouds I, and the rationale is that the instances of the first

---

server power is proportional to the amount of traffic processed, it can be readily incorporated into the traffic routing cost, which will be formulated later.

2 Here we use a dummy SF 0 to represent the input process at each source edge cloud s.

SF of each SFC s can be placed at multiples edge clouds for the purpose of service locality. (2) The cost of routing the intermediate traffic in each direct hop $m \to n$ of the SFC s ($h_{mn}^s = 1$), $\sum_{i \in I} \sum_{j \in J} h_{mn}^s y_{sij}^{mn}(t) d_{ij}$. (3) The routing cost of routing the output traffic from the edge clouds that host instances of the tail SF of SFC s back to the source edge cloud s ($h_{m0}^s = 1$), $\sum_{m \in M} \sum_{i \in I} h_{m0}^s \alpha_m^s z_{si}^m(t) d_{is}$. Summing up the above terms, the total cost of routing traffic of SFC s can be given by: $\sum_{m \in M} \sum_{i \in I} [h_{0m}^s z_{si}^m(t) d_{si} + h_{m0}^s \alpha_m^s z_{si}^m(t) d_{is}] + \sum_{m \in M} \sum_{i \in I} \sum_{n \in M} \sum_{j \in J} h_{mn}^s y_{sij}^{mn}(t) d_{ij}$. Summing over all the SFC s, the overall routing cost $C_R(t)$ of all the traffic in time slot t is given by $\sum_s \sum_m \sum_i h_{0m}^s z_{si}^m(t) d_{si} + \sum_s \sum_m \sum_i h_{m0}^s \alpha_m^s z_{si}^m(t) d_{is} + \sum_s \sum_m \sum_i \sum_n \sum_j h_{mn}^s y_{sij}^{mn}(t) d_{ij}$. For ease of presentation, we let $g_{si}^m = h_{0m}^s d_{si} + h_{m0}^s \alpha_m^s d_{is}$, then the above term can be simplified to:

$$C_R(t) = \sum_{s \in S} \sum_{m \in M} \sum_{i \in I} \left(g_{si}^m z_{si}^m(t) + \sum_{n \in M} \sum_{j \in I} h_{mn}^s d_{ij} y_{sij}^{mn}(t) \right).$$

4.1.3.4 The Cost Minimization Problem

In this chapter, we aim to develop a cost-efficient service function chaining framework to facilitate the cross-edge deployment of SFCs, towards the goal of minimizing the holistic cost of the cross-edge system. To this end, we formulate a joint optimization on instance provisioning and traffic routing, aiming at minimizing the overall cost over the long-term.

$$\mathsf{P}: \quad \min \sum_{t \in T} [C_I(t) + C_S(t) + C_R(t) + C_O(t)],$$

$$\text{s.t.} \sum_{s \in S} z_{si}^m(t) \le x_i^m(t) b_i^m, \forall t \in T, i \in I, m \in M, \tag{4.1a}$$

$$\sum_{i \in I} z_{si}^m(t) \ge \beta_s^m a_k^s(t), \forall t \in T, s \in I, m \in M, \tag{4.1b}$$

$$z_{si}^m(t) \ge \sum_{j \in I} \sum_{n \in M} h_{nm}^s y_{sji}^{nm}(t),$$
$$\forall t \in T, i, s \in I, m \in M/\{m_F^s\}, \tag{4.1c}$$

$$\alpha_s^m z_{si}^m(t) \le \sum_{j \in I} \sum_{n \in M} h_{mn}^s y_{sij}^{mn}(t),$$
$$\forall t \in T, i, s \in I, m \in M/\{m_L^s\}, \tag{4.1d}$$

$$a_s(t) + u_s(t) \ge A_s(t), \forall t \in T, s \in I, \tag{4.1e}$$

$$y_{sij}^{mn}(t) \ge 0, \forall t \in T, i, j, s \in I, m, n \in M, \tag{4.1f}$$

$$z_{si}^m(t) \ge 0, \forall t \in T, i, s \in I, m \in M, \tag{4.1g}$$

$$a_s(t) \ge 0, \forall t \in T, s \in I, \tag{4.1h}$$

$$u_s(t) \ge 0, \forall t \in T, s \in I, \tag{4.1i}$$

$$x_i^m(t) \in \{0, 1, ..., C_i^m\}, \forall t \in T, i \in I, m \in M. \tag{4.1j}$$

Here m_F^s and m_L^s represent the head SF and tail SF of SFC s, respectively. Equation (4.1a) is the capacity constraint that enforces that, for each SF m, the traffic routed to each edge cloud i does not exceed the provisioned processing capacity $x_i^m(t)b_i^m$. Equation (4.1b) is the load balancing constraint that indicates that the total incoming traffic of SFC s to SF m in all edge clouds should be no less than the aggregated traffic of SFC s at this SF m. Equations (4.1c) and (4.1d) are the traffic reservation constraints that guarantee that, for each *SF* in each edge cloud i, the outgoing traffic rate is no less than α_s^m times the incoming traffic rate. Equation (4.1e) ensures that all the incoming input traffic can be served by either the edge clouds or the remote central cloud. Equations (4.1f)–(4.1i) are the non-negative constraints for the decision variables. Finally, Equation (4.1j) is the integrality constraint for the number of deployed SF instances.

In the above problem formulation P, we only consider the computing resource constraint and omit the network constraint. The rationale is that, typically edge servers are inter-connected by high-speed local area network and computing resource sharing can be much more demanding Taleb et al. (2017), Jiao et al. (2018). Also, in practice, the network bottleneck can be the cross-edge link, or uplink/downlink of each edge node, or both. However, to the best of our knowledge, there is no empirical measurement study identifying the network bottleneck in collaborative edge computing environments. To avoid misleading assumptions, we do not consider the constraint of network bandwidth capacity in this chapter. We hope that our problem formulation will stimulate the research community to conduct empirical measurements to uncover the network bottleneck in collaborative edge computing environments, and we are glad to extend our model to incorporate this new constraint in our future work.

Solving the above optimization problem P is non-trivial due to the following dual challenges. First, the long-term cost minimization problem P is a time-coupling problem that involves further system information, as the instance switching cost $C_S(t)$ couples the decision of consecutive time slots. However, in a realistic cross-edge system, parameters such as traffic arrival rates typically fluctuate over time and thus cannot be readily predicted. Then, how can we minimize the long-term cost in an online manner, without knowing the future information as a priori knowledge? Second, even with an offline setting where all the future information is given as a priori knowledge, the corresponding cost minimization problem is NP-hard. Specifically, our problem can be reduced from the classical minimum knapsack problem (MKP) Vazirani (2013) which is known to be NP-hard.

4.1.4 Online Optimization for Long-term Cost Minimization

To address the dual challenges of the time-coupling effect and NP-hardness of the long-term cost minimization problem P, we blend the advantages of a regularization method for online algorithm design and a dependent rounding technique for approximation algorithm design to propose a provably efficient online algorithm. The key idea of the proposed online algorithm is two-fold: (1) by regularizing the time-coupling switching cost $C_S(t)$ and relaxing the integer variable $x_i^m(t)$, we decompose the long-term problem into a series of one-shot fractional problems that can be readily solved. (2) By rounding the fractional solution with a dependent rounding scheme, we obtain a near-optimal solution to the original problem, with a bounded optimality gap.

4.1.4.1 Problem Decomposition via Relaxation and Regularization

In response to the challenge of the NP-hardness of problem P, we first relax the integrality constraint Equation (4.1j), obtaining the fractional optimization problem P_R as follows:

$$P_R : \quad \min \sum_{t \in \mathcal{T}} C_I(t) + C_S(t) + C_O(t) + C_R(t),$$

$$\text{s.t. Constraint (4.1f) to (4.1i)},$$

$$x_i^m(t) \in [0, C_i^m], \quad \forall t \in \mathcal{T}, i \in \mathcal{I}, m \in \mathcal{M}.$$

For the above fractional problem P_R, the relaxed switching cost $C_S(t)$ still temporally couples $x_i^m(t)$ across the time span. To address this issue, a natural solution would be greedily adopting the best decision for the relaxed problem in each independent time slot. However, this naive solution does not necessarily reach the global optimum for the long-term, and may even lead to arbitrary bad results.

Towards a worst-case performance guarantee for the online algorithm of our problem, we exploit the algorithmic technique of regularization in online learning Buchbinder et al. (2014). The basic idea of regularization is to solve the relaxed problem P_R with regularized objective function to substitute the intractable $[x_i^m(t) - x_i^m(t-1)]^+$. Specifically, in this chapter, to approximate the term $[x_i^m(t) - x_i^m(t-1)]^+$, we employ the widely adopted convex regularizer relative entropy function Buchbinder et al. (2014) as follows:

$$\Delta(x_i^m(t) || x_i^m(t-1)) = x_i^m(t) \ln \frac{x_i^m(t)}{x_i^m(t-1)} + x_i^m(t) - x_i^m(t-1). \qquad (4.2)$$

Here we obtain the relative entropy function by summing the relative entropy term $x_i^m(t) \ln \frac{x_i^m(t)}{x_i^m(t-1)}$ and a linear term denoting the movement $x_i^m(t) - x_i^m(t-1)$. Due to the convexity of the above regularizer $\Delta(x_i^m(t) || x_i^m(t-1))$, it has been widely adopted to approximate optimization problems that involve L1-distance terms (e.g., the switching cost $C_S(t)$ in our problem) in online learning. To ensure that the fraction is still valid when no instance of SF m is deployed in edge cloud i at time slot $t-1$ (i.e. $x_i^m(t-1) = 0$), we add a positive constant term ϵ to both $x_i^m(t)$ and $x_i^m(t-1)$ in the relative entropy term in Equation (4.2). To normalize the switching cost $C_S(t)$ by regularization, we also define an approximation weight factor $\eta_i^m = \ln(1 + \frac{C_i^m}{\epsilon})$ and multiply the improved relative entropy function by $\frac{1}{\eta_i^m}$.

By using the enhanced regularizer $\Delta(x_i^m(t) || x_i^m(t-1))$ to approximate the time-coupling term $[x_i^m(t) - x_i^m(t-1)]^+$ in the switching cost $C_S(t)$, we further obtain the relaxed and regularized problem, which is denoted as P_{Re}. Although P_{Re} is still time-coupling, the convex, differentiable and logarithmic-based regularizer $\Delta(x_i^m(t) || x_i^m(t-1))$ enables us to temporally decouple P_{Re} into a series of one-shot convex programs P_{Re}^t, which can be solved in each individual time slot t based the solution obtained from the previous time slot $t-1$. These series of solutions generated in each time slot thus constitute a feasible yet near-optimal solution to our original problem, with provable performance guarantee even for the worst-case (to be analyzed in Section 4.5). Specifically, the decomposed sub-problem P_{Re}^t for each time slot t, $\forall t \in \mathcal{T}$ can be denoted as follows:

$$P_{Re}^t : \quad \min \ C_I(t) + C_O(t) + C_R(t) + \sum_{i \in \mathcal{I}} \sum_{m \in \mathcal{M}} \frac{q_i^m}{\eta_i^m} \left(\left(x_i^m(t) + \epsilon \right) \ln \frac{x_i^m(t) + \epsilon}{x_i^m(t-1) + \epsilon} \right.$$

$$\left. + x_i^m(t-1) - x_i^m(t) \right),$$

s.t. Constraint (4.1f) to (4.1i),

$$x_i^m(t) \in [0, C_i^m], \quad \forall t \in \mathcal{T}, i \in \mathcal{I}, m \in \mathcal{M}.$$

Since the problem P_{Re}^t is a standard convex optimization with linear constraints, it can be optimally solved in polynomial time, by taking existing convex optimization technique as exemplified by the classical interior point method Boyd and Vandenberghe (2004).

Note that at each time slot t, the variable $x_i^m(t-1)$, $\forall i \in \mathcal{I}, m \in \mathcal{M}$ has been obtained when solving P_{Re}^{t-1} at time slot $t-1$, and it is required as the input to solve P_{Re}^t at time slot t. In this regard, we develop an online regularization-based fractional algorithm (ORFA) as shown in Algorithm 5, which generates an optimal fractional solution $(\tilde{x}(t), \tilde{y}(t), \tilde{z}(t), \tilde{a}(t), \tilde{u}(t))$ at each time slot t by using the previous and current system information. It is obvious that this optimal solution of the relaxed and regularized problem P_{Re}^t, constitutes a feasible solution to the relaxed (but unregularized) problem P_R. Later, this feasibility will be leveraged to derive the competitive ratio of the ORFA algorithm.

Algorithm 5 An online regularization-based fractional algorithm – ORFA.

Input: $\mathcal{I}, \mathcal{M}, S, b, C, g, h, l, r, q, \alpha, \beta, \eta, \epsilon$

Output: x, y, z, u, a

1: initialization: $x = 0, y = 0, z = 0, u = 0, a = 0$

2: **for** each time slot $t \in \mathcal{T}$ **do**

3:     observe values of $A(t), q(t)$ and $x(t-1)$

4:     invoke the interior point method to solve the regularized problem P_{Re}^t

5:     **return** the optimal fractional solution $x(t), y(t), z(t)$

6: **end for**

4.1.4.2 A Randomized Dependent Rounding Scheme

The proposed online regularization-based fractional algorithm ORFA obtains a fractional solution of problem P_{Re}^t. In order to satisfy the integrality constraint Equation (4.1j) of the original problem P, we need to round the optimal fractional solution $\tilde{x}(t)$ to an integer solution $\bar{x}(t)$. To this end, a straightforward solution is the independent randomized rounding scheme Raghavan and Tompson (1987), whose basic idea is to round up each fractional $\tilde{x}(t)$ to the nearest integer $\bar{x}(t) = \lceil \tilde{x}(t) \rceil$ with a probability of $\tilde{x}(t) - \lfloor \tilde{x}(t) \rfloor$, i.e. $\Pr\{\bar{x}(t) = \lceil \tilde{x}(t) \rceil\} = \tilde{x}(t) - \lfloor \tilde{x}(t) \rfloor$, and round down $\tilde{x}(t)$ to the nearest integer $\bar{x}(t) = \lfloor \tilde{x}(t) \rfloor$ with a probability of $\lceil \tilde{x}(t) \rceil - \tilde{x}(t)$, i.e. $\Pr\{\bar{x}(t) = \lfloor \tilde{x}(t) \rfloor\} = \lceil \tilde{x}(t) \rceil - \tilde{x}(t)$.

While the above independent rounding policy can always generate a feasible solution (since the remote central cloud is able to cover all the unserved traffic incurred by rounding down $\tilde{x}_i^m(t)$), directly applying this policy to round $\tilde{x}_i^m(t)$ may incur high instance running costs or cloud outsourcing costs. That is, with certain probability, an excessive amount or even all the SF instances at each edge cloud are destroyed, leading to the situation that an enormous amount of input traffic is outsourced to the expensive central cloud. Similarly, an excessive amount of SF instances may also be launched due to aggressively rounding up $\tilde{x}_i^m(t)$), leading to a sharp growth of the instance running cost.

To address the above challenge, we therefore develop a randomized and dependent pairwise rounding scheme Rajiv et al. (2006) that can exploit the inherent dependence of the variables $\tilde{x}_i^m(t)$. The key idea is that a rounded-down variable will be compensated by another rounded-up variable, ensuring that the input traffic absorbed by the edge clouds could be fully processed by the edge clouds even after the rounding phase. With such a dependent rounding scheme, the variables would not be aggressively rounded up or down, reducing the cost of using the expensive central cloud or launching an excessive amount of SF instances at the edge clouds.

For each SF $m \in \mathcal{M}$, we introduce two sets $\mathcal{I}_{mt}^+ = \{i | \tilde{x}_i^m(t) \in \mathbb{Z}\}$ and $\mathcal{I}_{mt}^- = \{i | \tilde{x}_i^m(t) \in \mathbb{R}^+\}$. Intuitively, $\mathcal{I}_{mt}^+$ denotes the set of edge clouds with integral $\tilde{x}_i^m(t)$ while $\mathcal{I}_{mt}^-$ denotes the set of edge clouds with fractional $\tilde{x}_i^m(t)$ and thus should be rounded. According to the above definitions, we have $\mathcal{I}_{mt}^+ \bigcup \mathcal{I}_{mt}^- = \mathcal{I}$ at each time slot t. For each element $i \in \mathcal{I}_{mt}^-$, we further introduce a probability coefficient p_i^m and a weight coefficient ω_i^m associated with it. Here we define $p_i^m = \tilde{x}_i^m(t) - \lfloor \tilde{x}_i^m(t) \rfloor$, and $\omega_i^m = b_i^m, \forall m \in \mathcal{M}, i \in \mathcal{I}_{mt}^-$.

The detailed randomized dependent instance provisioning (RDIP) algorithm, which rounds the fractional solution, is shown in Algorithm 6.Specifically, for each SF $m \in \mathcal{M}$, to round the elements in $\mathcal{I}_{jt}^-$ with fractional $x_{ij}(t)$, the proposed RDIP algorithm runs a series of rounding iterations. At each iteration, we randomly select two elements i_1 and i_2 from $\mathcal{I}_{mt}^-$, and let the probability of one of these two elements round to 0 or 1, decided by the coupled coefficient γ_1 and γ_2. By doing so, at each iteration, the number of elements in $\mathcal{I}_{mt}^-$ would decrease at least by 1. Finally, when $\mathcal{I}_{mt}^-$ has only one element in the last iteration, we directly round it up with its current probability.

The proposed rounding scheme is cost-efficient, in terms of that it would not aggressively launch new SF instances or outsource unserved input traffic to the central cloud. This cost-efficiency is achieved by maintaining three desirable properties in the main loop of each iteration.

Firstly, the **continuous reduction property**. At least one of the two selected variables $\tilde{x}_{i_1}^m(t)$ and $\tilde{x}_{i_2}^m(t)$ is rounded into integer. For example, if $\varphi_1 = 1 - p_{i_1}^m$ and $\varphi_2 = p_{i_1}^m$ (line 11), then $p_{i_1}^m = 1$ if line 13 is executed and $p_{i_1}^m = 0$ if line 15 is executed. In both cases, $\tilde{x}_{i_1}^m(t)$ will be rounded to a integer.

Secondly, the **weight conservation property**. That is, after the main loop of each iteration, the total weighted resource capacity of the selected two elements (i.e. SF instance) remains unchanged, i.e. the sum $\tilde{x}_{i_1}^m(t)b_{i_1}^m + \tilde{x}_{i_2}^m(t)b_{i_2}^m$ stays constant. For example, if line 13 is executed, we have $[\tilde{x}_{i_1}^m(t) + \varphi_1]b_{i_1}^m + [\tilde{x}_{i_2}^m(t) - \frac{b_{i_1}^m}{b_{i_2}^m}\varphi_1]b_{i_2}^m = \tilde{x}_{i_1}^m(t)b_{i_1}^m + \tilde{x}_{i_2}^m(t)b_{i_2}^m$. Similarly, if line 15 is executed, we can also prove that this equation still holds.

Thirdly, the **marginal distribution property**. That is, the probability of rounding up or down each element $i \in \mathcal{I}_{mt}^-, \forall m \in \mathcal{M}$ after the main loop is determined by the fractional part $\tilde{x}_i^m(t) - \lfloor \tilde{x}_i^m(t) \rfloor$ of the fractional solution $\tilde{x}_i^m(t)$. More specifically, $\Pr\{\overline{x}_i^m(t) = \lceil \tilde{x}_i^m(t) \rceil\} = \tilde{x}_i^m(t) - \lfloor \tilde{x}_i^m(t) \rfloor, \Pr\{\overline{x}_i^m(t) = \lfloor \tilde{x}_i^m(t) \rfloor\} = 1 - (\tilde{x}_i^m(t) - \lfloor \tilde{x}_i^m(t) \rfloor)$. Based on this marginal distribution property, which has been proven in Rajiv et al. (2006), we

---

Algorithm 6 Randomized dependent instance provision – RDIP.

Input: $\mathcal{I}, \mathcal{M}, \widetilde{\boldsymbol{x}}(t-1), b$

Output: $\bar{x}(t)$

1: **for** each SF $m \in \mathcal{M}$ **do**

2:    let $\mathcal{I}_{mt}^{+} = \{i | \widetilde{x}_i^m(t) \in \mathbb{Z}\}, \mathcal{I}_{mt}^{-} = \{i | \widetilde{x}_i^m(t) \in \mathbb{R}^{+}\}$

3:    **for** each edge cloud $i \in \mathcal{I}_{it}^{+}$ **do**

4:      Set $\bar{x}_i^m(t) = \widetilde{x}_i^m(t)$

5:    **end for**

6:    **for** each edge cloud $i \in \mathcal{I}_{it}^{-}$ **do**

7:      let $p_i^m = \widetilde{x}_i^m(t) - \lfloor \widetilde{x}_i^m(t) \rfloor, \omega_i^m = b_i^m$

8:    **end for**

9:    **while** $|\mathcal{I}_{mt}^{-}| > 1$ **do**

10:      randomly select two elements i_1, i_2 from $\mathcal{I}_{mt}^{-}$

11:      define $\varphi_1 = \min\{1 - p_{i_1}^m, \frac{\omega_{i_2}^m}{\omega_{i_1}^m} p_{i_2}^m\}, \varphi_2 = \min\{p_{i_1}^m, \frac{\omega_{i_2}^m}{\omega_{i_1}^m}(1 - p_{i_2}^m)\}$

12:      with the probability $\frac{\varphi_2}{\varphi_1 + \varphi_2}$ set

13:      $p_{i_1}^m = p_{i_1}^m + \varphi_1, p_{i_2}^m = p_{i_2}^m - \frac{\omega_{i_1 J}}{\omega_{i_2 J}} \varphi_1;$

14:      With the probability $\frac{\varphi_1}{\varphi_1 + \varphi_2}$ set

15:      $p_{i_1}^m = p_{i_1}^m - \varphi_2, p_{i_2}^m = p_{i_2}^m + \frac{\omega_{i_1}^m}{\omega_{i_2}^m} \varphi_2$

16:      if $p_{i_1}^m \in \{0, 1\}$, then set $\bar{x}_{i_1}^m(t) = \lfloor \widetilde{x}_{i_1}^m(t) \rfloor + p_{i_1}^m$

17:      $\mathcal{I}_{mt}^{+} = \mathcal{I}_{mt}^{+} \cup \{i_1\}, \mathcal{I}_{mt}^{-} = \mathcal{I}_{mt}^{-} \setminus \{i_1\}$

18:      if $p_{i_2}^m \in \{0, 1\}$, then set $\bar{x}_{i_2}^m(t) = \lfloor \widetilde{x}_{i_2}^m(t) \rfloor + p_{i_2}^m$

19:      $\mathcal{I}_{mt}^{+} = \mathcal{I}_{mt}^{+} \cup \{i_2\}, \mathcal{I}_{mt}^{-} = \mathcal{I}_{mt}^{-} \setminus \{i_2\}$

20:    **end while**

21:    **if** $|\mathcal{I}_{mt}^{-}| = 1$ **then**

22:      set $\bar{x}_i^m(t) = \lceil \widetilde{x}_i^m(t) \rceil$ for the only element $i \in \mathcal{I}_{mt}^{-}$

23:    **end if**

24: **end for**

---

can further derive:

$$\mathbb{E}\{\bar{x}_i^m(t)\} = (\widetilde{x}_i^m(t) - \lfloor \widetilde{x}_i^m(t) \rfloor) \lceil \widetilde{x}_i^m(t) \rceil +$$
$$[1 - (\widetilde{x}_i^m(t) - \lfloor \widetilde{x}_i^m(t) \rfloor)] \lfloor \widetilde{x}_i^m(t) \rfloor$$
$$= (\widetilde{x}_i^m(t) - \lfloor \widetilde{x}_i^m(t) \rfloor)(\lfloor \widetilde{x}_i^m(t) \rfloor + 1) +$$
$$[1 - (\widetilde{x}_i^m(t) - \lfloor \widetilde{x}_i^m(t) \rfloor)] \lfloor \widetilde{x}_i^m(t) \rfloor$$
$$= \widetilde{x}_i^m(t).$$

The above equation shows that the expectation of each rounded solution is exactly the value of the original fractional solution. This property indicates that new SF instances would not be aggressively launched when rounding the fractional solution. As a result, the optimality gap between the expected total cost incurred by the rounded solution and that incurred by the optimal fractional solution is bounded, as we will prove in Section 4.1.5.3.

4.1.4.3 Traffic Re-routing

After performing the randomized and dependent rounding scheme RDIP at each time slot t, the instance provisioning decision $\bar{x}(t)$ produced by the RDIP algorithm together with the traffic routing decision $(\tilde{y}(t), \tilde{z}(t), \tilde{u}(t), \tilde{a}(t))$ may well not be a feasible solution of the original problem P. This means that we further need to modify the fractional traffic routing decision $(\tilde{y}(t), \tilde{z}(t), \tilde{u}(t), \tilde{a}(t))$ accordingly, in order to maintain the solution feasibility.

To the above end, a naive approach is to bring the feasible integral solution $\bar{x}(t)$ back to the original problem P, and solve the degraded traffic re-routing problem to obtain a complete feasible solution $(\bar{x}(t), \bar{y}(t), \bar{z}(t), \bar{u}(t), \bar{a}(t))$. However, if we further consider the weight conservation property maintained by the randomized dependent rounding scheme in Section 4.1.4.2, we may obtain a complete feasible solution in a simpler manner. Specifically, the weight conservation property of the RDIP algorithm ensures that for each SF $m \in \mathcal{M}$, the total resource capacity $\sum_i b_i^m \bar{x}_i^m$ of the rounded solution is no smaller than $\sum_i b_i^m \tilde{x}_i^m$ of the fractional solution. This further indicates that the resource capacity of the rounded solution is sufficient to cover the total traffic absorbed by the edge clouds when performing the ORFA algorithm. Then, an intuition is to let $\bar{u}(t) = \tilde{u}(t), \bar{a}(t)) = \tilde{a}(t)$ and solve the following simplified traffic re-routing problem to obtain $\bar{y}(t)$ and $\bar{z}(t)$:

$$\min \ C_R(t)$$
$$\text{s.t. Constraints (1a)–(1d), (1f) and (1g).}$$

Since this traffic re-routing problem is a linear programming, therefore its optimal solution can be readily computed in polynomial time, by applying linear programming techniques as exemplified by interior point method.

Remark: in terms of optimality, we acknowledge that the weight conservation property based traffic re-routing scheme is not as better as the aforementioned naive approach, since the latter has a larger feasible space. However, in a realistic cross-edge system, the cost of cloud outsourcing is far more expensive than edge-based processing, the optimality gap between those two approaches is small actually. In addition, here we study the former one due to the following three reasons: (1) it has lower time complexity; (2) it is more straightforward for performance analysis (to be detailed in Section 4.1.5); and (3) it can work as a bridge to analyze the performance bound of the aforementioned naive approach for traffic re-routing, as we will discuss later in Section 4.1.5.3.

4.1.5 Performance Analysis

In this section, we rigorously analyze the theoretical performance of the proposed online algorithm via competitive analysis. As a standard approach to quantifying the performance of online algorithms, the basic idea of competitive analysis is to compare the performance of the online algorithm to the theoretical optimal solution in the offline case, where all the future information is given as a priori knowledge. In particular, we prove that the proposed online algorithm has guaranteed performance, which is quantified by a worst-case parameterized competitive ratio.

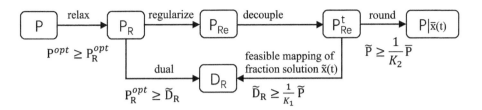

Figure 4.3 An illustration of the basic idea of the performance analysis.

4.1.5.1 The Basic Idea

Essentially, we derive the competitive ratio based on the well-established primal-dual framework for convex optimization. To this end, as illustrated in Figure 4.3, we introduce the dual problem of the relaxed problem D_R to act as the bridge that connects the original problem and regularized problem through the following chain of inequalities.

$$P^{opt} \geq P_R^{opt} \geq \tilde{D}_R \geq \frac{1}{K_1}\tilde{P} \geq \frac{1}{K_1 K_2}\overline{P}. \tag{4.4}$$

Here P^{opt} and P_R^{opt} denote the objective values achieved by the optimal solutions of the original problem P and relaxed problem P_R, respectively. $\tilde{D}_R$ denotes the objective value of P_R's dual problem D_R, achieved by a feasible solution mapped from the optimal fractional solution of the regularized problem P_{Re}. Finally, $\tilde{P}$ and $\overline{P}$ are the objective values of the original problem P, achieved by the optimal fractional solution and the rounded solution, respectively.

Since P_R is a relaxed problem of the original minimization problem P, we thus have $P^{opt} \geq P_R^{opt}$. The inequality $P_R^{opt} \geq D_R$ holds due to the celebrated weak duality theorem in convex optimization theory. In addition, to connect the dual problem D_R to the regularized problem P_{Re}, we construct a feasible solution for D_R mapped from the optimal fractional solution for P_{Re}. Based on such mapping, we can derive the competitive ratio K_1 by applying Karush–Kuhn–Tucker (KKT) conditions, i.e. the first-order necessary conditions characterizing the optimal solution to exploit the inherent common structures shared by P_{Re} and D_R. Finally, based on the three desirable properties maintained by the dependent rounding scheme, we can further characterize the competitive ratio K_2 between the optimal fractional solution and the rounded solution.

4.1.5.2 Competitive Ratio of ORFA

We establish $\tilde{D}_R \geq \frac{1}{K_1}\tilde{P}$ and derive the competitive ration K_1 of ORFA in this subsection. Specifically, we first drive the Lagrange dual problem D_R of the relaxed problem P_R. Then, we characterize the the optimality conditions of the regularized problem P_{Re}. Finally, we construct a feasible solution for D_R mapped from the primal and dual optimal solutions for P_{Re}. Due to space limits, interested readers are referred to Zhou et al. (2019b) for the detailed mathematical analysis, and the competitive ratio of the proposed ORFA algorithm is given in the following Theorem 4.1.

Theorem 4.1 *For the proposed ORFA algorithm, it achieves a competitive ratio of K_1. That is, the objective value of problem P achieved by the optimal fractional solution, denoted by $\tilde{P}$, is no larger than K_1 times the offline optimum P^{opt}, where K_1 is given by:*

$$K_1 = \ln\left(1 + \frac{C_{max}}{\epsilon}\right) + \frac{C_{max}}{\delta} + 1.$$

4.1.5.3 Rounding Gap of RDIP

We next study the rounding gap incurred by the randomized dependent rounding scheme, in terms of competitive ratio K_2 of the cost $\overline{P}$ achieved by the final rounded solution to the cost $\tilde{P}$ achieved by the fractional solution. The basic idea is to leverage the relationship between $\tilde{x}(t)$ and $x(t)$, which has been characterized by the three desirable properties in Section 4.1.4.2, to establish the connection between their instance running costs $\sum_i\sum_m b_i^m(t)\tilde{x}_i^m(t)$ and $\sum_i\sum_m b_i^m(t)\overline{x}_i^m(t)$. Then, we further take this connection as a bridge to bound the cost terms in the objective function of the original problem P. Due to space limits, interested reader are referred to Zhou et al. (2019b) for the detailed mathematical analysis. Based on the intermediate results that maps the cost terms of the rounded solution to that of the optimal fractional solution, the competitive ratio of the proposed RDIP algorithm is now given in the following Theorem 4.2.

Theorem 4.2 *For the proposed RDIP algorithm, it achieves a competitive ratio of K_2. That is, the objective value $\overline{P}$ of problem P achieved by the rounded solution is no larger than K_2 times of $\tilde{P}$ achieved by the optimal fractional solution, where $K_2 = \xi_1 + \xi_2 + \xi_3$, and*

$$\xi_1 = (1 + \kappa)\max_{t,i,m} \frac{p_i^m(t)}{b_i^m} \max_{t,i,m} \frac{b_i^m}{p_i^m(t)},$$

$$\xi_2 = (1 + \kappa)\max_{t,i,m} \frac{b_i^m}{p_i^m(t)} \max_{i,m} \frac{q_i^m}{b_i^m},$$

$$\xi_3 = (1 + \kappa)\max_{t,i,m} \frac{b_i^m}{p_i^m(t)}(\max_{s,i,m} g_{si}^m + \max_{ij} l_{ij}).$$

4.1.5.4 The Overall Competitive Ratio

Based on the competitive ratio of the proposed ORFA and RDIP algorithms, given in Theorems 4.1 and 4.2, respectively, we give the overall competitive ratio of the online algorithm for joint instance provisioning and traffic routing in the following Theorem 4.3.

Theorem 4.3 *For the proposed online algorithm for joint instance provisioning and traffic routing, it achieves a competitive ratio of $K_1 K_2$. That is, the objective value $\overline{P}$ of problem P achieved by the rounded solution is no larger than $K_1 K_2$ times of the offline optimum, where K_1 and K_2 are derived in Theorems 4.1 and 4.2, respectively.*

Theorem 4.3 follows immediately the chain of inequalities in Equation (4.4). We now discuss some insights into the final competitive ratio $K_1 K_2$.

Firstly, the final competitive ratio decreases with increases in the tunable parameter ϵ. By increasing ϵ to be large enough, we can push the competitive ratio K_1 given in Theorem 4.2

arbitrarily close to $\frac{C_{\max}}{\delta} + 1$. However, overly aggressive increases in the control parameter ϵ can also increase the time complexity of the regularized problem P_{Re}, since the convexity of P_{Re} deteriorates as ϵ grows (according to the decreasing of the second-order derivative of the objective function). Therefore, the control parameter ϵ works as a knob to balance the performance-complexity trade-off.

Secondly, recall that when performing traffic re-routing based on $\bar{x}(t)$ to obtain a complete feasible solution, we exploit the weight conservation property to modify $\tilde{y}(t)$ and $\tilde{z}(t)$, while keeping $\tilde{u}(t)$ and $\tilde{a}(t)$ unchanged. Compared to the naive approach that directly brings $\bar{x}(t)$ back to the original problem P to obtain the complete feasible solution, the above applied approach has lower time complexity but also reduced optimality. Therefore, the derived ratio K_1K_2 can also serve as an upper bound of the competitive ratio of the scheme that applies the above naive approach to traffic re-routing.

Finally, and interestingly, we observe that the derived competitive ratio is deterministic, although the proposed rounding scheme is random in nature. The rationale is that it is derived based on the weight conservation property, which is actually deterministic rather than random. Furthermore, if we leverage the marginal distributed property that ensures $\mathbb{E}\{\bar{x}_i^m(t)\} = \tilde{x}_i^m(t)$ after the main loop of the RDIP algorithm, we may derive a random competitive ratio. We leave this horizontal route as an extension to be addressed in further work.

4.1.6 Performance Evaluation

In this section, we conduct trace-driven simulations to evaluate the practical benefits of the proposed online orchestration framework. The simulations are based on real-world workload traces and electricity prices.

4.1.6.1 Experimental Setup

Infrastructure: We simulate a cross-edge system that deploys $M = 4$ edge clouds in the geographical center of four regions in New York city: Manhattan, Brooklyn, Queues and Bronx. The input traffic at each edge cloud requires a SFC that contains three SFs randomly chosen from the four extensively studied virtual network functions (VNFs) in Table 4.1.

Workload trace: Since edge computing is still in a very early stage, there is no public accessible workload trace from an edge cloud. In response, we adopt the workload trace of mega-scale data centers to simulate the traffic arrival at each edge cloud. Specifically, we associate to the one-week hourly traces of the data centers of Google Google, Facebook Chen et al. (2011), HP Liu et al. (2011), and Microsoft Narayanan et al. (2008) to each of the four edge clouds, respectively. The normalized traffic arrival of the traces is shown in

Table 4.1 Configuration of four required SFs.

SF	Change ratio α	Instance capacity b
Firewall	0.8–1.0	30
Proxy	1.0	30
NAT	0.6	20
IDS	0.8–1.0	20

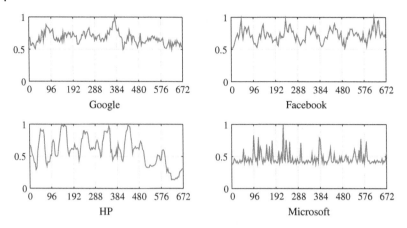

Figure 4.4 The workload trace of Google, Facebook, HP, and Microsoft data centers.

Figure 4.4. In the following simulations, we proportionally scale the normalized trace up to 1000 times to represent the actual traffic arrival at each edge cloud.

Real cost data: To incorporate the spatial and temporal diversities of the instance running cost $p_i^m(t)$, here we use the product of the instance capacity and the regional electricity price to simulate the instance running cost $p_i^m(t)$. Specifically, we download the hourly locational marginal prices (LMP) in the real-time electricity markets of the four regions from the website of NYISO (New York Independent System Operator) ISO. The time period of this data is October 10–16, 2018, including one week or 168 h.

System parameters: We set C_i^m, the maximal number of available instances of SF m at edge cloud i, to be the minimal integer that can serve 80% of the peak-arrival at each edge cloud locally. For the cross-edge traffic routing cost parameter d_{ij}, we assume it is proportional to the geographical distance, and the total traffic routing cost $C_R(t)$ has a similar order of magnitude to the instance running cost $C_I(t)$. For the switching cost parameter q_i^m, we also set it to make the switching cost $C_S(t)$ has a similar order of magnitude to $C_I(t)$. For the cloud outsourcing cost parameter r_s, we set it is sufficient large, such that the central cloud would only be used in the presence of flash crowd.

Benchmarks: to empirically evaluate the competitive ratio of the proposed online framework, we adopt the state-of-the-art MILP solver Gurobi to obtain the offline optimum of the long-term cost minimization problem P. To demonstrate the efficacy of the online algorithm ORFA, we compare it to the lazy instance provisioning (LIP) that has been extensively applied to decouple time-coupling term in the literature Lin et al. (2011), Zhang et al. (2013). Furthermore, to demonstrate the efficacy of the rounding scheme RDIP, we compare it to another three benchmarks: (1) the randomized independent instance provisioning (RIIP) Raghavan and Tompson (1987), as we have discussed in the first paragraph of Section 4.1.4.2. (2) The EC-Greedy approach, which directly rounds-up all the fractional solutions to process the absorbed traffic by the edge clouds. (3) The CC-Greedy approach, which directly rounds-down all the fractional solutions, and using the central cloud to serve the traffic which can not be covered by the edge clouds.

4.1.6.2 Evaluation Results

Efficiency of the online algorithm. We first examine the competitive ratio achieved by the online algorithm ORFA (e.g., Algorithm 4.1) without rounding the factional solution, which is the ratio of the total cost achieved by ORFA to that achieved by the offline minimum cost. For comparison, we plot the competitive ratio of ORFA as well as the benchmark LIP in Figure 4.5, under varying number of total time slots. From this figure, we observe that: (1) ORFA always outperforms the LIP that has been widely applied in literature Lin et al. (2011), Zhang et al. (2013), demonstrating the effectiveness of our proposed regularization-based online algorithm. (2) As the number of total time slots varies, the competitive ratio of ORFA only changes very slightly, indicating that ORFA has stable performance against varying time span.

Effect of the switching cost. We continue to investigate the effect of the instance switching cost (i.e. q_i^m in our formulation) on the competitive ratio of ORFA and LIP, by multiplying the switching cost of each SF instance with a various scaling ratios. The results are plotted in Figure 4.6, which shows that as the switching cost increases the competitive ratios of both ORFA and LIP increase dramatically. The rationale is that, by increasing the switching cost q_i^m, we would increasingly focus on minimizing the total switching cost. However, since the switching cost term is time-coupling and involves future stochastic information, the optimality gap incurred by any online algorithm will increase

Figure 4.5 The competitive ratio of different online algorithms.

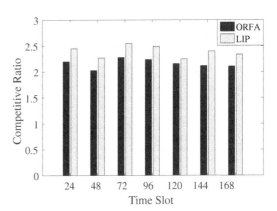

Figure 4.6 The effect of the switching cost on the competitive ratio.

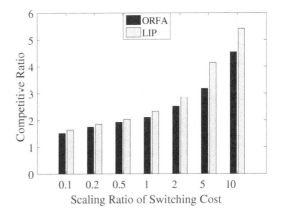

as the switching cost q_i^m grows. Since the weights of other cost terms remains unchanged, the optimality gap of the switching cost term will deteriorate the overall competitive ratio of the online algorithm.

Sensibility to varying parameters. Recall that in Theorem 4.1, the worst-case competitive ratio of ORFA is parameterized by the control parameter ϵ and $C_{\max}$, the maximum number of available instance for each SF m at each edge cloud i. We now examine the effect of parameters ϵ and $C_{\max}$ on the actual competitive ratio of ORFA. Figure 4.7 depicts the actual competitive ratio under various ϵ and $C_{\max}$, from which we observe that: (1) as the control parameter ϵ increases, the competitive ratio descents. This quantitatively corroborates the insights we unfolded in Section 4.1.5.4. However, we should also note that while ϵ varies significantly, the competitive ratio only changes slightly, meaning that the effect of ϵ on the competitive ratio is very limited. (2) As $C_{\max}$ changes, the competitive ratio remains relatively stable. This suggest that in practice, the impact of $C_{\max}$ is not obvious.

Competitive ratio of the rounding scheme. After executing the online algorithm ORFA to obtain the fractional solution, we need to round it to an integer solution for instance provisioning. Here we compare the competitive ratio of different rounding schemes in Figure 4.8, which demonstrates that the competitive ratio of our proposed RDIP is relatively stable and substantially outperforms those of the three benchmarks.

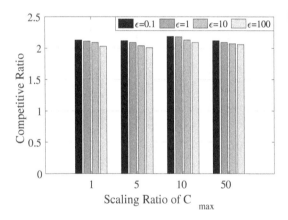

Figure 4.7 The effect of the control parameter ϵ on the competitive ratio.

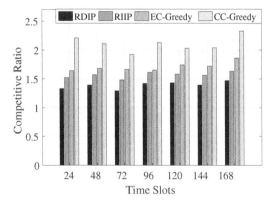

Figure 4.8 The competitive ratio of various rounding schemes.

Figure 4.9 The overall competitive ratio of different algorithm combinations.

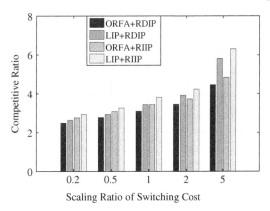

Here we also note that the independent rounding scheme RDIP outperforms the other two greedy strategies, due to the fact that the optimality introduced by the rounding can be compensated for by the traffic-rerouting conducted afterwards.

Examining the overall competitive ratio. We finally verify the performance of our complete online algorithm for joint instance provisioning and traffic routing, by comparing the overall competitive ratio of different combinations of the online algorithm and the rounding scheme. The results in Figure 4.9 show that the combination of our proposed online algorithm and rounding scheme performs the best. While in sharp contrast, the combination of the benchmarks leads to the worst performance. This confirms the efficacy of both the proposed ORFA algorithm and the RDIP rounding scheme. We also observe that, similar to Figure 4.6, the competitive ratio rises as the switching cost q_i^m increases, due to the aforementioned fact that the optimality gap of the switching cost term rises as the switching cost q_i^m increases.

4.1.7 Future Directions

To shed light on future potential research on cost-efficient service function chaining across dispersed edge computing nodes, in this section we discuss several research directions to extend the proposed online optimization framework in this chapter.

- **Levering the power of prediction.** The proposed online orchestration framework in this chapter can make near-optimal and online decisions without assuming any future information as a priori. However, with the recent advancement in deep learning, the classical challenging problem of time series prediction has witnessed renewed research interest and progress, as highlighted by some highly effective methods exemplified by the recurrent neural network (RNN) and long short term memory (LSTM). How to incorporate these emerging deep learning-based prediction methods into our optimization framework, and thus to make predictive online decisions to improve the optimality could be a promising direction.
- **Incorporating AI-tailored optimization strategies.** In this chapter, we assume that the resource demand of each traffic flow is fixed and untunable by default. However, in practice, for some AI-based applications as exemplified by video analytics and image recognition, the resource demand of the application workload can be tunable. For

example, for video analytics, both the frame rate and resource of each video frame can be adjusted to tune the resource consumption. Intuitively, such AI-tailored optimization strategies can be exploited to further reduce the system-wide cost of service function chaining across dispersed edge computing nodes. However, as such AI-tailored optimization strategy typically degrades the accuracy of the AI model, how to balance the resource-accuracy trade-off should be carefully considered.

- **Addressing DAG-style function graph.** In this chapter, we assume that each traffic flow is served by a chain of consecutive service functions. However, for some complex applications and also for the benefit of better parallelism, the service functions invoked by the application are represented by a directed cyclical graph (DAG). However, extending our optimization framework to address such a DAG-style function graph could also be an interesting future research direction.

4.2 Dynamic Network Slicing for High-quality Services

This chapter studies fog computing systems in which cloud data centers can be supplemented by a large number of fog nodes deployed in a wide geographical area. Each node relies on harvested energy from the surrounding environment to provide computational services to local users. We propose the mechanism of *dynamic network slicing* in which a regional orchestrator coordinates workload distribution among local fog nodes, providing partitions/slices of energy and computational resources to support a specific type of service with certain quality-of-service (QoS) guarantees. The resources allocated to each slice can be dynamically adjusted according to service demands and energy availability. A stochastic overlapping coalition-formation game is developed to investigate distributed cooperation and joint network slicing between fog nodes under randomly fluctuating energy harvesting and workload arrival processes. We observe that the overall processing capacity of the fog computing network can be improved by allowing fog nodes to maintain a belief function about the unknown state and the private information of other nodes. An algorithm based on a belief-state partially observable Markov decision process (B-POMDP) is proposed to achieve the optimal resource slicing structure among all fog nodes. We describe how to implement our proposed dynamic network slicing within the 3GPP network sharing architecture, and evaluate the performance of our proposed framework using the real BS location data of a real cellular system with over 200 BSs deployed in the city of Dublin. Our numerical results show that our framework can significantly improve the workload processing capability of fog computing networks. In particular, even when each fog node can coordinate only with its closest neighbor, the total amount of workload processed by fog nodes can be almost doubled under certain scenarios.

4.2.1 Service and User Requirements

With the widespread proliferation of intelligent systems, IoT devices, and smart infrastructure, computation-intensive mobile applications that require low delay and fast processing are becoming quite popular. Next-generation mobile networks (e.g., 5G and beyond) are expected to serve over 50 billion mobile devices, most of which are smart devices requiring as low as 1 ms latency and very little energy consumption (ITU-T, Aug. 2014, Huawei, Andrews et al., June 2014). Major IT service providers, such as Google, Yahoo, Amazon, etc., are heavily investing in large-scale data centers to meet the demand for future data services. However, these data centers are expensive and often built in remote areas to save costs. This makes it difficult to provide the QoS requirements of end users, especially for users located at the edge of a coverage area. To provide low-latency services to end users, a new framework referred to as fog computing has emerged (Vaquero and Rodero-Merino, 2014), in which a large number of wired/wireless, closely located, and often decentralized devices, commonly referred to as fog nodes, can communicate and potentially cooperate with each other to perform certain computational tasks. Fog computing complements existing cloud services by distributing computing, communication, and control tasks closer to end users. Fog nodes include a variety of devices between end users and data centers, such as routers, smart gateways, access points (APs), base stations (BSs), and set-top boxes. According to the Next-Generation Mobile Network (NGMN) Alliance (NGMN Alliance, Feb. 2015), fog computing will be an important component of 5G systems, providing support for computation-intensive applications that require low latency, high reliability, and secure services. Examples of these applications include intelligent transportation, smart infrastructure, e-healthcare, and augmented/virtual reality (AR/VR). The success of fog computing heavily relies on the ubiquity and intelligence of low-cost fog nodes to reduce the latency and relieve network congestion (Vahid Dastjerdi et al., 2016, Yi et al., Hangzhou, China, June 2015, Yannuzzi et al., Athens, December 2014).

Over the last decade, there has been a significant interest in climate change and energy sustainability for information and communication technologies (Cloud, Times, August 2016, Microsoft, Fortune.com, 2017). The telecommunication network infrastructure is already one of the leading sources of global carbon dioxide emissions (Chamola and Sikdar, 2016). In addition, the unavailability of a reliable energy supply from electricity grids in some areas is forcing mobile network operators (MNOs) to use sources like diesel generators for power, which will not only increase operating costs but also contribute to pollution. Energy harvesting is a technology that allows electronic devices to be powered by the energy converted from the environment, such as sunlight, wind power, and tides. It has recently attracted significant interest due to its potential to provide a sustainable energy source for electronic devices with zero carbon emission (Ulukus et al., 2015, Xiao et al., 2015b, Lu et al., 2015, Xiao et al., London, UK, June 2015, Ge et al., 2015). Major IT providers including Apple, Facebook, and Google have already upgraded all their cloud computing servers to be fully supported by renewable energy (Fortune.com, 2017, Microsoft, Cloud). Allowing fog nodes to utilize the energy harvested from nature can provide ubiquitous computational resources anywhere at any time. For example, fog nodes deployed inside an edge network can rely on renewable energy sources to support low-latency, real-time computation for applications such as environmental control, traffic monitoring and congestion avoidance, automated real-time vehicle guidance systems, and AR/VR assisted manufacturing.

Incorporating energy harvesting into the design of the fog computing infrastructure is still relatively unexplored. In contrast to data centers that can be supported by massive photovoltaic solar panels or wind turbines, fog nodes are often limited in size and location. In addition, it is generally difficult to have a global resource manager that coordinates resource distribution among fog nodes in a centralized fashion. Developing a simple and effective method for fog nodes to optimize their energy and computational resources, enabling autonomous resource management according to the time-varying energy availability and user demands, is still an open problem.

Enabled by software-defined networking (SDN) and network function virtualization (NFV) technologies, the mechanism of *network slicing* has recently been introduced by 3GPP to further improve the flexibility and scalability of fog computing for 5G systems (3GPP, June 2016a, J, Vaezi and Zhang, 2017, NGMN Alliance, September 2016).

Network slicing allows a fog node to support multiple types of service (use cases) by partitioning its resources, such as spectrum, infrastructure, network functionality, and computing power among these types. Resource partitions, commonly referred to as *slices*, can be tailored and orchestrated according to different QoS requirements of different service types (e.g., real-time audio, image/video processing with various levels of delay tolerance, etc.) (Richart et al., 2016). Multiple SDN-based network slicing architectures have been proposed by 3GPP (Samdanis et al., 2016), The NGMN Alliance (NGMN Alliance, Feb. 2015,S), and the Open Networking Foundation (ONF) (ONF, Apr. 2016, 2016). However, these architectures are all based on a centralized control plane and cannot be directly applied to large-scale network systems.

In this chapter, we introduce a new *dynamic network slicing* architecture for large-scale energy-harvesting fog computing networks. This architecture embodies a new network entity, the *regional SDN-based orchestrator*, that coordinates the workload processed by multiple closely located fog nodes and creates slices of energy and computational resources for various types of service requested by end users. To minimize the coordination cost, the workload of each user is first sent to the closest fog node. Fog nodes will then make autonomous decisions on how much energy resource is spent on activating computational resources and how to partition the activated computational resources according to time-varying energy availability, user demands, and QoS requirements. If a fog node decides that it needs help from its neighboring fog nodes to process a part of its received workload, or if it has surplus resource to help other fog nodes in proximity, it will coordinate with these fog nodes through the regional SDN-based orchestrator. Our main objective is to develop a simple distributed network slicing policy that can maximize the utilization efficiency of available resources and balance the workloads among fog nodes over a wide geographical area. The distributed and autonomous decision making process at each fog node makes game theory a suitable tool to analyze the interactions among fog nodes. In this chapter, we develop a stochastic overlapping coalition-formation game-based framework, called *dynamic network slicing game*, to analyze such interactions. In contrast to the traditional partition-based coalition formation game, in our game, players are allowed to interact with each other across multiple coalitions, which has the potential to further improve the resource utilization efficiency and increase the outcome for players. Unfortunately, finding a stable coalition structure in this game is known to be notoriously difficult. Because each player can allocate a fraction of its resources to each coalition, there can be infinitely many possible coalitions among players. It has already been proved that an overlapping coalition game may not always have a stable coalition structure. Even it does, there is no general method that can converge to such a structure. We propose a distributed algorithm based on a belief-state partially observable Markov decision process (B-POMDP) for each fog node to sequentially learn from its past experience and update its belief function about the state and offloading capabilities of other nodes. We prove that our proposed algorithm can achieve the optimal resource slicing policy without requiring back-and-forth communication between fog nodes. Finally, we evaluate the performance of our proposed framework by simulations, using actual BS topological deployment of a large-scale cellular network in the city of Dublin. Results show that our proposed framework can significantly improve the workload offloading capability of fog nodes. In particular, even when each fog node can only cooperate with its closest neighbor, the total amount of workload offloaded by the fog nodes can almost be doubled especially for densely deployed fog nodes in urban areas.

4.2.2 Related Work

A key challenge for fog computing is to provide QoS-guaranteed computational services to end users while optimizing the utilization of local resources owned by fog nodes (Sarkar et al., Bonomi et al., 2014, Datta et al., Madrid, Spain, June 2015). In (Do et al., Siem Reap, Cambodia, January 2015), the joint optimization of allocated resources while minimizing the carbon footprint was studied for video streaming services over fog nodes. A service-oriented resource estimation and management model was proposed in (Aazam and Huh, St. Louis, MO, Mar. 2015) for fog computing systems. In (Xiao and Krunz, November 2018), a distributed optimization algorithm was proposed for fog computing-supported tactile internet applications requiring ultra-low latency services.

SDN and NFV have been considered as key enablers for fog computing (Gupta et al., 2016, Khan et al., Las Vegas, NV, January 2018). In particular, popular SDN protocols such as ONF's OpenFlow have already been extended into fog computing networks (Hakiri et al., Hammamet, Tunisia, October 2017). To further improve the scalability and flexibility of OpenFlow when extending to wireless systems, the authors in (Tootoonchian and Ganjali, San Jose, CA, Apr. 2010) introduced a hybrid SDN control plane that combines the optimized link state routing protocol (OLSR), a popular IP routing protocol for mobile ad hoc networks, with OpenFlow to perform path searching and selection as well as network monitoring. Recently, a new framework, referred to as the hierarchical SDN, has been introduced to reduce the implementation complexity of SDN by organizing the network components such as controllers and switchers in a layered structure (Fang et al., June 2015). In particular, the authors in (Hakiri et al., Hammamet, Tunisia, October 2017) proposed a hierarchical framework, called hyperflow, in which groups of switches have been assigned to controllers to keep the decision making within individual controllers. The authors in (Hassas Yeganeh and Ganjali, Aug. 2012) developed a new control plane consisting of two layers of controllers: bottom-layer and top-layer controllers. The former runs only locally controlled applications without inter-connections or the knowledge of the network-wide state. And the latter corresponds to a logically centralized controller that maintains the network-wide state. In (Liu et al., London, UK, June 2015), the authors investigated the optimization of the hierarchical organization for a set of given network scales. It shows that using a four-layer SDN is sufficient for most practical network scales.

Recently, game theory has been shown to be a promising tool to analyze the performance and optimize fog computing networks. Specifically, in (Zhang et al., 2016), a hierarchical game-based model was applied to analyze the interactions between cloud data centers (CDCs) and fog nodes. An optimal pricing mechanism was proposed for CDCs to control resource utilization at fog nodes. To the best of our knowledge, this section is the first work to study the distributed workload offloading and resource allocation problem for energy harvesting fog computing networks with fog node cooperation.

4.2.3 System Model and Problem Formulation

4.2.3.1 Fog Computing

A generic fog computing architecture consists of the following elements (Chiang et al., 2017, Zhang et al., 2017, 2020, Tong et al., San Francisco, CA, Apr. 2016):

(1) *Cloud computing service provider (CSP)*: The CSP owns and manages large-scale CDCs that can provide abundant hardware and software resources with low processing delay. CDCs are often built in low-cost remote areas and therefore services processed at the CDC are expected to experience high transmission latency.

(2) *Fog computing service provider (FSP)*: The FSP controls a large number of low-cost fog nodes (e.g., mini-servers), deployed in a wide geographical area. Typically, fog nodes do not have high-performance processing units. However, they are much cheaper to deploy and require much less energy to operate. In this chapter, we focus on energy-harvesting fog computing networks in which computational resources that can be activated are time-varying and solely depend on the harvested energy.

(3) *Networking service provider (NSP)*: The NSP deploys a large wired or wireless network infrastructure that connects users to fog nodes and/or remote CDCs.

(4) *Tenants*: Tenants can correspond to virtual network operators (VNOs) that lack network infrastructure or with limited capacity and/or coverage, and have to lease resources from other service/infrastructure providers. They can also be over-the-top (OTT) service/content providers, such as Netflix, Spotify, Skype, etc., operating on top of the hardware/software infrastructures deployed by CSP, FSP, and/or NSP. In this chapter, we assume each tenant always requests networking and/or computational resources (e.g., slices) from one or more providers to serve the needs for the users.

(5) *Users*: Users are mobile devices or user applications that consume the services offered by tenants. Users can locate in a wide geographical area and can request different types of services with different QoS requirements.

Note that the above elements may not always be physically separated from one another. For example, a cellular network operator (an NSP) with insufficient computational resources can rent computational resources (e.g., server/CPU times) from an FSP to support computational intensive service (e.g., AR/VR-based services, online gaming, etc.) requested by its subscribers (Hadžić et al., 2017). In this case, the NSP will also be considered as a tenant of the computational infrastructure of the FSP. Similarly, CSP/FSP can also rent networking infrastructure of an NSP to reduce the service delay of its users. In this case, the CSP/FSP will be the tenant of the networking infrastructure of the NSP. The interaction between tenants and service providers (e.g., NSP/CSP/FSP) is closely related to the ownership as well as availability of the shared resources. In this chapter, we consider a decentralized architecture and assume the tenants, CSP, NSP, and FSP are associated with different providers/operators.

We also mainly focus on the resource slicing/partitioning of the computational resources of the FSP and assume each tenant can always obtain sufficient networking resources from the NSP to deliver the requests of users to the intended fog servers.

4.2.3.2 Existing Network Slicing Architectures

A comprehensive SDN architecture was introduced by ONF in (ONF, 2016). In this architecture, an intermediate control plane is used to deliver tailored services to users in the application plane by configuring and abstracting the physical resources. The proposed SDN architecture can naturally support network slicing (ONF, Apr. 2016). In particular, the SDN controller is supposed to collect all the information needed to communicate with each user and create a complete abstract set of resources (as resource groups) to support control logic that constitutes a slice, including the complete collection of related service attributes of users. Although ONF's SDN architecture provides a comprehensive view of the control plane functionalities that enable network slicing, its centralized nature cannot support scalable deployment.

In (Sciancalepore et al., Atlanta, GA, May 2017, Caballero et al., Atlanta, GA, May 2017, Bega et al., Atlanta, GA, May 2017, Samdanis et al., 2016), the authors introduced the mechanism of a 5G network slicing broker in 3GPP service working group SA 1. The proposed mechanism is based on the 3GPP's network sharing management architecture(3GPP, June 2016a, J). In this mechanism, each tenant can acquire slices from service providers to run network functions. In contrast to the ONF's network slicing architecture in which the slicing is created by exchanging resource usage information between tenants and service providers, the 5G network slicing broker allows the service providers to directly create network slices according to the requirement of tenants. It can therefore support on-demand resource allocation and admission control. However, the network slicing broker only supports slicing of networking resources that are centrally controlled by a master operator-network manager (MO-NM)[3].

There are many other architectures that can also be extended to support network slicing. However, in terms of the entities (either each tenant directly requests the service slices from the FSP, or tenants and FSP must negotiate and jointly decide the slicing/partitioning of the resource) that need to be involved when making network slicing decisions, these architectures can be considered as the special cases of ONF and 3GPP architectures.

4.2.3.3 Regional SDN-based Orchestrator

We propose a dynamic network slicing architecture that supports large-scale fog computing network on a new entity, the regional orchestrator. In our architecture, each fog node i coordinates with a subset of its neighboring fog nodes C_i via a regional orchestrator to create network slices for a common set of services requested by the local users. More specifically, each tenant sends the resource request together with the location information of each user. The workload request of each user is first assigned to the closest fog node. Each fog node will then partition its own resources according to the total received requests. If a fog node receives requests that exceed its available resources, it will coordinate with the regional orchestrator to outsource a part of its load to one or more neighboring fog nodes. Similarly, if a fog node has surplus resources, it will report this surplus to the regional orchestrator, who will then coordinate with other fog nodes to forward the appropriate amounts of their workload to the nodes with surplus resource. Our proposed architecture is illustrated in Figure 4.10.

---

3 According to 3GPP's network sharing management architecture (3GPP, June 2016a, J), to ensure optimized and secure resource allocation, a single master operator must be assigned as the only entity to centrally monitor and control shared network resources.

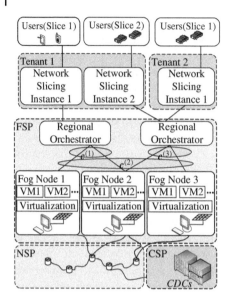

Figure 4.10 Dynamic network slicing architecture.

4.2.3.4 Problem Formulation

We consider a fog computing network consisting of a set of N fog nodes, labeled as $\mathcal{F} = \{1, 2, \dots, N\}$. Each fog node i serves a set $\mathcal{B}_i$ of tenants (e.g., VNO or service/content providers) located in its coverage area. Different types of service can have different QoS requirements, here measured by the service delay. Each tenant can support at least one type of service. Examples of service types that require ultra-low response-time (<10 ms) and high computational resources include AR/VR as well as traffic guidance and planning services for high-speed (self-driving) vehicles(ATT, IETF, Mar. 2017). Other services that are more latency-tolerant (e.g., ≈100 ms) include speech recognition and language translation. Let $\mathcal{V}$ be the set of K types of service supported by the FSP. Each tenant can request a subset of the service types in $\mathcal{V}$. We use superscript k to denote the parameters related to the kth service type. Let $\theta^{(k)}$ be the maximum tolerable service delay for type k service. We assume the tenants can always obtain sufficient networking resources from the NSP to deliver their workload to fog nodes and receive feedback results. Our methods can be directly extended to support slicing of both communication and computational resources, e.g., if network resources obtained by each tenant can only support a portion of the users' workload, then the tenant will only request slices of computational resources to process this particular portion of workload.

4.2.4 Implementation and Numerical Results

4.2.4.1 Dynamic Network Slicing in 5G Networks

The dynamic network slicing among multiple fog nodes can be supported by the network sharing management architecture recently introduced by 3GPP (3GPP, June 2016a, J). In particular, a *network slice* consists of a set of isolated computational and networking resources (e.g., processing units, network infrastructure, and bandwidth) orchestrated according to a specific type of service. Popular wireless services that can be supported by fog computing include voice processing (e.g., voice recognition applications such as Apple's Siri and Amazon's Alexa services) and image processing (e.g., image recognition applications such as the traffic sign recognition in automotive devices). To facilitate on-demand resource allocation, admission control, workload distribution and monitoring, the regional SDN-based orchestrator can be deployed at the evolved packet core (EPC) with accessibility to the core network elements such as a mobility management entity (MME) and packet data gateways (P-GW) via the S1 interface. Each fog node can be deployed inside of the eNB. The regional SDN-based orchestrator can coordinate the workload distribution among connected fog nodes via the X2 interface.

The regional SDN-based orchestrator can also connect with the network element manager (NEM) of each eNB to evaluate the received workload for every type of service and decide the amount of workload to be offloaded by each connected fog node. It also performs workload monitoring, information exchange and control of the computational resources belonging to different fog nodes through the NEMs of their associated eNBs.

4.2.4.2 Numerical Results

We simulate the possible implementation of the fog computing infrastructure in over 200 BS locations (including GSM and UMTS BSs) deployed by a primary MNO in the city of Dublin(Francesco et al., 2018, Kibilda et al., 2016). The locations and coverage areas of the BSs are presented in Figure 4.11(a). To compare the workload offloading performance with different deployment densities of BSs, we consider five areas from the city center to the rural areas as shown in Figure 4.11(b). We assume a mini-server consisting of 100 processing units that can be activated or deactivated according to the energy availability has been built inside of each BS. We assume each fog node can support two types of services: image and voice recognition with maximum tolerable response times of 50 ms and 100 ms, respectively. These values are typical for today's voice and image recognition-based applications(Assefi et al., Las Vegas, NV, Aug. 2015, Kyllonen and Zu, Apr. 2016). Each processing unit can process 10 image or 40 voice recognition requests per second. We assume the number of energy units that can be harvested by each fog node follows a uniform distribution between 0 and a given value. We consider two settings of offload forwarding. In the first setting, each fog node can only cooperate with its closest fog node. In the second one, each fog node can forward its workload to any neighboring fog nodes within a limited distance called the (offload) forwarding distance. We assume the round-trip time between two fog nodes within the workload forwarding distance can be regarded as a constant given by $\tau_{ij} = 20$ ms. We investigate the fog computing system under backlogged traffic conditions.

To evaluate the performance improvement that can be achieved by offload forwarding, we compare the number of requests that can be processed by each BS in the five considered areas shown in Figure 4.12. We can observe that by allowing each BS to cooperate with all the neighboring BSs within a 500 m of the forwarding distance can significantly improve the numbers of offloaded requests for both supported services. We can also observe that even when each fog node can only cooperate with its closest fog node, the workload processing capability in terms of the number of offloaded requests can almost double compared to the case without offload forwarding. The workload processing capability can be further improved if each fog node can cooperate with more neighboring nodes. Note that in Figure 4.12, we observe that in areas 4 and 5, allowing each fog node to cooperate with

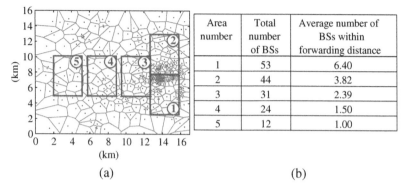

Area number	Total number of BSs	Average number of BSs within forwarding distance
1	53	6.40
2	44	3.82
3	31	2.39
4	24	1.50
5	12	1.00

(a)   (b)

Figure 4.11 Distribution of BSs and considered areas.

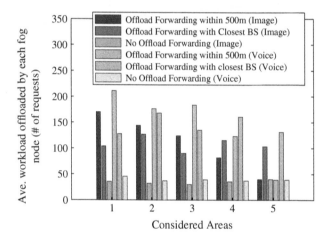

Figure 4.12 Offloaded workload for both types of service in different areas in Figure 4.11.

other fog nodes within 500 m cannot achieve the highest workload offloading performance. This is because in these two rural areas, some fog nodes cannot have any other fog nodes located in the 500 m.

It can be observed that the round-trip workload transmission latency between neighboring fog nodes directly affect the performance of the offload forwarding. In Figure 4.13, we compare the average number of offloaded service requests over all the considered areas when fog nodes within the workload forwarding distance have different round-trip times between each other. We can again observe that allowing each fog node to cooperate with all the neighboring nodes within the forwarding distance achieves the best offloading performance compared to other strategies, even when the transmission latency between fog nodes becomes large. This is because in our proposed dynamic network slicing, each fog node can carefully schedule the energy consumed for activating the computational resources at each time slot. In this case, when the transmission latency between fog nodes is small, each fog node is more willing to spend energy in processing the workload for others or forwarding workload to others. On the other hand, when the round-trip time between fog nodes

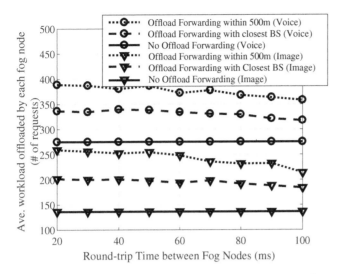

Figure 4.13 Offloaded workload under different RTT between fog nodes.

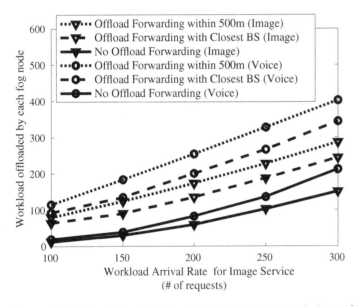

Figure 4.14 Offloaded workload under different workload arrival rates for image service.

becomes large, always forwarding workload to other neighboring fog nodes will introduce higher transmission latency. In this case, fog nodes will reserve more energy for its own use instead of helping others.

In Figure 4.14, we fix the workload arrival rate of the voice service for each fog node and compare the workload that can be processed by the fog nodes when the arrival rate for the image services changes. We can observe that both offload forwarding settings can significantly improve the offloading performance for the video service. This is because when each fog node receives more workload for the image processing service, it will be more

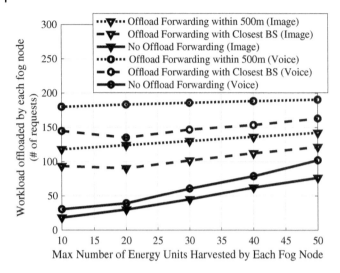

Figure 4.15 Offloaded workload under different amounts of harvested energy.

likely to seek help from other fog nodes and jointly offload this highly loaded service with others. We can also observe that each fog node is always willing to offload more workload for the voice service than the image service. This is because the voice service require less computational resources for each activated processing unit.

We consider the effect of energy harvesting process on the workload offloading performance of fog nodes in Figure 4.15 where we investigate the workload offloading performance of the fog layer when the maximum energy that can be harvested by each fog node is changed. Note that we assume the number of energy units that can be harvested by each fog node in each time slot is upper bounded by a maximum value. Therefore, the maximum amount of energy that can be harvested by each fog node also reflects the average energy that can be obtained by fog nodes as well as the energy available to activate the local computational resources. We observe that when the harvested energy for each fog node is limited, allowing two or more fog nodes to help each other can significantly improve the workload offloading performance of the fog layer. As the amount of harvested energy increases, more computational resources can be activated and the queuing delay of fog nodes will be further reduced.

4.3 Collaboration of Multiple Network Operators

Recent studies predict that by 2020 up to 50 billion IoT devices will be connected to the internet, straining the capacity of wireless network that has already been overloaded with data-hungry mobile applications, such as high-definition video streaming and virtual reality (VR)/augmented reality (AR). How to accommodate the demand for both the massive scale of IoT devices and high-speed cellular services in the physically limited spectrum without significantly increasing the operational and infrastructure costs is one of the main challenges for operators. In this chapter, we introduce a new multi-operator network sharing framework that supports the coexistence of IoT and high-speed cellular services. Our framework is based on the radio access network (RAN) sharing architecture recently introduced by 3GPP as a promising solution for operators to improve their resource utilization and reduce the system roll-out cost. We evaluate the performance of our proposed framework using the real base station location data in the city of Dublin collected from two major operators in Ireland. Numerical results show that our proposed framework can almost double the total number of IoT devices that can be supported and coexist with other cellular services compared with the case without network sharing.

4.3.1 Service and User Requirements

The IoT is a holistic framework for supporting the communication of intelligent devices and services that are employed in diverse verticals including e-health, environment control, smart city, and autonomous vehicles. It has been considered as the key technology to fulfill the 5G vision of ubiquitous connectivity. The fast proliferation of IoT applications has been driven by continuous decreases in cost, size, and power consumption of IoT devices, and fast growing demands for intelligent services. According to Cisco, *by* 2020, up to 50 billion IoT devices will be connected to the internet via cellular networks, generating over $1.9 trillion total added value across a wide variety of industries including retail, healthcare, manufacturing, transportation, agriculture, and others (Cisco, November 2018).

Since there are no frequency bands that are exclusively allocated to IoT services, IoT devices must share a spectrum with other technologies. 3GPP Release 13 introduces multiple solutions that enable the coexistence of IoT services and regular cellular services. The main challenge for operators is therefore to accommodate the traffic generated by both IoT and fast-growing high-speed cellular services without significantly increasing their operational and infrastructure costs. Recent 3GPP LTE standards promote the idea of network sharing, i.e. allowing operators to share radio access network (RAN) resources, including network infrastructure and spectrum, to improve the utilization of individual operator's resource and reduce the system roll-out cost/delay. Recent studies reported that network sharing has the potential to save at least half the infrastructure cost in 5G deployment for a typical European cellular operator (Samdanis et al., 2016).

Despite its great potential, it is known that network sharing between multiple operators could significantly increase the implementation complexity of wireless systems. In addition, 3GPP's network sharing architecture is mainly introduced to support high-speed data service in which a single operator can temporally access a much wider frequency band to support the high-throughput service requested by a single piece of User Equipment (UE). However, IoT devices typically generate low-throughput data traffic and their data transmission can be intermittent. How to quickly establish a large number of data connections and allocate the required frequency bands for a massive scale of IoT devices that can be associated with multiple operators is still an open problem. In this chapter, we propose a novel network sharing framework that allows the coexistence of IoT and high-speed data services across the network and spectrum of multiple operators. Our proposed framework is based on the active RAN sharing architecture recently introduced in 3GPP Release 13. We introduce multiple new design solutions that could reduce the implementation complexity of the network sharing for IoT applications. We simulate a multi-operator cellular system using the actual BS location information in the city of Dublin obtained from two major telecommunication operators in Ireland. Such trace-driving simulations are used to evaluate the performance of our proposed framework under various practical scenarios. The rest of this chapter is organized as follows: we provide an overview of recent 3GPP solutions on IoT and discuss the challenges for a massive deployment of IoT services in cellular networks. We then introduce our proposed framework and discuss various design issues. Finally, we present numerical results to demonstrate the promising potential of our proposed framework. To the best of our knowledge, this is the first work that discusses the deployment of IoT solutions on the multi-operator network sharing architecture.

4.3.2 System Model and Problem Formulation

4.3.2.1 IoT Solutions in 3GPP Release 13

In 3GPP Release 13, three solutions have been standardized for cellular IoT deployment: extended coverage GSM IoT (EC-GSM-IoT), Narrowband IoT (NB-IoT) and enhanced machine-type communication (eMTC) (Lin et al., 2017, 3GPP, V1.0.0, December 2016). EC-GSM-IoT operates on legacy GSM bands and can support up to 240 kbps peak data rate with 200 kHz bandwidth per channel. It applies advanced repetition and signal combining techniques to further extend the service coverage. NB-IoT is a new radio added to the LTE focusing on the low-end IoT applications. For example, T-mobile recently announced plans to provide the NB-IoT service at a rate of $6 per year per device with up to 12 MB of data. It can achieve up to 250 kbps peak data rate with 180 kHz bandwidth of spectrum on a regular GSM or LTE band, or an LTE guard-band. eMTC extends from LTE by introducing new power saving functions that can support up to 10 years of operation with a 5 W h battery. Due to its low transmit power, eMTC can coexist with high-speed LTE services. eMTC devices can support up to a 1 Mbps data rate in both uplink and downlink with 1.08 MHz bandwidth. We summarize the main specifications of 3GPP IoT solutions in Table 4.2.

To further improve the battery life of IoT devices, all IoT solutions adopt discontinuous reception (DRX) cycle similar to LTE. In this setting, each device will only periodically check the system information broadcast in the control channel according to the DRX cycle and only request a channel connection if it identifies a service request (e.g., receiving calls, messages and connection requests). A typical LTE device can have up to 2.56 s of DRX cycles. 3GPP further extends the mechanism of DRX by introducing new extended discontinuous reception (eDRX) power saving modes for all three IoT solutions. In

Table 4.2 IoT Solutions in 3GPP Release 13 (Shirvanimoghaddam et al., September 2017).

	EC-GSM-IoT	NB-IoT	eMTC
Frequency	850–900 MHz and 1800–1900 MHz GSM bands	2G/3G/4G spectrum between 450 MHz and 3.5 GHz; Sub-2 GHz bands are preferred for applications requiring good coverage	Legacy LTE between 450 MHz and 3.5 GHz
Bandwidth	200 kHz	180 kHz	1.08 MHz
Transmit Power	33 dBm, 23 dBm	23 dBm, 20 dBm	23 dBm, 20 dBm

particular, two modes have been introduced for NB-IoT and eMTC: connected mode (C-eDRX) and idle mode (I-eDRX). C-eDRX supports 5.12 s and 10.24 s of DRX cycles for eMTC and NB-IoT, respectively. In I-eDRX, the DRX cycle can be further extended to 44 min and 3 h for eMTC and NB-IoT, respectively. EC-GSM-IoT supports up to 52 min of DRX cycles.

4.3.2.2 Challenges for Massive IoT Deployment

In spite of the strong push from industry and standardization organizations, many challenges remain to be addressed for massive deployment of IoT.

- **Coexistence mechanism design for massive IoT and high-speed cellular services**. Motivated by the fact that IoT devices require lower transmit powers and narrower bandwidth, most existing works focus on developing optimal power control, channel allocation, as well as transmission scheduling algorithms for IoT services to adapt to the dynamics of the coexisting cellular traffic. However, IoT devices are usually low-cost with limited or no computational power to calculate and instantaneously adjust their transmit powers and channel usage. Some recent works have suggested deploying edge/nano-computing servers at the edge of the network, e.g., the BSs, to collect the necessary information and make decision for IoT devices in proximity (Cisco, 2015).

  These solutions make optimal resource allocation and instantaneous interference control possible for IoT devices. However, deploying new infrastructure such as edge servers, enhanced/upgraded base stations (BSs), and new interface supporting coordination and information exchange between BSs and edge servers requires extra investment from operators. For example, recent announcements from AT&T and Verizon revealed that billions of dollars of investment are required to upgrade their infrastructure for supporting the IoT-based 5G networks. These investment will be eventually reflected on the prices/bills charged to the IoT service users operated on the 5G network infrastructure.

- **Excessive overhead and inefficiency of the random access channel procedure**. Another issue is that the Random Access Channel (RACH) procedure currently used in LTE and GSM is energy-consuming and requires a significant amount of signaling overhead to establish connection between devices and network infrastructure. Directly extending this procedure into IoT systems is uneconomic and unrealistic.

  In particular, it has been reported that in a typical cellular system, 100 bytes of data transmission for a mobile device may generate up to 59 bytes and 136 bytes of overhead on uplink and downlink, respectively (Shariatmadari et al., 2015).

  In addition, the RACH procedure was originally designed to support only a limited number of mobile devices (around 100 mobile devices per cell). For example, if a device tries to establish a connection, it must randomly choose a preamble signal sent to the BS over the Physical Random Access Channel (PRACH). In existing LTE systems, each device can only choose one preamble from a set of 64 pre-defined preamble signals. If two or more devices choose the same preamble, a conflict will happen which will result in retransmission and further delay in resource allocation.

- **Diverse QoS requirements**. Another problem is that existing IoT solutions ignore the diverse requirements of the IoT services and treat all the data generated by different IoT services as the same. In particular, for some massive-type IoT applications such as long-term environmental monitoring and parcel tracking, certain amounts of data loss and data delivery latency can be tolerated. However, in mission-critical IoT applications such as fire/gas alarm, health monitoring and traffic safety, the data delivery must be instant and highly reliable.

How to differentiate the service requirements for different applications and distribute appropriate resources to meet the needs for various IoT services is still an open problem.

- **Mobility management and traffic dynamic control**. Due to the mobility of UEs and IoT devices as well as time-varying traffic of different services, the resource demand and LTE/IoT coexisting topologies can be dynamic. Most existing solutions are focusing on optimizing the long-term performance based on a priori knowledge and/or prediction results. For example, an IoT device can predict the future change of its movement, change of data traffic as well as activities of other UEs in proximity so it can prepare for the future (e.g., scheduling/reserving a certain amount of bandwidth for future use if it predicts that these resources will soon be limited). However, always relying on each IoT device to calculate the prediction and resource allocation results is uneconomic due to the limited processing capability. Currently, there is still a lack of a simple and economic solution that allows each IoT device to instantaneously adapt to the environmental dynamics without sacrificing the device's cost and battery life.

4.3.2.3 Inter-operator Network Sharing Architecture

The mechanism of network sharing was first introduced in 3GPP Release 10 to allow multiple operators to share their physical networks. Early development of network sharing mainly focused on the infrastructure sharing, also referred to as *passive RAN sharing* (3GPP, June 2016a, J). In this scenario, operators share the same site locations and site supporting infrastructures such as power supply, shelters and antenna masts. However, each operator still needs to install its own antennas and backhaul equipment for individual usage. To this end, 3GPP Release 14 introduced the *active RAN sharing* architecture. Operators can now share their spectrum resources as well as core network equipment including BSs (i.e. eNBs) based on a network sharing agreement, which can include mutual agreement on legal, finance, and joint operations. To ensure efficient and secure resource management, a master operator (MOP) will be designated as the only entity to manage the resource shared among participating operators (POPs). The MOP can be a third-party manager designated by POPs. It can also be one of the POPs that has been agreed by others to act as the MOP to control and supervise the allocation of the shared resource. In the 3GPP architecture, the MOP can charge POPs according to the requested data volume and the required QoS.

According to the entities shared among POPs, active RAN sharing architectures can be further divided into two categories:

- **RAN-only sharing**, also called multi-operator core network (MOCN), where a set of BSs sharing the same spectrum can be accessed by all the POPs. Each POP however maintains its own core network elements including mobility management entity (MME) and serving and packet gateways (S/P-GW). Each POP can connect its core network elements to the shared RAN via the S1 interface.
- **Gateway core network** (GWCN) in which, in addition to sharing the same set of BSs, POPs can also share a common MME to further reduce costs.

To simplify our description, in the rest of this section, we assume each POP corresponds to a cellular operator that can divide its network infrastructure and licensed spectrum into two parts: *exclusive use part* that is reserved and exclusively used by itself and a *shared part* that can be accessed by other operators. The shared part of the network as well as the spectrum of all the POPs will be combined and managed by the MOP. Each IoT device or UE has already been assigned to a POP. The BSs of each POP need to calculate the channel reuse structure between the low-power narrow-band IoT devices and regular UEs so the cross-interference between both channel-sharing devices is below a tolerable threshold. In LTE, for example, the interference threshold for each UE is -62 dBm. If the exclusive use part of spectrum is insufficient to support the traffic generated by the associated IoT and cellular services, the POP can temporally request a portion of shared spectrum from the MOP. If the spectrum requests of a POP have been approved, the POP can assign any of its traffic (IoT or cellular) to the shared spectrum without consulting the MOP. If the spectrum requested by all the POPs exceeds the total amount of shared spectrum, MOP will partition the shared spectrum and assign the divided spectrum to each POP according to a pre-determined mutual agreement.

Both active RAN sharing architectures can be extended to our framework to support spectrum sharing between IoT and high-speed cellular services. In particular, RAN-only sharing allows each POP to adjust the traffic traversed through the shared network or its exclusive network resources according to the mobility of IoT devices (e.g., IoT services in wearable devices and vehicle networks). In this case, each POP needs to keep track of traffic dynamics and the required QoS requirements for both IoT services and its regular cellular services. The POP can then adjust the traffic sent through shared infrastructure and its own exclusive infrastructure accordingly. GWCN further reduces the cost for each POP, by sharing a common MME among all POPs. It however cannot provide the same flexibility as RAN-only sharing for each POP because in this case mobility for devices is restricted for inter-RAN scenarios. In other words, each POP cannot adjust the data traffic sent through the shared infrastructure and its own exclusive infrastructure by itself. In this case, POPs will need to predict the traffic from IoT services and cellular services and reserve resources for each service accordingly.

In active RAN sharing, different POPs can access/rent different part of the shared infrastructure (e.g., a set of BSs that can be accessed by all POPs). However, the BSs in the shared RAN must operate on the same spectrum. Based on the spectrum used by the shared RAN, the multi-operator network sharing architecture can be further divided into the following two sub-categories, as illustrated in Figure 4.16:

- **Spectrum pooling**. POPs can merge their licensed (GSM and/or LTE) bands to form a common pool to be used by the shared RAN as shown in Figure 4.16(a) (Xiao et al., 2014). Allowing the shared BSs to operate on the pooled spectrum can significantly reduce the complexity of spectrum management, i.e. it is uneconomic and too complex to allow each BS to switch its operational bands when it has been rented by different POPs. Spectrum pooling has been considered as one of the main use cases for the network sharing architecture in the 3GPP technical specification. In this architecture, each POP will need to coordinate with the MOP network management controller for channel assignment to avoid inter-cell interference between the BSs in the exclusive-use RAN and those in the shared RAN.

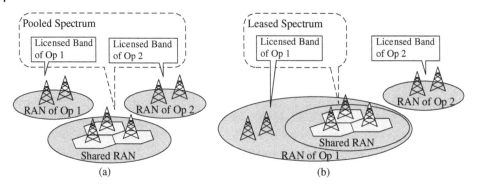

Figure 4.16 Inter-operator network sharing: (a) a spectrum pool and (b) spectrum leasing.

- *Spectrum Leasing* 3GPP architecture allows one of the POPs to serve as the MOP to manage and control the resource allocation of the shared RAN, as shown in Figure 4.16(b). In this case, it is possible for one POP to lease a part of its BSs and the licensed band to be shared with other POPs. In spectrum leasing, to maintain the required QoS for the MOP, IoT, and UE associated with MOP can have the priority to access the shared spectrum. The POPs can only offload a limited traffic to the shared RAN if the resulting impact (e.g., throughput degradation) to the existing traffic of the MOP is below a tolerable level. If two or more POPs can lease their network infrastructure and licensed bands to each other at different time periods according to their traffic demands and resource availability, the spectrum leasing becomes equivalent to the *mutual renting* introduced in METIS' future spectrum system mechanism (Xiao et al., 2016, Singh et al., 2015, Xiao et al., 2015a).

4.3.2.4 Design Issues

There are several important issues when deploying IoT services using our proposed multi-operator network sharing framework:

- **Fair revenue division among operators for spectrum pooling**. In 3GPP's network sharing architecture, MOP can charge services (e.g., IoT services) using the shared resource according to the data usage and required QoS profiles. One intrinsic problem is then how to divide the revenue obtained by MOP from serving IoT among all the resource-sharing POPs. This revenue division determines each POP's perception on the fairness of the sharing, and will in turn affect its willingness to share the licensed band with others. In other words, the revenue allocation must be fair in the sense that it needs to protect the interests of all the contributing operators and, more importantly, incentivize POPs to contribute their resources to the pool.

  In addition, to encourage operators with higher investment and more licensed spectrum resources to contribute, it must also take into consideration the contributions of different operators. In other words, operators that contribute more resources should have a larger share of the revenue from the pool. Various fairness criteria have been investigated for the spectrum pooling. In particular, in our previous work (Xiao et al., 2014), we consider the scenarios that multiple operators form a spectrum pool and allow coexistence of their cellular service and other low-power services (e.g., IoT services) in the same band as long as the resulting interference is less than a tolerable threshold. We prove that operators can use the price charged to the spectrum access of low-power services to control the admission of devices. We also investigate the fair revenue division between resource sharing operators. This framework can be directly extended to analyze the coexistence of IoT (e.g., eMTC) and cellular services. In this case, the IoT traffic admitted to the spectrum pool will be controlled by the price of the MOP.

- **NOMA for coexistence between cellular UEs and massive IoT**. As mentioned earlier, the existing RA-based resource allocation approach cannot be applied to IoT devices due to the physical limit of the licensed band and the inefficient design of the protocol. One possible solution is to apply Non-orthogonal Multiple Access (NOMA). In particular, NOMA improves the utilization of the cellular spectrum by exploiting power and code domain reuse. It provides the operators with more flexibility to increase the number of channel sharing devices, e.g., each BS can carefully choose different numbers of low-power IoT devices and high-power UEs at different locations to share the same channel. Furthermore, NOMA does not require IoT devices to perform the RACH procedure for data transmission. In particular, in NOMA, the random access and data communication can be combined (Ding et al., 2017). For example, each IoT device can randomly pick up a narrowband and start data transmission without waiting for the channel assignment from the BS. The BS can then perform successive interference cancellation to decode the message of each IoT device received in each frequency band. The authors of (Shirvanimoghaddam et al., September 2017) suggested applying rateless raptor codes to generate as many coded symbols as required by each BS, so each BS can differentiate the message sent by different IoT devices. It has been observed that the more difference in channel gains between IoT/UE and the BS, the higher performance improvement can be achieved by the NOMA.

- *Network slicing for diverse IoT services*. Network slicing is a mechanism recently introduced by 3GPP to further improve the flexibility and scalability of 5G. The main idea is to create logical partitions of a common resource (e.g., spectrum, antenna and network infrastructure), known as the slices, to be orchestrated and customized according to different service requirements. Network slicing has the potential to significantly improve spectrum efficiency and enable more flexible and novel services that cannot otherwise be supported by the existing network architecture. In our previous work, we proposed an inter-operator network slicing framework to support different services with different requirements on a commonly shared resource pool formed by multiple operators (Xiao et al., 2018). In this framework, a software-defined mobile network controller will be deployed in the MOP's network infrastructure that can isolate and reserve a certain amount of resource for each type of IoT services (e.g., wearable IoT devices, machine-type IoT and smart infrastructure). The controller will predict the possible future traffic of all the supported IoT services and can adjust the portion of the resource reserved for each service.

4.3.3 Performance Analysis

To evaluate the performance improvement that can be achieved by our framework, we simulate a multi-operator network sharing architecture using over 200 real BS locations in the

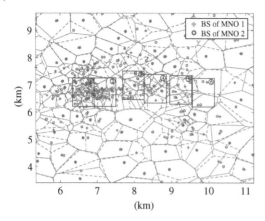

Figure 4.17 Locations of BSs deployed by two major cellular operators in the city of Dublin.

city of Dublin deployed by two major telecom operators in Ireland. The actual distribution and the deployment densities of BSs are shown in Figure 4.17. We consider saturated traffic for both UEs and IoT devices and evaluate the possible coexistence of IoT (e.g., eMTC) devices and cellular UEs for uplink data communication in the same LTE band. Our results can be regarded as the maximum performance improvement that can be achieved by multi-operator network sharing architecture. The transmit powers of each IoT device and cellular UE are set to 20 dBm and 25 dBm, respectively. We assume 20 pieces of UE and 50,000 IoT devices are uniformly randomly located in each cell. Each piece of UE occupies a 5 MHz bandwidth. Each IoT device is randomly allocated with a 1 MHz bandwidth channel and can only send data with 20 dBm of transmit power. IoT devices can only be supported when the interference to the UE is lower than the LTE tolerable interference threshold (-62 dBm).

In Figure 4.18, we carefully select six areas from the city center to suburban areas (representing different sizes and deployment density of cells) and compare the maximum number of IoT devices that can simultaneously transmit data with the UE in the same LTE bands when each piece of UE can tolerate 10% of throughput degradation. We observe that when the size of the cell is small, the number of IoT devices that can share the same spectrum as the UE is limited due to the high cross-interference between IoT devices and cellular UE. However, as the size of the cell increases, the total number of coexisting IoT devices can increase significantly. In addition, allowing both operators to share their spectrum via pooling can almost double the total number of IoT devices when the deployment density of BSs is low. This result complements the existing efforts of 3GPP on promoting the network sharing for 5G networks and could have the potential to influence the future practical implementation of the network sharing architecture between major operators.

In Figure 4.19, we compare the maximum number of IoT devices that can share the same channel with UE when throughput degradations that can be tolerated by the each piece of UE are different. We observe that the number of IoT devices increases when the UEs can tolerate a higher degradation for their throughput. In addition, network sharing provides more improvement in coexisting IoT traffic when the UEs can only tolerate a small throughput degradation, i.e. network sharing can almost double the maximum number of coexisting IoT devices when each UE can tolerate 20% throughput degradation. However, when

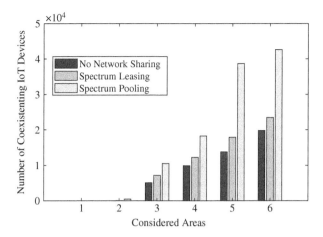

Figure 4.18 Maximum number of IoT (eMTC) devices that can coexist with cellular UE in different considered areas.

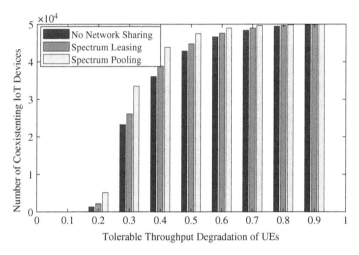

Figure 4.19 Maximum number of IoT (eMTC) devices that can coexist with cellular UE under various tolerable throughput degradation of UE.

the tolerable throughput degradation of UE increases to 90%, the total number of coexisting IoT devices approaches the maximum values even without network sharing. In other words, network sharing can provide more performance improvement when the UE requires a stringent QoS guarantee with a limited interference tolerance.

4.4 Conclusion

This chapter presents an online orchestration framework for cross-edge service function chaining. It jointly optimizes the resource provisioning and traffic routing to maximize the holistic cost-efficiency. Since the formulated optimization problem is NP-hard and involves future uncertain information, a joint optimization framework is designed that carefully blends the advantages of an online optimization technique and an approximate optimization method. Then, we propose the mechanism of dynamic network slicing for energy-harvesting fog computing networks consisting of a set of fog nodes that can offload and/or forward its received workload to CDCs using the energy harvested from the natural environment. In this mechanism, each fog node can offload multiple types of service with guaranteed QoS using the computational resources activated by the harvested energy. The limited resources and uncertainty of the energy harvesting process restrict the total amount of workload that can be processed by each individual fog node. Finally, we review the current IoT solutions introduced by the 3GPP. We have introduced a multi-operator network sharing framework based on 3GPP's network sharing architecture to support coexistence of massive IoT and regular cellular services offered by multiple operators. We simulate a multi-operator network sharing scenario using the real BS location data provided by two major operators in the city of Dublin. Our numerical results show that our proposed framework can almost double the coexisting IoT traffic under certain scenarios.

References

The cloud comes to you: AT&T to power self-driving cars, AR/VR and other future 5G applications through edge computing.

Intelligence at the Edge: How Machine Learning is Driving Edge Computing. URL https://vilmate.com/blog/intelligent-edge-supercharge-your-analytics-with-ml/.

3GPP. Telecommunication management; network sharing; concepts and requirements. 3GPP TS 32.130, June 2016a.

3GPP. Network sharing; artechecture and functional description. 3GPP TR 23.251, June 2016b.

3GPP. Service requirements for next generation new services and markets. 3GPP TS 22.261, V1.0.0, December 2016.

M. Aazam and Eui-Nam Huh. Dynamic resource provisioning through fog micro datacenter. In *Proc. of the IEEE PerCom Workshops*, pages 105–110, St. Louis, MO, Mar. 2015. doi: 10.1109/PERCOMW.2015.7134002.

Mohammad Abu-Lebdeh, Diala Naboulsi, Roch Glitho, and Constant Wette Tchouati. On the placement of vnf managers in large-scale and distributed nfv systems. *IEEE Transactions on Network and Service Management*, 14 (4):875–889, 2017.

Ganesh Ananthanarayanan, Paramvir Bahl, Peter Bodík, Krishna Chintalapudi, Matthai Philipose, Lenin Ravindranath, and Sudipta Sinha. Real-time video analytics: The killer app for edge computing. *Computer*, 50 (10):58–67, 2017.

J. G. Andrews, S. Buzzi, W. Choi, S. V. Hanly, A. Lozano, A. C. K. Soong, and J. C. Zhang. What will 5G be? *IEEE Journal on Selected Areas in Communications*, 32(6):1065–1082, June 2014.

M. Assefi, M. Wittie, and A. Knight. Impact of network performance on cloud speech recognition. In *Proc of the ICCCN Conference*, pages 1–6, Las Vegas, NV, Aug. 2015. doi: 10.1109/ICCCN.2015.7288417.

Dario Bega, Marco Gramaglia, Albert Banchs, Vincenzo Sciancalepore, Konstantinos Samdanis, and Xavier Costa-Perez. Optimising 5g infrastructure markets: The business of network slicing. In *Proc. of the IEEE INFOCOM Conference*, Atlanta, GA, May 2017.

Ilias Benkacem, Tarik Taleb, Miloud Bagaa, and Hannu Flinck. Optimal vnfs placement in cdn slicing over multi-cloud environment. *IEEE Journal on Selected Areas in Communications*, 36(3):616–627, 2018.

Flavio Bonomi, Rodolfo Milito, Preethi Natarajan, and Jiang Zhu. *Big Data and Internet of Things: A Roadmap for Smart Environments*, chapter Fog Computing: A Platform for Internet of Things and Analytics, pages 169–186. Springer, 2014. ISBN 978-3-319-05029-4. doi: 10.1007/978-3-319-05029-4_7.

Stephen Boyd and Lieven Vandenberghe. *Convex optimization*. Cambridge university press, 2004.

Niv Buchbinder, Shahar Chen, and Joseph Naor. Competitive analysis via regularization. In *Proc. of ACM/SIAM SODA*, 2014.

Pablo Caballero, Albert Banchs, Gustavo de Veciana, and Xavier Costa-Perez. Network slicing games: Enabling customization in multi-tenant networks. In *Proc. of the IEEE INFOCOM Conference*, Atlanta, GA, May 2017.

V. Chamola and B. Sikdar. Solar powered cellular base stations: current scenario, issues and proposed solutions. *IEEE Communications Magazine*, 54(5): 108–114, May 2016. ISSN 0163-6804. doi: 10.1109/MCOM.2016.7470944.

N. Chen, Y. Yang, T. Zhang, M. Zhou, X. Luo, and J. Zao. Fog as a service technology. *IEEE Communications Magazine*, 56(11):95–101, 2018a.

Xu Chen, Wenzhong Li, Sanglu Lu, Zhi Zhou, and Xiaoming Fu. Efficient resource allocation for on-demand mobile-edge cloud computing. *IEEE Transactions on Vehicular Technology*, 67(9):8769–8780, 2018b.

Xu Chen, Zhi Zhou, Weigang Wu, Di Wu, and Junshan Zhang. Socially-motivated cooperative mobile edge computing. *IEEE Network*, (99): 12–18, 2018c.

Yanpei Chen, Archana Ganapathi, Rean Griffith, and Randy Katz. The case for evaluating mapreduce performance using workload suites. In *Proc. of IEEE MASCOTS*, 2011.

Mung Chiang, Bharath Balasubramanian, and Flavio Bonomi. *Fog for 5G and IoT*. John Wiley & Sons, 2017.

Cisco. *Cisco Fog Computing Solutions: Unleash the Power of the Internet of Things*. 2015.

Cisco. *Visual Networking Index: Forecast and Trends, 2017-C2022*. November 2018. URL https://www.cisco.com/c/en/us/solutions/collateral/service-provider/visual-networking-index-vni/white-paper-c11-741490.html.

Google Cloud. Google cloud and the environment. URL https://cloud.google.com/environment/.

S.K. Datta, C. Bonnet, and J. Haerri. Fog computing architecture to enable consumer centric internet of things services. In *Proc. of IEEE ISCE Conference*, pages 1–2, Madrid, Spain, June 2015. doi: 10.1109/ISCE.2015.7177778.

Z. Ding, X. Lei, G. K. Karagiannidis, R. Schober, J. Yuan, and V. K. Bhargava. A survey on non-orthogonal multiple access for 5G networks: Research challenges and future trends. *IEEE J. Sel. Area Commun.*, 35(10):2181–2195, Oct 2017. ISSN 0733-8716. doi: 10.1109/JSAC.2017.2725519.

Lam Dinh-Xuan, Michael Seufert, Florian Wamser, Phuoc Tran-Gia, Constantinos Vassilakis, and Anastasios Zafeiropoulos. Performance evaluation of service functions chain placement algorithms in edge cloud. In *International Teletraffic Congress (ITC 30)*, 2018.

C.T. Do, N.H. Tran, Chuan Pham, M.G.R. Alam, Jae Hyeok Son, and Choong Seon Hong. A proximal algorithm for joint resource allocation and minimizing carbon footprint in geo-distributed fog computing. In *Proc. of the IEEE ICOIN Conference*, pages 324–329, Jan Siem Reap, Cambodia, January 2015. doi: 10.1109/ICOIN.2015.7057905.

Luyuan Fang, Fabio Chiussi, Deepak Bansal, Vijay Gill, Tony Lin, Jeff Cox, and Gary Ratterree. Hierarchical sdn for the hyper-scale, hyper-elastic data center and cloud. In *Proceedings of the 1st ACM SIGCOMM Symposium on Software Defined Networking Research*, pages 7:1–7:13, New York, NY, USA, June 2015. ACM. ISBN 978-1-4503-3451-8. doi: 10.1145/2774993.2775009. URL http://doi.acm.org/10.1145/2774993.2775009.

Xincai Fei, Fangming Liu, Hong Xu, and Hai Jin. Towards load-balanced vnf assignment in geo-distributed nfv infrastructure. In *Proc. of IEEE/ACM IWQoS*, 2017.

Fortune.com. Apple, facebook, and google top greenpeace energy report card, 2017.

P. Di Francesco, F. Malandrino, and L. A. DaSilva. Assembling and using a cellular dataset for mobile network analysis and planning. *IEEE Transactions on Big Data*, 4(4), 2018.

X. Ge, B. Yang, J. Ye, G. Mao, C. Wang, and T. Han. Spatial spectrum and energy efficiency of random cellular networks. *IEEE Trans. Commun.*, 63(3): 1019–1030, Mar 2015. ISSN 0090-6778. doi: 10.1109/TCOMM.2015.2394386.

Google. Google Cluster Data. URL \protect\LY1\textbraceleft https://code.google.com/p/googleclusterdata/\protect\LY1\textbraceright.

Racha Gouareb, Vasilis Friderikos, and A Hamid Aghvami. Virtual network functions routing and placement for edge cloud latency minimization. *IEEE Journal on Selected Areas in Communications*, 36, 2018.

Abhishek Gupta, Brigitte Jaumard, Massimo Tornatore, and Biswanath Mukherjee. A scalable approach for service chain mapping with multiple sc instances in a wide-area network. *IEEE Journal on Selected Areas in Communications*, 36(3):529–541, 2018.

Harshit Gupta, Shubha Brata Nath, Sandip Chakraborty, and Soumya K. Ghosh. Sdfog: A software defined computing architecture for qos aware service orchestration over edge devices. *CoRR*, abs/1609.01190, 2016. URL http://arxiv.org/abs/1609.01190.

Ilija Hadžić, Yoshihisa Abe, and Hans C. Woithe. Edge computing in the epc: A reality check. In *Proc. of the ACM/IEEE Symposium on Edge Computing*, pages 13:1–13:10, New York, NY, USA, 2017. ACM. ISBN 978-1-4503-5087-7. doi: 10.1145/3132211.3134449. URL http://doi.acm.org/10.1145/3132211.3134449.

A. Hakiri, B. Sellami, P. Patil, P. Berthou, and A. Gokhale. Managing wireless fog networks using software-defined networking. In *Proc. of IEEE/ACS AICCSA Conference*, pages 1149–1156, Hammamet, Tunisia, October 2017.

Hajar Hantouti, Nabil Benamar, Tarik Taleb, and Abdelquoddous Laghrissi. Traffic steering for service function chaining. *IEEE Communications Surveys & Tutorials*, 21(1):487–507, 2018.

Soheil Hassas Yeganeh and Yashar Ganjali. Kandoo: A framework for efficient and scalable offloading of control applications. In *Proceedings of the 1st Workshop on Hot Topics in Software Defined Networks*, pages 19–24, Aug. 2012. ISBN 978-1-4503-1477-0. doi: 10.1145/2342441.2342446. URL http://doi.acm.org/10.1145/2342441.2342446.

Huawei. 5G vision: 100 billion connections, 1 ms latency, and 10 gbps throughput. URL https://www.huawei.com/minisite/5g/en/defining-5g.html.

IETF. Problem statement: Transport support for augmented and virtual reality applications. Internet-Draft, Mar. 2017. URL https://tools.ietf.org/id/draft-han-iccrg-arvr-transport-problem-00.xml.

New York ISO. Energy Market & Operational Data. URL https://www.nyiso.com/energy-market-operational-data.

ITU-T. The Tactile Internet. ITU-T technology watch report, Aug. 2014. URL https://www.itu.int/dms_pub/itu-t/opb/gen/T-GEN-TWATCH-2014-1-PDF-E.pdf.

Yongzheng Jia, Chuan Wu, Zongpeng Li, Franck Le, and Alex Liu. Online scaling of nfv service chains across geo-distributed datacenters. *IEEE/ACM Transactions on Networking*, 26(2):699–710, 2018.

L. Jiao, L. Pu, L. Wang, X. Lin, and J. Li. Multiple granularity online control of cloudlet networks for edge computing. In *Proc. of IEEE SECON*, 2018.

Lei Jiao, Antonia Maria Tulino, Jaime Llorca, Yue Jin, and Alessandra Sala. Smoothed online resource allocation in multi-tier distributed cloud networks. *IEEE/ACM Transactions on Networking (TON)*, 25(4):2556–2570, 2017.

A. A. Khan, M. Abolhasan, and W. Ni. 5g next generation vanets using sdn and fog computing framework. In *Proc of IEEE CCNC Conference*, pages 1–6, Las Vegas, NV, January 2018. doi: 10.1109/CCNC.2018.8319192.

J. Kibilda, B. Galkin, and L. A. DaSilva. Modelling multi-operator base station deployment patterns in cellular networks. *IEEE Transactions on Mobile Computing*, 15(12):3087–3099, Dec 2016. ISSN 1536-1233. doi: 10.1109/TMC.2015.2506583.

Patrick C Kyllonen and Jiyun Zu. Use of response time for measuring cognitive ability. *Journal of Intelligence*, 4(4):14, Apr. 2016.

Abdelquoddouss Laghrissi, Tarik Taleb, Miloud Bagaa, and Hannu Flinck. Towards edge slicing: Vnf placement algorithms for a dynamic & realistic edge cloud environment. In *Proc. of IEEE Globecom*, 2017.

M. Lin, A. Wierman, L. L. H. Andrew, and E. Thereska. Dynamic Right-Sizing for Power-Proportional Data Centers. In *Proc. of IEEE INFOCOM*, 2011.

X. Lin, J. Bergman, F. Gunnarsson, O. Liberg, S. M. Razavi, H. S. Razaghi, H. Rydn, and Y. Sui. Positioning for the internet of things: A 3GPP perspective. *IEEE Communications Magazine*, 55(12):179–185, December 2017. ISSN 0163-6804. doi: 10.1109/MCOM.2017.1700269.

Y. Liu, A. Hecker, R. Guerzoni, Z. Despotovic, and S. Beker. On optimal hierarchical sdn. In *Proc of IEEE ICC Conference*, pages 5374–5379, June London, UK, June 2015.

Z. Liu, M. Lin, A. Wierman, S. Low, and L. L. H. Andrew. Geographical Load Balancing with Renewables. In *Proc. of ACM GreenMetrics*, 2011.

Xiao Lu, Ping Wang, D. Niyato, Dong In Kim, and Zhu Han. Wireless networks with RF energy harvesting: A contemporary survey. *IEEE Communications Surveys Tutorials*, 17(2):757–789, 2015. ISSN 1553-877X. doi: 10.1109/COMST.2014.2368999.

Ahmed M Medhat, Tarik Taleb, Asma Elmangoush, Giuseppe A Carella, Stefan Covaci, and Thomas Magedanz. Service function chaining in next generation networks: State of the art and research challenges. *IEEE Communications Magazine*, 55(2):216–223, 2017.

Microsoft. Microsoft environment: Enabling a sustainable future. URL https://www.microsoft.com/en-us/environment/default.aspx.

Carla Mouradian, Somayeh Kianpisheh, Mohammad Abu-Lebdeh, Fereshteh Ebrahimnezhad, Narjes Tahghigh Jahromi, and Roch H Glitho. Application component placement in nfv-based hybrid cloud/fog systems with mobile fog nodes. *IEEE Journal on Selected Areas in Communications*, 37(5):1130–1143, 2019.

D. Narayanan, A. Donnelly, and A. Rowstron. Write off-loading: Pratical power management for enterprise storage. In *Proc. of USENIX Conference on File and Storage Technologies (FAST)*, 2008.

NGMN Alliance. 5G white paper, Feb. 2015. URL https://www.ngmn.org/uploads/media/ NGMN_5G_White_Paper_V1_0.pdf.

NGMN Alliance. Description of network slicing concept, September 2016. URL https://www .ngmn.org/uploads/media/161010_NGMN_Network_Slicing_framework_v1.0.8.pdf.

ONF. SDN architecture issue 1.1. ONF TR-521, 2016.

ONF. Applying SDN architecture to 5G slicing issue 1. ONF TR-526, Apr. 2016.

Long Qu, Maurice Khabbaz, and Chadi Assi. Reliability-aware service chaining in carrier-grade softwarized networks. *IEEE Journal on Selected Areas in Communications*, 36(3):558–573, 2018.

Prabhakar Raghavan and Clark D Tompson. Randomized rounding: a technique for provably good algorithms and algorithmic proofs. *Combinatorica*, 7(4): 365–374, 1987.

Rajiv, Khuller, Samir, Parthasarathy, Srinivasan, Srinivasan, and Aravind. Dependent rounding and its applications to approximation algorithms. *Journal of the Acm*, 53(3):324–360, 2006.

M. Richart, J. Baliosian, J. Serrat, and J. L. Gorricho. Resource slicing in virtual wireless networks: A survey. *IEEE Transactions on Network and Service Management*, 13(3):462–476, September 2016. ISSN 1932-4537. doi: 10.1109/TNSM.2016.2597295.

K. Samdanis, X. Costa-Perez, and V. Sciancalepore. From network sharing to multi-tenancy: The 5g network slice broker. *IEEE Communications Magazine*, 54(7):32–39, July 2016. ISSN 0163-6804. doi: 10.1109/MCOM.2016.7514161.

S. Sarkar, S. Chatterjee, and S. Misra. Assessment of the suitability of fog computing in the context of internet of things. *to appear at IEEE Transactions on Cloud Computing*. ISSN 2168-7161. doi: 10.1109/TCC.2015.2485206. URL http://ieeexplore.ieee.org/document/ 7286781/.

Vincenzo Sciancalepore, Konstantinos Samdanis, and Xavier Costa-Perez. Mobile traffic forecasting for maximizing 5g network slicing resource utilization. In *Proc. of the IEEE INFOCOM Conference*, Atlanta, GA, May 2017.

H. Shariatmadari, R. Ratasuk, S. Iraji, A. Laya, T. Taleb, R. Jantti, and A. Ghosh. Machine-type communications: current status and future perspectives toward 5g systems. *IEEE Communications Magazine*, 53(9):10–17, September 2015. ISSN 0163-6804. doi: 10.1109/MCOM.2015.7263367.

Mahyar Shirvanimoghaddam, Mischa Dohler, and Sarah J Johnson. Massive non-orthogonal multiple access for cellular IoT: Potentials and limitations. *IEEE Communications Magazine*, 55(9):55–61, September 2017.

B. Singh, S. Hailu, K. Koufos, A. Dowhuszko, O. Tirkkonen, R. Jantti, and R. Berry. Coordination protocol for inter-operator spectrum sharing in co-primary 5g small cell networks. *IEEE Communications Magazine*, 53(7): 34–40, July 2015. ISSN 0163-6804. doi: 10.1109/MCOM.2015.7158263.

Tarik Taleb, Konstantinos Samdanis, Badr Mada, Hannu Flinck, Sunny Dutta, and Dario Sabella. On multi-access edge computing: A survey of the emerging 5g network edge cloud architecture and orchestration. *IEEE Communications Surveys & Tutorials*, 19(3):1657–1681, 2017.

Tarik Taleb, Pantelis Frangoudis, Ilias Benkacem, and Adlen Ksentini. Cdn slicing over a multi-domain edge cloud. *IEEE Transactions on Mobile Computing*, 19(9), 2020.

New York Times. Apple becomes a green energy supplier, with itself as customer, August 2016. URL https://www.nytimes.com/2016/08/24/business/energy-environment/as-energy-use-rises-corporations-turn-to-their-own-green-utility-sources.html.

L. Tong, Y. Li, and W. Gao. A hierarchical edge cloud architecture for mobile computing. In *Proc. of IEEE INFOCOM Conf.*, San Francisco, CA, Apr. 2016.

Amin Tootoonchian and Yashar Ganjali. Hyperflow: A distributed control plane for OpenFlow. In *Proceedings of the 2010 Internet Network Management Workshop/Workshop on Research on enterprise networking*, pages 1–6, San Jose, CA, Apr. 2010.

S. Ulukus, A. Yener, E. Erkip, O. Simeone, M. Zorzi, P. Grover, and K. Huang. Energy harvesting wireless communications: A review of recent advances. *IEEE J. Sel. Areas in Commun.*, 33(3):360–381, Mar. 2015. ISSN 0733-8716. doi: 10.1109/JSAC.2015.2391531.

Mojtaba Vaezi and Ying Zhang. *Virtualization and Cloud Computing*, pages 11–31. Springer, 2017. ISBN 978-3-319-54496-0. doi: 10.1007/978-3-319-54496-0_2. URL https://doi.org/10.1007/978-3-319-54496-0_2.

A. Vahid Dastjerdi, H. Gupta, R. N. Calheiros, S. K. Ghosh, and R. Buyya. Fog Computing: Principals, Architectures, and Applications. *ArXiv e-prints*, January 2016.

Luis Vaquero and Luis Rodero-Merino. Finding your way in the fog: Towards a comprehensive definition of fog computing. *Proc. of ACM SIGCOMM Comput. Commun. Rev.*, 44(5):27–32, October 2014. ISSN 0146-4833. doi: 10.1145/2677046.2677052.

Vijay V Vazirani. *Approximation algorithms*. Springer Science & Business Media, 2013.

Y. Xiao, D. Niyato, Z. Han, and K.C. Chen. Secondary users entering the pool: A joint optimization framework for spectrum pooling. *IEEE J. Sel. Area Commun.*, 32(3):572–588, Mar. 2014.

Y. Xiao, Z. Han, C. Yuen, and L. A. DaSilva. Carrier aggregation between operators in next generation cellular networks: A stable roommate market. *IEEE Trans. Wireless Commun.*, 15(1):633–650, Jan 2016. ISSN 1536-1276. doi: 10.1109/TWC.2015.2477077.

Y. Xiao, M. Hirzallah, and M. Krunz. Distributed resource allocation for network slicing over licensed and unlicensed bands. *IEEE J Sel. Area. Commun.*, 36 (10):2260–2274, Oct 2018. ISSN 0733-8716. doi: 10.1109/JSAC.2018.2869964.

Y. Xiao, Z. Han, D. Niyato, and C. Yuen. Bayesian reinforcement learning for energy harvesting communication systems with uncertainty. In *Proc. of the IEEE ICC Conference*, London, UK, June 2015.

Yong Xiao and Marwan Krunz. Distributed optimization for energy-efficient fog computing in the tactile internet. *IEEE J. Sel. Areas Commun.*, 36(11): 2390–2400, November 2018.

Yong Xiao, Kwang-Cheng Chen, Chau Yuen, Zhu Han, and Luiz DaSilva. A Bayesian overlapping coalition formation game for device-to-device spectrum sharing in cellular networks. *IEEE Trans. Wireless Commun.*, 14 (7):4034–4051, July 2015a.

Yong Xiao, D. Niyato, Zhu Han, and L.A. DaSilva. Dynamic energy trading for energy harvesting communication networks: A stochastic energy trading game. *IEEE J. Sel. Areas Commun.*, 33(12):2718–2734, December 2015b. ISSN 0733-8716. doi: 10.1109/JSAC.2015.2481204.

M. Yannuzzi, R. Milito, R. Serral-Gracia, D. Montero, and M. Nemirovsky. Key ingredients in an iot recipe: Fog computing, cloud computing, and more fog computing. In *Proc. of the IEEE International Workshop on Computer Aided Modeling and Design of Communication Links and Networks*, pages 325–329, Athens, December 2014. doi: 10.1109/CAMAD.2014.7033259.

Shanhe Yi, Cheng Li, and Qun Li. A survey of fog computing: Concepts, applications and issues. In *Proc. of ACM Workshop on Mobile Big Data*, pages 37–42, Hangzhou, China, June 2015. ISBN 978-1-4503-3524-9.

H. Zhang, Y. Xiao, S. Bu, D. Niyato, F. R. Yu, and Z. Han. Computing resource allocation in three-tier iot fog networks: a joint optimization approach combining stackelberg game and matching. *IEEE Internet of Things Journal*, 4(5):1204–1215, 2017. ISSN 2327-4662. doi: 10.1109/JIOT.2017.2688925.

H. Zhang, Y. Xiao, S. Bu, R. Yu, D. Niyato, and Z. Han. Distributed resource allocation for data center networks: A hierarchical game approach. *IEEE Transactions on Cloud Computing*, 8(3), 2020.

Huaqing Zhang, Yong Xiao, Shengrong Bu, Dusit Niyato, Richard Yu, and Zhu Han. Fog computing in multi-tier data center networks: A hierarchical game approach. In *Proc. of the IEEE ICC Conference*, Kuala Lumpur, Malaysia, May 2016.

Linquan Zhang, Chuan Wu, Zongpeng Li, Chuanxiong Guo, Minghua Chen, and Francis CM Lau. Moving big data to the cloud: An online cost-minimizing approach. *IEEE Journal on Selected Areas in Communications*, 31(12): 2710–2721, 2013.

Qixia Zhang, Fangming Liu, and Chaobing Zeng. Adaptive interference-aware vnf placement for service-customized 5g network slices. In *IEEE INFOCOM*, 2019.

R. Zhou. An online placement scheme for vnf chains in geo-distributed clouds. In *Proc. of IEEE IWQoS, Poster*, 2018.

Zhi Zhou, Fangming Liu, Shutong Chen, and Zongpeng Li. A truthful and efficient incentive mechanism for demand response in green datacenters. *IEEE Transactions on Parallel and Distributed Systems*, 2018.

Zhi Zhou, Xu Chen, En Li, Liekang Zeng, Ke Luo, and Junshan Zhang. Edge intelligence: Paving the last mile of artificial intelligence with edge computing. *Proceedings of the IEEE*, 2019a.

Zhi Zhou, Qiong Wu, and Xu Chen. Online orchestration of cross-edge service function chaining for cost-efficient edge computing. *IEEE Journal on Selected Areas in Communications*, 37(8):1866–1880, 2019b.

5

Lightweight Privacy-Preserving Learning Schemes*

5.1 Introduction

With the advances of sensing and communication technologies, the internet of things (IoT) will become a main data generation infrastructure in the future. The drastically increasing amount of data generated by IoT will create unprecedented opportunities for various novel applications powered by machine learning (ML). However, various system challenges need to be addressed to implement the envisaged intelligent IoT.

IoT in nature is a distributed system consisting of heterogeneous nodes with distinct sensing, computing, and communication capabilities. Specifically, it consists of massive mote-class sensors deeply embedded in the physical world, personal devices that move with people, widespread network fog nodes such as wireless access points, as well as the cloud backend. Implementing the fabric of IoT and ML faces the following two key challenges:

- **Separation of data sources and ML computing power.** Most IoT data will be generated by the end devices that often have limited computation resources, while the computing power needed by ML model training and execution will be located at the fog nodes and in the cloud. In addition, the communication channels between the end devices and the edge/cloud are often constrained, in that they are limited in bandwidth, intermittency, and have long delays.
- **Privacy preservation.** As the end devices can be deeply embedded in people's private space and time, the data generated by them will contain privacy-sensitive information. To gain wide acceptance, the IoT-ML fabric must respect the human users' privacy. The lack of privacy preservation may even go against the recent legislation such as the General Data Protection Regulation in the European Union. However, privacy preservation often presents substantial challenges to the system design.

Privacy-preserving ML has received extensive research in the context of cloud computing. Thus, it is of great interest to investigate whether the existing solutions can be applied in the context of IoT. To this end, this section provides a brief review of the existing privacy-preserving ML approaches, which are classified into two categories: *privacy-preserving training* and *privacy-preserving inference*. The nodes in a privacy-preserving ML system often have two roles: *participant* and *coordinator*. The participants

---

* This chapter is based on the authors' conference papers Jiang et al. (2019a), Jiang et al. (2019b) and journal paper Xu et al. (2020).

Intelligent IoT for the Digital World: Incorporating 5G Communications and Fog/Edge Computing Technologies, First Edition. Yang Yang, Xu Chen, Rui Tan, and Yong Xiao.

are often the data generators (e.g., smartphones), whereas the coordinator (e.g., a cloud server) orchestrates the ML process. The *privacy-preserving training* schemes aim to learn a global ML model or multiple local ML models from disjoint local datasets which, if aggregated, would provide more useful/precise knowledge. Thus, the primary objective of privacy protection is to preserve the privacy of the data used for building an ML model in the training phase. Differently, *privacy-preserving inference* schemes focus on the scenario where a global ML model at the coordinator has been trained and the participants transmit the unlabeled data to the coordinator for inference. The aim is to protect the privacy of the input data in the inference phase and maintain the inference accuracy. Since the computation and communication overheads are the key considerations in the design of IoT systems, ML for IoT should be of low-overhead and with privacy preservation.

In a privacy-preserving training process orchestrated by the coordinator, the participants collaboratively train a global model from their disjoint training datasets while the privacy of the training datasets is preserved. Distributed machine learning (DML) Hamm et al. (2015b), Shokri and Shmatikov (2015), McMahan et al. (2017), Dean et al. (2012), Zinkevich et al. (2010), Boyd et al. (2011) is a typical scheme of this category, in which only model weights are exchanged among the nodes. However, the local model training and the iterative weight exchanges are compute- and communication-intensive. Recently, the federated learning scheme Yang et al. (2019) has received wide research interest. Federated learning is a type of DML. During the training process of federated learning, the training data possessed by each participant is not exchanged. Instead, only the weights of the model trained by each participant are transmitted to the coordinator. Thus, federated learning is considered promising for privacy preservation. However, it requires each participant to train a local model based on the training data it possesses. The intensive computation of the local training renders federated learning ill-suited for IoT devices acting as participants. Therefore, federating learning is mostly studied under the context where each participant is an enterprise that has sufficient computing power.

If the training data samples are to be transmitted to the coordinator, they can be obfuscated or encrypted for data privacy protection. Obfuscation is often achieved via additive perturbation and multiplicative projection. Additive perturbation implemented via Laplacian Dwork et al. (2006b), exponential McSherry and Talwar (2007), and median Roth and Roughgarden (2010) mechanisms can provide differential privacy Dwork (2011). Multiplicative projection Liu et al. (2012), Shen et al. (2018) protects the confidentiality of the raw forms of the original data. In Liu et al. (2012), the participants use distinct secret projection matrices, where the Euclidean distances among the projected data samples are no longer preserved. This can degrade the performance of distance-based ML algorithms. To address this issue, in Liu et al. (2012), the participants need to project a number of public data vectors and return the results to the coordinator which will learn a regress function to preserve Euclidean distances. However, this approach is only applicable to distance-based classifier. The conventional classifier does not scale well with the volume of the training data and the complexity of the data patterns. ML can be also performed based on homomorphically encrypted data samples DeMillo (1978), Graepel et al. (2012), Zhan et al. (2005), Qi and Atallah (2008). However, homomorphic encryption incurs high compute overhead compared with additive perturbation and multiplicative projection.

Privacy-preserving inference approaches assume that the ML model at the coordinator has been previously trained using public plaintext data. They aim to protect the privacy contained in test data vectors while maintaining the inference accuracy. Additive perturbation is generally not advisable for deep models because the inference accuracy of deep models can be significantly degraded by small perturbations on input data Zheng et al. (2016). In order to achieve privacy preservation in the inference phase against an honest-but-curious coordinator running the ML model, CryptoNets Gilad-Bachrach et al. (2016) and multi-party computation (MPC) Barni et al. (2006) are proposed. Gilad-Bachrach et al. (2016) adjust the feed-forward neural network trained with plaintext data so that it can be applied to the homomorphically encrypted data to make encrypted inference. Unfortunately, the high computational complexity of HE renders CryptoNets unpractical for IoT devices. Moreover, although CryptoNets does not need to support training over ciphertext, the neural network still needs to satisfy certain conditions. MPC Goldreich (1998) enables the parties involved to jointly compute a function over their inputs while keeping those inputs private. Barni et al. (2006) apply MPC in *privacy-preserving inference*. However, MPC requires many rounds of communication between the participant and the coordinator, representing considerable communication overhead.

To address the computation and communication overheads with privacy preservation approaches, in this chapter, we propose three novel privacy-preserving ML schemes on the training or the inference in the context of IoT. The first scheme is a lightweight privacy-preserving collaborative learning (PPCL) scheme. It applies independent Gaussian random projection at each IoT object to obfuscate data and trains a deep neural network at the coordinator based on the projected data from the IoT objects. This approach introduces light computation overhead to the IoT objects and moves most workload to the coordinator that can have sufficient computing resources. The second scheme is also a PPCL approach, in which the fog nodes and the cloud train different stages of a deep neural network, and the data transmitted from an fog node to the cloud is perturbed by Laplacian random noises to achieve ϵ-differential privacy. The last scheme is designed as a privacy-preserving inference approach. It is a lightweight and unobtrusive approach to obfuscate the inference data by executing a small-scale well-designed neural network at the fog nodes for data obfuscation. The three proposed approaches are all designed for an IoT system consisting of distributed IoT nodes with limited computing powers.

The remainder of the chapter is as follows. Section 5.2 describes the system model and state the mutual problem. Section 5.3 presents the three proposed schemes and their evaluation. Section 5.4 concludes the chapter.

5.2 System Model and Problem Formulation

In this chapter, we consider a decentralized ML system consisting of multiple participants and a coordinator. The participants are resource-constrained data generators with plenty of training and inference data samples. The coordinator is an honest-but-curious IoT backend with sufficient computing power to orchestrate the ML process. The participants and coordinator are collaborative to realize a classification system in the context of the IoT. As the learning process is often compute-intensive, most of the learning computation should be accomplished by the coordinator. In this chapter, we focusing on the problem of a decentralized machine learning scheme while protecting certain privacy contained in the data samples.

The privacy concern regarding the data vectors is primarily due to the fact that the data vectors may contain information beyond the classification objective in question. For example, consider a PPCL system for training a classifier to recognize human body activity (e.g., sitting, walking, climbing stairs, etc.). The recognition is based on various body signals (e.g., motion, heart rate, breath rate, etc.) that are captured by wearable sensors. However, the raw body signals can also be used to infer the health status of the participants and even pinpoint the patients of certain diseases. In this chapter, we adopt the following threat and privacy models.

- **Honest-but-curious coordinator:** We assume that the coordinator will honestly coordinate the decentralized learning process and inference process, aiming to realize the best ML performance. Thus, it will not tamper with any data or parameters collected from or transmitted to the participants. However, the coordinator is curious about the participants' privacy contained in the data vectors.
- **Potential collusion between participants and coordinator:** We assume that the participants are not trustworthy in that they may collude with the coordinator in finding out other participants' details contained in the data vectors. The colluding participants are also honest, i.e. they will faithfully contribute their training data to improve the supervised classifier. The design of the decentralized system should keep the privacy preservation for a participant when any or all other participants are colluding with the coordinator.

Label privacy: The class labels may also contain information about the participant. In this chapter, we do not consider label privacy because the participant willingly contributes the labeled data vectors and should have no expectation of privacy regarding labels.

Thus, the problem in this chapter is how to design privacy-preserving ML schemes against the above threat, and privacy models in the decentralized ML system in the context of the IoT. Based on the categories of privacy-preserving training and privacy-preserving inference, we propose three different novel privacy-preserving ML schemes below.

5.3 Solutions and Results

In this section we describe the details of three novel schemes to realize privacy-preserving ML for the IoT and show the respective evaluation and implementation results on the three schemes.

5.3.1 A Lightweight Privacy-preserving Collaborative Learning Scheme

In this subsection, we study the design and implementation of a PPCL approach that is lightweight for resource-constrained participants, while maintaining privacy preservation against an honest-but-curious learning coordinator. The coordinator can be a cloud server or a resource-rich fog node, e.g., access points, base stations, network routers, etc. We propose applying (1) multiplicative *Gaussian random projection* (GRP) at the resource-constrained IoT objects to obfuscate the contributed training data and (2) *deep learning* at the coordinator to address the much increased complexity of the data patterns due to the GRP. Specifically, each participant uses a private, time-invariant but randomly generated Gaussian matrix to project each plaintext training data vector and transmits the result to the coordinator. GRP gives several privacy preservation properties of (1) the computational difficulty for the coordinator to reconstruct the plaintext without knowing the Gaussian matrix Liu et al. (2006), Rachlin and Baron (2008), and (2) quantifiable plaintext reconstruction error bounds even if the coordinator obtains the Gaussian matrix Liu et al. (2006). From a system perspective, GRP is computationally lightweight and does not increase the data volume. Thus, GRP is a practical privacy protection method suitable for resource-constrained IoT objects. Regarding GRP's impact on the design of the machine learning algorithms, the random projection can be viewed as a process of mapping the original data vectors to some domain in which the data vectors in different classes are less separable. If the original data vectors are readily separable (that is, they are features), the inverse of the Gaussian matrix can be considered as a linear feature extraction matrix. With the deep learning's unsupervised feature learning capability, this inverse matrix can be implicitly captured by the trained deep model. Thus, we conjecture that the randomly projected training samples can still be used by the coordinator to build the deep model for classification.

To achieve robustness of the privacy preservation against the collusion between any single participant and the curious learning coordinator, each participant should generate its own Gaussian matrix independently. However, this presents the main challenge to the PPCL system's scalability with respect to the number of participants (denoted by N). Specifically, assuming that the training data samples for each class are horizontally distributed among the participants, the number of data patterns for a class will increase from one in the plaintext domain to N in the projection data domain. This increased pattern complexity is to be addressed by the strong learning capability of deep learning. Thus, in the proposed PPCL approach, most of the computational workload is offloaded to the resourceful coordinator at the edge or in the cloud, unlike the existing DML and homomorphic encryption approaches that introduce significant or prohibitive compute overhead to the smart objects beneath the IoT edge.

To understand the effectiveness of the GRP approach and its scalability with the number of participants and the pattern complexity of the training data, we conduct extensive evaluation to compare GRP with several other lightweight PPCL approaches. The evaluation is based on two example applications with low and moderate pattern complexities, i.e. handwritten digit recognition and spam e-mail detection. The baseline approaches include various combinations between (1) multiplicative GRP versus additive noisification for differential privacy (DP) at the participants, and (2) deep neural networks (DNNs), including the multi-layer perceptron (MLP) and convolutional neural network (CNN), versus support vector machines (SVMs) at the coordinator. The results show that, for the two example applications, the proposed GRP-DNN approach can support up to hundreds of participants without sacrificing the learning performance much, whereas the GRP-SVM approach may fail to capture the projected data patterns and the performance of the DP-DNN approach is susceptible to additive noisification. The results of this chapter suggest that GRP-DNN is a practical PPCL approach for resource-constrained IoT objects observing data with low- or moderate-complexity patterns.

We also implement GRP-DNN, Crowd-ML Hamm et al. (2015a) (a federated learning approach based on shallow learning), and CryptoNets Gilad-Bachrach et al. (2016)

(a homomorphic encryption approach) on a testbed of 14 IoT devices. Experiments show that, compared with GRP-DNN, Crowd-ML incurs 350× computation overhead and 3.5× communication overhead to each IoT device. Deep federated learning will only incur more computation overhead. CryptoNets incurs 2.6 million times higher computation overhead to the IoT device, compared with GRP.

The remainder of this section is organized as follows. Section 5.3.1.1 introduces preliminaries about random Gaussian projection. Section 5.3.1.2 overviews our approach. Section 5.3.1.3 present two toy examples under our approach. Section 5.3.1.4 presents the learning performance evaluation for various lightweight PPCL approaches. Section 5.3.1.8 presents the benchmark results of GRP-DNN, Crowd-ML, and CryptoNets on the testbed.

5.3.1.1 Random Gaussian Projection (GRP)

Let $\mathbf{R} \in \mathbb{R}^{k \times d}$ represent a random Gaussian matrix, i.e. each element in $\mathbf{R}$ is drawn independently from the normal distribution $\mathcal{N}(0, \sigma^2)$. GRP has the following two properties Liu et al. (2006):

Property 5.1 For data vectors $\mathbf{x}_1$, $\mathbf{x}_2$ and their projections $\mathbf{y}_1 = \frac{1}{\sqrt{k\sigma}}\mathbf{R}\mathbf{x}_1$, $\mathbf{y}_2 = \frac{1}{\sqrt{k\sigma}}\mathbf{R}\mathbf{x}_2$, the dot product and Euclidean distance between $\mathbf{y}_1$ and $\mathbf{y}_2$ are unbiased estimates of those between $\mathbf{x}_1$ and $\mathbf{x}_2$, i.e. $\mathbb{E}[\mathbf{y}_1^\mathsf{T}\mathbf{y}_2] = \mathbf{x}_1^\mathsf{T}\mathbf{x}_2$ and $\mathbb{E}[\|\mathbf{y}_1 - \mathbf{y}_2\|_2^2] = \|\mathbf{x}_1 - \mathbf{x}_2\|_2^2$. The estimation error bounds are $\mathrm{Var}[\mathbf{y}_1^\mathsf{T}\mathbf{y}_2] \leq \frac{2}{k}$ and $\mathrm{Var}[\|\mathbf{y}_1 - \mathbf{y}_2\|_2^2] \leq \frac{32}{k}$.

Property 5.2 Given a Gaussian matrix instance $\mathbf{R} \in \mathbb{R}^{k \times d}$ where $k < d$ and the projection $\mathbf{y} = \frac{1}{\sqrt{k\sigma}}\mathbf{R}\mathbf{x}$, the minimum norm estimate of $\mathbf{x}$, denoted by $\hat{\mathbf{x}}$, is an unbiased estimate of $\mathbf{x}$, i.e. $\mathbb{E}[\hat{\mathbf{x}}] = \mathbf{x}$. The estimation error for the ith element of $\mathbf{x}$ is $\mathrm{Var}[x_i] = \frac{2}{k}x_i^2 + \frac{1}{k}\sum_{j,j\neq i}x_j^2$.

The estimation error given by Property 5.2 will be used in the later sections of this chapter to measure the degree of privacy protection provided by our proposed approach.

5.3.1.2 Gaussian Random Projection Approach

Figure 5.1 illustrates the PPCL system. We propose a GRP-based approach that is computationally lightweight and communication efficient for the participants. The overview of our approach is presented as follows.

At the system initialization, each participant i independently generates a random Gaussian matrix $\mathbf{R}_i \in \mathbb{R}^{k \times d}$, where d is the dimension of the data vector. During the learning phase, the participant i keeps $\mathbf{R}_i$ secret and uses it to project all the training data vectors. The participant i transmits the projected training dataset $\mathcal{D}_i = \{\mathbf{R}_i\mathbf{x}_{i,j}, y_{i,j} | j \in [1, M_i], y_{i,j} \in C\}$ to the coordinator. After collecting all projected training datasets $\mathcal{D}_i$, $i = 1, \ldots, N$,

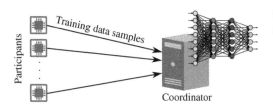

Participants · · · Training data samples → Coordinator

Figure 5.1 A collaborative learning system.

the coordinator applies deep learning algorithms to train the classifier $h(\cdot|\theta^*)$. During the classification phase, the participant i still uses $\mathbf{R}_i$ to project the test data vector $\mathbf{x}$ and obtains the classification result $h(\mathbf{R}_i\mathbf{x}|\theta^*)$. The classification computation can be carried out at the participant or the coordinator, depending on whether the participant is capable of executing the trained deep model. In our approach, each participant independently generates its random projection matrix to counteract the collusion between participants and coordinator.

We adopt Gaussian matrices. Specifically, each element of $\mathbf{R}_i$ is sampled independently of the standard normal distribution Ailon and Chazelle (2009). We set the row dimension of $\mathbf{R}_i$ to be smaller than or equal to its column dimension, i.e. $k \leq d$. Thus, the GRP can also compress the data vector. We define the compression ratio as $\rho = d/k$. The understanding regarding the admission of compression into the training data projection is as follows. From the compressive sensing theory Candès and Wakin (2008), a sparse signal can be represented by a small number of linear projections of the original signal and recovered faithfully. Therefore, in the compressively projected data vector, the feature information still exists, provided that the adopted compression ratio is within an analytic bound Candès and Wakin (2008). With GRP, if $\mathbf{R}_i$ is kept confidential to the coordinator, it is computationally difficult (practically impossible) for the coordinator to generate a meaningful reconstruction of the original data vector from the projected data vector Liu et al. (2006), Rachlin and Baron (2008). Thus, GRP protects the form of the original data. In the worst case where the coordinator obtains $\mathbf{R}_i$, the estimation error given by Property 5.2 can be used as a measure of privacy protection. Random projection has been used as a lightweight approach to protect data from confidentiality in various contexts Li et al. (2013), Tan et al. (2017), Wang et al. (2013), Xue et al. (2017).

Feature extraction is a critical step of supervised learning. With the traditional *shallow learning*, the classification system designer needs to handcraft the feature. The emerging deep learning method LeCun et al. (2015) automates the design of feature extraction by *unsupervised feature learning*, which is often based on a neural network consisting of a large number of parameters. Thus, the deep model is often a tandem of the feature extraction stage and the classification stage. For example, a CNN for image classification consists of convolutional layers and dense layers, which are often considered performing the feature extraction and classification, respectively.

Our approach leverages on the unsupervised feature learning capability of deep learning to address the data distortion introduced by the GRP. We now illustrate this using a simple example system, in which there is only one participant and the projection matrix $\mathbf{R}$ is a square invertible matrix. Moreover, we make the following two assumptions to simplify our discussion. First, we assume that a linear transform $\mathbf{\Psi} \in \mathbb{R}^{f \times d}$ gives effective features of the data vectors, where f is the feature dimension. That is, $\mathbf{f} = \mathbf{\Psi}\mathbf{x}$ is an effective representation of the data vector $\mathbf{x}$ for classification. Second, we assume that $\mathbf{\Psi}$ can be learned in the form of a neural network by the unsupervised feature learning. Now, we discuss the impact of the random projection on the unsupervised feature learning. After the projection, the data vector becomes $\mathbf{R}\mathbf{x}$. Moreover, the linear transform $\mathbf{\Psi}\mathbf{R}^{-1}$ will be an effective feature extraction method, since $\mathbf{f} = (\mathbf{\Psi}\mathbf{R}^{-1})(\mathbf{R}\mathbf{x})$. It is reasonable to expect that the unsupervised feature learning can also build a neural network to capture the linear transform $\mathbf{\Psi}\mathbf{R}^{-1}$, similar to the unsupervised feature learning to capture the $\mathbf{\Psi}$ based on the plaintext training data **x**. As a result, the deep model trained using the projected data can still classify future projected data vectors.

The above discussion based on linear features provides a basis for us to understand how the unsupervised feature learning helps address the distortion caused by the GRP. In practice, effective feature extractions are generally nonlinear mappings. Neural network-based deep learning has shown strong capability in capturing sophisticated features beyond the above ideal linear features.

As discussed earlier, each participant independently generates a Gaussian matrix to counteract the potential collusion between participants and the coordinator. However, this introduces a challenge to deep learning, because the pattern for a class of projected data vectors from N participants will be a composite of N different patterns. Thus, intuitively, a deeper neural network and a larger volume of training data will be needed to well capture the data patterns with increased complexity due to the participants' independence in generating their projection matrices. We note that, the participants' independence also engenders the following possible situation that undermines the learning performance and leads to classification errors: $\mathbf{R}_u\mathbf{x}_u = \mathbf{R}_v\mathbf{x}_v$, where $\mathbf{x}_u$ and $\mathbf{x}_v$ are generated by participants u and v and belong to different classes respectively. However, for high-dimensional data vectors, the probability of the above situation is low. The more complex data patterns due to the independent projection matrix generation will be the major challenge. In this approach, we conduct extensive experiments to assess how well deep learning can scale with the number of participants, compared with the traditional learning approaches.

5.3.1.3 Illustrating Examples

We use two examples to illustrate the intuitions. First, we consider a PPCL system with four participants (i.e. $N = 4$) to build a two-class classifier. The original data vectors in the two classes follow two two-dimensional Gaussian distributions with means of $[-2, -2]^\mathsf{T}$ and $[2, 2]^\mathsf{T}$, and the same covariance matrix of $[1, 0; 0, 1]$. Figure 5.2(a) shows the plaintext data vectors generated by the four participants. From the figure, the plaintext data vectors

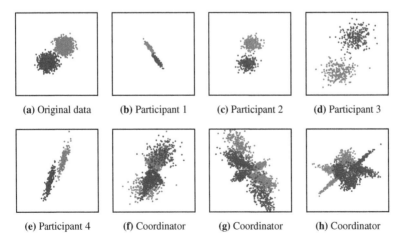

| (a) Original data | (b) Participant 1 | (c) Participant 2 | (d) Participant 3 |

| (e) Participant 4 | (f) Coordinator | (g) Coordinator | (h) Coordinator |

Figure 5.2 Two-dimensional example. Original data vectors and projected data vectors (light points: class 0; dark points: class 1). The ranges for the x and y axes are $[-10, 10]$.

of the two classes can be easily separated using a simple hyperplane. Each participant independently generates a Gaussian random matrix. Figures 5.2(b)–5.2(e) show the projected data vectors of each participant. We can see that the patterns of the projected data vectors are different across the participants. Figure 5.2(f) shows the mixed projected data vectors received from all participants. Compared with Figure 5.2(a), the pattern of the mixed projected data from all participants is highly complex. Moreover, no simple hyperplane can well divide the two classes. We also generate two other sets of the random projection matrices for all participants. Figures 5.2(g) and 5.2(h) show the mixes of all participants' projected data vectors with the two sets of random projection matrices, respectively. Similarly, the pattern of the mixed projected data from all participants is highly complex.

Now, we use another example system to understand the effect of deep learning's unsupervised feature learning capability in addressing the data distortion caused by the random projection. This example is a PPCL system with only one participant (i.e. $N = 1$). The original data vectors in two classes follow two 10-dimensional Gaussian distributions, with $[-2, -2, \ldots, -2]^\mathsf{T}$ and $[2, 2, \ldots, 2]^\mathsf{T}$ as the respective mean vectors, and the 10-dimensional identity matrix as their identical covariance matrix. We assume that the projection matrix $\mathbf{R}$ is invertible and the unsupervised feature learning tend to capture $\mathbf{\Psi R}^{-1}$. As learning algorithms are based on numerical computation on the training data, an ill-conditioned matrix $\mathbf{R}$ will impede efficient fitting of $\mathbf{\Psi R}^{-1}$. We verify this intuition by assessing the learning performance of the single-participant PPCL system using different $\mathbf{R}$ matrices with varying condition numbers. Specifically, by following a method described in Bierlaire et al. (1991), the participant generates a random square matrix $\mathbf{R}$ that has a certain condition number value. The condition number is defined as $\|\mathbf{R}\|_F \|\mathbf{R}^+\|_F$ Paige and Saunders (1982), where $\mathbf{R}^+$ denotes the pseudoinverse of $\mathbf{R}$ and $\| \cdot \|_F$ represents the Frobenius norm. Figure 5.4 shows the test accuracy of the MLP and SVM classifiers trained using data projected by $\mathbf{R}$ versus the condition number of $\mathbf{R}$. Note that a larger condition number means that the matrix is more ill-conditioned. We can see that the test accuracy decreases with the condition number, consistent with the intuition.

The study Chen and Dongarra (2005) analyzes the distribution of the condition numbers of Gaussian random matrices. The results show that a Gaussian random matrix is well-conditioned with a high probability. For instance, it is shown in Chen and Dongarra (2005) that for a 10×5 Gaussian random matrix, the probability that its condition number

Figure 5.3 Test accuracy based on projected data versus the number of participants.

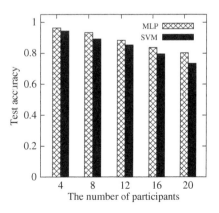

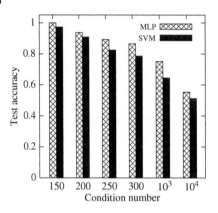

Figure 5.4 Test accuracy based on projected data versus the condition number.

is larger than 100 is less than 6×10^{-7}. This is a basis for our choice of using Gaussian random matrices to project data.

In the following section, we extensively compare the accuracy achieved by various approaches. The computation and communication overhead of these approaches will be profiled in Section 5.3.1.8 based on their implementations on a testbed.

5.3.1.4 Evaluation Methodology and Datasets

We conduct extensive evaluation to compare several approaches:

- **GRP-DNN:** This is the proposed approach consisting of GRP at the participants and collaborative learning based on a DNN at the coordinator. The design or choice of the DNN model will be application specific. The DNN models and training algorithms are implemented based on PyTorch pyt (2018).
- **GRP-SVM:** This baseline approach applies GRP at the participants and trains an SVM-based classifier at the coordinator. The SVM-based classifier is implemented using LIBSVM Chang and Lin (2018). The classifier uses an RBF kernel with two configurable parameters C and λ. During the training phase, we apply a grid search to determine the best settings for C and λ. This grid search is often lengthy in time (e.g., several days).
- **GRP-NCL:** This is the non-collaborative learning (NCL) baseline approach. It runs GRP at the participants and trains a separate DNN for each participant at the coordinator. Compared with other approaches, this approach additionally requires the identity of the participant for each training sample.
- **ϵ-DP-DNN:** This approach implements ϵ-DP by adding Laplacian noise vectors to the data vectors and performs collaborative deep learning based on a DNN at the coordinator.
- **ϵ-DP-SVM:** This approach implements ϵ-DP by adding Laplacian noise vectors to the data vectors and performs collaborative learning based on SVM at the coordinator.
- **CNN, SVM, MLP:** These are the plain learning approaches based on the CNN, SVM, and MLP, respectively. They do not protect any privacy.

The performance evaluation is performed based on two datasets, i.e. MNIST LeCun et al. (2018) and spambase Hopkins et al. (2018). The MNIST dataset consists of 60,000 training samples and 10,000 testing samples. Each sample is a 28×28 grayscale image showing a handwritten digits from 0 to 9. Figure 5.5(a) shows an instance of each digit. The spambase dataset consists of 4601 samples. Each sample consists of (i) a 57-dimensional feature

Figure 5.5 Example images from the MNIST dataset.

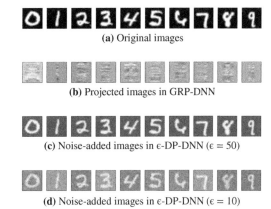

(a) Original images

(b) Projected images in GRP-DNN

(c) Noise-added images in ε-DP-DNN (ε = 50)

(d) Noise-added images in ε-DP-DNN (ε = 10)

vector that is extracted from an e-mail message and (ii) a class label indicating whether the e-mail message is an unsolicited commercial e-mail. The details of the feature vector can be found in Hopkins et al. (2018). As the data volume of this spambase dataset is limited, we apply data augmentation to the spambase by adding zero-mean Gaussian noises, resulting in 40,000 training samples and 400 testing samples. We choose these two datasets because the small sizes of the data vectors are commensurate with the limited computing and transmission capabilities of IoT end devices.

Training a spam detector based on user-contributed samples (e.g., e-mails) may cause privacy concerns. Thus, our proposed approach fits well in this case. The choice of the vision-based character recognition task with the MNIST dataset allows us to leverage on the learning capabilities of the deep models that are often designed for image classification. Moreover, by using images as the data vectors, the effect of the distortion caused by noise adding or random projection can be visualized for intuitive understanding. Although the recognition task in MNIST is not privacy-sensitive, its results will provide understanding of other image classification-based privacy-sensitive applications, such as collaboratively training a mood classifier using the photos in the album of the users' smartphones.

For a PPCL system with N participants, we divide both the training and testing samples into N disjoint sets evenly. Each set is assigned to a participant. Under GRP-DNN, GRP-SVM, and GRP-NCL, each participant independently generates its random Gaussian matrix and uses the matrix to project its plaintext data vectors. The deep models and SVM are trained by the coordinator based on the projected or noise-added training data vectors from the participants. The trained deep models and SVM are used to classify the projected or noise-added testing data vectors to measure the test accuracy as the evaluation results.

5.3.1.5 Evaluation Results with the MNIST Dataset

We design a CNN that is used in the GRP-DNN, GRP-NCL, and ε-DP-DNN approaches. The CNN consists of two convolutional layers and three dense layers of ReLUs. We apply max pooling after each convolutional layer to reduce the dimension of data after convolution. The max pooling controls overfitting effectively and improves the CNN's robustness to small spatial distortions in the input image. The last dense layer has ten ReLUs corresponding to the ten classes of MNIST. A softmax function is used to make the classification decision based on the outputs of the last dense layer. Figure 5.6 illustrates the design of the

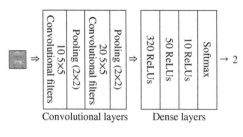

Figure 5.6 CNN with a projected MNIST image as input.

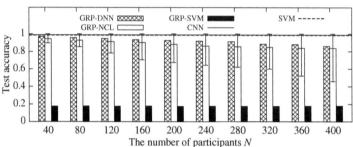

Figure 5.7 Impact of the number of participants (MNIST). The error bars for GRP-NCL represent minimum and maximum.

CNN. Note that, without random projection, the CNN and the SVM with grid search for kernel parameters can achieve test accuracy of 98.7% and 98.52%. This shows that the CNN and SVM can well capture the patterns of MNIST.

First, we evaluate the impact of the number of participants N on the learning performance of GRP-DNN, GRP-NCL, and GRP-SVM. Figure 5.7 shows the results. The two horizontal lines in Figure 5.7 represent the test accuracy of the plain CNN and SVM without any privacy protection. The two lines overlap. When N increases from 40 to 400, the test accuracy of GRP-DNN decreases from 96.87% to 86.18%. If N is no greater than 280, GRP-DNN can maintain a test accuracy greater than 90%. The drop of accuracy with increased N is consistent with the understanding that distinct random projection matrices increase the pattern complexity of the aggregated data. However, for MNIST data with light pattern complexities, the GRP-DNN approach can support up to 280 IoT objects for a satisfactory classification accuracy of 90%.

Under the GRP-NCL approach, the deep models corresponding to the participants having different test accuracy values. The histogram and error bars in Figure 5.7 represent the average, minimum, and maximum of the test accuracy values across all trained deep models. Under each setting of N, the maximum test accuracy is 100%. However, the average test accuracy is consistently lower than that of GRP-DNN. This shows that, the GRP-NCL that needs to compromise data anonymity yields inferior average learning performance compared with GRP-DNN. This result shows the advantage of collaborative learning. Lastly, the GRP-SVM approach gives poor test accuracy of around 17.5%. This is because no efficient RBF kernels can be found to create proper hyperplanes for classification. This suggests that DNNs are more efficient in coping with the distortions caused by projections.

Second, we evaluate the impact of GRP's data compression on the learning performance. Figure 5.8 shows the results when $N = 100$. When the compression ratio increases from 1

Figure 5.8 Impact of data compression on learning performance (MNIST, $N = 100$).

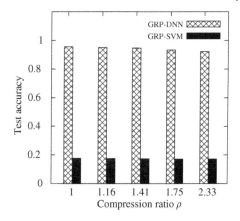

Figure 5.9 Impact of differential privacy loss on learning performance (MNIST).

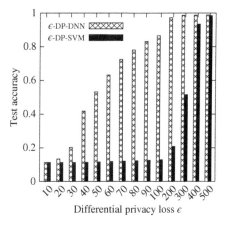

(i.e. no compression) to 2.33 (i.e. 43% of data volume is retained), the test accuracy of GRP-DNN decreases from 95.52% to 92.85% only. The good tolerance of GRP-DNN against data compression is due to the high sparsity of the MNIST images. In contrast, the GRP-SVM approach performs poorly under all compression ratio settings.

Then, we evaluate the impact of adding Laplacian noises to implement ϵ-DP on the learning performance. Figure 5.9 shows the test accuracy of ϵ-DP-DNN versus the privacy loss level ϵ. When $\epsilon = 100$ (small Laplacian noises and large differential privacy loss), the ϵ-DP-DNN achieves a test accuracy of 86.6%, lower than those achieved by GRP-DNN when N is up to 400. When $\epsilon = 10$, the performance of ϵ-DP-DNN drops to 11.4%, close to the performance of random guessing. For comparison, we visualize the projected and noise-added images with two ϵ settings in Figure 5.5. From Figure 5.5(b), we cannot visually interpret the projected images. However, from Figures 5.5(c) and 5.5(d), the noise-added images are easily interpreted when ϵ is down to 10. Note that in our evaluation, we use the same CNN model as shown in Figure 5.6 for the GRP-DNN, GRP-NCL, and ϵ-DP-DNN approaches. We do not spend special effort to improve the CNN design in favor of any approach; we only make sure the CNN fed with the original MNIST images achieves satisfactory performance. The poor performance of ϵ-DP-DNN is consistent with the understanding that the performance of deep learning can be susceptible to small perturbations to the data vectors Zheng et al. (2016). There are also systematic approaches to generating adversary examples with small differences from the training samples to yield wrong classification results Goodfellow et al. (2015), Bose and Aarabi (2018). Special care is needed in the deep model design to improve robustness against human-indiscernible perturbations Zheng et al. (2016). Significant noise, which is required to achieve good DP protection, is still an open challenge to deep learning. Thus, under the ϵ-DP framework, it is challenging to achieve a desirable trade-off between the privacy protection strength and learning performance.

The additive noisification for ϵ-DP is ineffective in achieving a good trade-off between learning performance and protecting the confidentiality of the raw forms of the training data. Now, we compare the results of GRP-DNN ($N = 1$, $k = d - 1$) and ϵ-DP-DNN. We consider the worst case for GRP-DNN, i.e. the projection matrix **R** is revealed to the curious coordinator. From Property 5.2, the minimum norm estimate of the original data vector by the coordinator will have a per-element variance of about 410 for any MNIST image. Under this setting, GRP-DNN can achieve a test accuracy of 94.82%. To achieve the same per-element variance of 410, the ϵ value adopted by the ϵ-DP-DNN should be 18.89. Under this ϵ setting, the test accuracy of ϵ-DP-DNN is 12.86% only.

Figure 5.9 also shows the test accuracy of the ϵ-DP-SVM approach. It performs poorly when $\epsilon \leq 100$. Only when the added noises are very small under the settings of $\epsilon = 400$ and $\epsilon = 500$, can this approach achieve good test accuracy.

5.3.1.6 Evaluation Results with a Spambase Dataset
We design a five-layer MLP classifier to detect spam. The numbers of ReLUs in the five layers are 57, 100, 50, 10, and 2, respectively. A softmax function is used lastly to make the final detection decision. Dropout is used during training to suppress overfitting. Without random projection, the MLP and the SVM with grid research for kernel parameters can achieve test accuracy of 96.52% and 96.25%, respectively. This shows that the MLP and SVM can capture well the patterns of the spambase.

We evaluate the impact of the number of participants N on the learning performance of GRP-DNN, GRP-NCL, and GRP-SVM. Figure 5.10 shows the results. The two horizontal lines in Figure 5.10 represent the test accuracy of the plain MLP and SVM without any privacy protection. When N increases from 1 to 200, the test accuracy of GRP-DNN decreases from 96% to 83.25%. If N is no greater 100, GRP-DNN can maintain a test accuracy of about

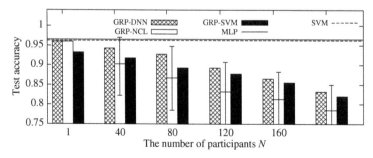

Figure 5.10 Impact of the number of participants (spambase). The error bars for GRP-NCL represent minimum and maximum.

90%. The average test accuracy of GRP-NCL is about 5% lower than that of the GRP-DNN, because GRP-NCL misses the advantages of collaborative learning. The test accuracy of the GRP-SVM is about 1.25% to 2.75% lower than that of the GRP-DNN. Thus, the GRP-SVM performs satisfactorily for this spambase dataset. The reasons are two-fold. First, in this spambase dataset, the classifiers operate on the e-mail features, rather than the raw data. Second, the RBF kernel is effective in capturing the features. In fact, the nature of this spambase dataset is similar to that of the 2-dimensional and 10-dimensional generated feature datasets used in Section 5.3.1.3, on which the GRP-DNN and GRP-SVM perform similarly.

5.3.1.7 Summary

We have several observations from the results in Sections 5.3.1.5 and 5.3.1.6:

- Compared with SVM, deep learning can better adapt to the complexity introduced by the multiplicative projections.
- Although the GRP-NCL approach additionally uses the identities of the participants, it gives inferior performance compared with the collaborative GRP-DNN. This shows the advantage of collaborative learning even with the privacy preservation requirement.
- Compared with GRP-DNN, the additive noisification for ϵ-DP achieves inferior trade-off between learning performance and protecting confidentiality of raw forms of training data.
- GRP-DNN shows promising scalability with the number of participants observing low-complexity data patterns. For the MNIST and spambase datasets, the GRP-DNN can well support 100 participants with a few percent test accuracy drop. For large-scale PPCL systems involving more participants, we envision a two-tier system architecture as follows. The participants are divided into groups. At the first tier, our GRP-DNN is applied within each group; at the second tier, the DML approach is applied among the group coordinators.

5.3.1.8 Implementation and Benchmark

In this section, we measure the overhead of two PPCL approaches (i.e. our GRP-DNN and Crowd-ML Hamm et al. (2015a)) and a privacy-preserving classification outsourcing approach (i.e. CryptoNets Gilad-Bachrach et al. (2016)) on a testbed of 14 Raspberry Pi 2 Model B nodes pi (2018) and a powerful workstation computer. The Raspberry Pi nodes act as PPCL participants and the workstation acts as the coordinator. They are interconnected using a 24-port network switch. We benchmark these approaches using the MNIST dataset. The training and testing samples are evenly allocated to the participants, resulting in 4285 training samples and 714 testing samples on each participant. The implementation of the three approaches (GRP-DNN, Crowd-ML, CryptoNets) on the same platform, i.e. Raspberry Pi, allows fair comparison. The participant part of our GRP-DNN can be implemented on mote-class platforms. Our previous work Tan et al. (2017) has implemented Gaussian matrix generation and GRP on the MSP430-based Kmote platform. However, it is difficult/impossible to implement Crowd-ML and CryptoNets on mote-class platforms.

We implement our GRP-DNN approach on the testbed. The compression ratio $\rho = 1$ (i.e. no compression). Table 5.1 shows the benchmark results. During the training phase, each GRP-DNN participant needs to transmit a total of 33.6 MB projected data.

Table 5.1 The overhead of various approaches.

Overhead	GRP-DNN	Crowd-ML	CryptoNets
Participant communication volume (training)	33.6 MB	117.2 MB	n/a [a)]
Participant compute time (training)	0.96 s	367.24 s	n/a[a)]
Coordinator compute time (training)	928.34 s	1.04 s	n/a[a)]
Participant communication volume (testing)	5.6 MB	n/a[a)]	15.0 MB
Participant compute time (testing)	0.16 s	4.67 s	116 h
Coordinator compute time (testing)	40.88 s	n/a[a)]	

a) n/a represents "not applicable."

A participant can complete projecting all the 4285 training images within 0.96 s. The coordinator needs 928.34 s to train the CNN. In our GRP-DNN implementation, the testing phase is performed on the coordinator. During the testing phase, each participant completes projecting all the 714 testing images within 0.16 s and transmits a total of 5.6 MB data to the coordinator. The coordinator needs 40.88 s to classify all projected testing images from the participants. Note that GPU acceleration is not used in this benchmark for GRP-DNN during both the training and testing phases.

The Crowd-ML Hamm et al. (2015a) is a DML approach. In Crowd-ML, a participant checks out the global classifier parameters from the coordinator and computes the gradients using its own training data. Then, the participants transmit the gradients to the coordinator that will update the global classifier parameters. Thus, during the training phase, the participants and the coordinator repeatedly exchange parameters. We apply an existing implementation of Crowd-ML Hamm (2018) on our testbed.

Our measurement shows that, during the training phase, each participant needs to upload and download a total of 117.2 MB data, which is 3.5× our GRP-DNN. The participant compute time is more than 350× that under GRP-DNN. Despite the larger volume of data exchanges, Crowd-ML achieves 91.28% test accuracy only, which is lower than the 95.58% test accuracy achieved by GRP-DNN. This is because Crowd-ML uses a simple multi-class logistic classifier, which is inferior compared with the CNN used by GRP-DNN in terms of learning performance. Note that during the testing phase of Crowd-ML, the participants execute their local classifiers. Thus, they do not need to transmit the testing samples to the coordinator for classification.

CryptoNets Gilad-Bachrach et al. (2016) uses a homomorphic encryption algorithm to encrypt a testing sample during the classification phase and transmits the encrypted sample to the coordinator. Then, the coordinator uses a neural network trained with plaintext data to classify the encrypted testing sample. We have implemented the homomorphic encryption part of CrytoNets that runs on the Raspberry Pi nodes. The volume of the 714 encrypted testing images is 15 MB, almost 3× the data volume generated by random projection. In particular, a Raspberry Pi node takes about 10 min and a total of 116 h to encrypt an image and all the testing images, respectively. This is 2.6 million times slower than the random projection computation. This result clearly shows that the high computation complexity of the homomorphic encryption makes CryptoNets ill-suited for resource-constrained devices.

5.3.2 A Differentially Private Collaborative Learning Scheme

In this section, we consider a honest-but-curious cloud that aims to infer private information from the data uploaded by the fog nodes during the training phase. We adopt the ϵ-DP Dwork (2006) as our privacy definition, which gives quantifiable indistinguishability of different data vectors yielded by the fog nodes against the honest-but-curious cloud. To implement ϵ-DP, a Laplacian random noise vector is added to the data vector generated by the front layers before being transmitted to the cloud. In our design, we apply batch normalization to the data vector generated by the front layers at the fog node to attain an analytic upper bound of the normalized data. The bound is used as the global sensitivity in setting the Laplacian noise generator parameters to guarantee ϵ-DP.

We apply our proposed approach to a case study of collaboratively training a convolutional neural network (CNN) for image classification. We use MNIST LeCun et al. (2018), an image dataset of handwritten digits, to train the CNN. Two convolutional layers with maximum pooling are trained by the fog nodes, while six dense layers are trained by the cloud. Results show that our approach maintains 99% and 96% classification accuracy in implementing privacy loss levels of $\epsilon = 5$ and $\epsilon = 2$, respectively. Note that, to provide good DP protection, the typical privacy loss level, i.e. ϵ, is often set to a value below 10. For example, in Hamm et al. (2015a), to obtain the balance between system performance and data privacy, the ϵ is set to be 10. In Abadi et al. (2016), the ϵ is set to 0.5, 2, or 3. Thus, the case study based on MNIST shows that our approach can achieve good DP protection while maintaining satisfactory classification performance.

The remainder of this section is organized as follows. Section 5.3.2.1 presents the design of the differentially private collaborative learning approach. Section 5.3.2.2 discusses the challenges to achieving differential privacy. Section 5.3.2.3 presents the performance evaluation results.

5.3.2.1 Approach Overview

In this section, we propose an approach that can protect the privacy of the extracted features before being transmitted to the coordinator to preserve the privacy contained in the original data. Perturbing original data directly to protect privacy may lead to significant learning performance degradation, which will be shown in Section 5.3.2.3.

From our analysis in Section 5.3.2.2, we can perturb the results computed from the original data before being transmitted to the coordinator to protect the privacy contained in the original data.

To realize the advantages of collaborative learning, the classification computation during the learning phase is performed on the coordinator to make good use of various data from different participants. In this chapter, we consider a convolutional neural network (CNN) to design collaborative learning system, since a CNN is an effective machine learning model. In a CNN, convolutional layers fold data in several channels to extract features with specific pooling layers and activation layers. The dense layers (i.e. fully connected layers) classify the extracted features to yield class labels. In our collaborative learning system, each participant runs convolutional layers to extract features that will be transmitted to the coordinator. The coordinator maintains the dense layers and forward-propagates them with the received features during the learning phase. Moreover, the participants will perturb the features before transmitting them to the coordinator.

Smart data
providers

DL service
provider

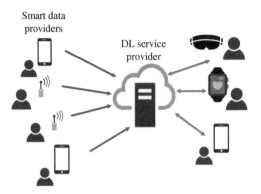

Figure 5.11 Overview of our proposed privacy-preserving collaborative learning approach.

Figure 5.11 illustrates the system architecture. Each participant collects data and extracts features locally. Under the privacy-preserving mechanism that will be presented in Section 5.3.2.2, the participant sends privacy-preserving features and original labels to the coordinator such that the coordinator can train the fully connected layers. The coordinator uses the back-propagation algorithm to update the fully connected layer parameters and meanwhile sends back the propagated loss to the participants which will update convolutional layers accordingly. In the above process, the convolutional layers of all the participants are updated based on each contributed training data sample. Thus, the participants enjoy the advantages of collaborative learning, which help them better extract features.

We now discuss several design issues.

- During the classification phase after the completion of the collaborative learning, the participant can send testing data features to the cloud, which will then perform classification using the dense layers. Alternatively, on the completion of the collaborative learning, the coordinator can disseminate the dense layers to all participants. Then, each participant can run the full CNN to perform classification without transmitting testing data.
- In order to utilize the large volume of training data to improve the effectiveness of the convolutional layers, it is desirable to maintain the same convolution layers at all the participants. We adopt the following method to keep convolutional layer consistency among participants. After the coordinator updates dense layer parameters, it broadcasts propagated loss to all the participants. Thus, all the participants can update their own convolutional layers simultaneously. Since we can configure the same hyper parameters for the convolutional layers at all the participants, we can maintain the convolutional layers at all the participants consistent.
- The system will have significant overhead if each participant immediately sends new extracted features once it generates new data. To solve this issue, in our design, if the data exceeds a specified value, the participant starts to process data to extract feature and transmit it. This method matches well with our privacy-preserving approach that adopts batch normalization and Laplacian noisification, which will be presented in the next subsection.

5.3.2.2 Achieving ϵ-Differential Privacy

Differential privacy is an information-theoretic approach to protecting data privacy. It aims to confound the query results based on adjacent datasets. In our approach, we adopt ϵ-differential privacy Dwork (2006) as our privacy definition. The ϵ-differential privacy (ϵ-DP) is formally defined as follows: *A randomized algorithm $\mathcal{A} : \mathbb{D} \to \mathbb{R}^t$ gives ϵ-DP if for all adjacent datasets $D_1 \in \mathbb{D}$ and $D_2 \in \mathbb{D}$ differing on at most one element, and all $S \subseteq Range(\mathcal{A})$, $\Pr(\mathcal{A}(D_1) \in S) \leq \exp(\epsilon) \times \Pr(\mathcal{A}(D_2) \in S)$.* Here, the differential privacy level ϵ, is a positive number which measures privacy loss. Smaller ϵ always means better protection: when ϵ is very small, $\Pr(\mathcal{A}(D_1) \in S) \approx \Pr(\mathcal{A}(D_2) \in S)$ for all $S \subseteq Range(\mathcal{A})$. Thus, the query results $\mathcal{A}(D_1)$ and $\mathcal{A}(D_2)$ are nearly indistinguishable, which prevents the attackers from recognizing the original dataset. An approach to implementing ϵ-DP is to add Laplacian noise. Concretely, for all function $\mathcal{F} : D \to \mathbb{R}^t$, the randomized algorithm $\mathcal{A}(D) = \mathcal{F}(D) + [n_1, n_2, \dots, n_t]^\mathsf{T}$ gives ϵ-DP, where each n_i is drawn independently from a Laplace distribution $\mathrm{Lap}(S(\mathcal{F})/\epsilon)$ and $S(\mathcal{F})$ denotes the global sensitivity of $\mathcal{F}$. Note that the global sensitivity $S(\mathcal{F})$ is $S(\mathcal{F}) = \max_{D,D' \in \mathbb{D}} ||\mathcal{F}(D) - \mathcal{F}(D')||_1$ while $\mathrm{Lap}(\lambda)$ is a zero-mean Laplace distribution with a probability density function of $f(x|\lambda) = \frac{1}{2\lambda} e^{-\frac{|x|}{\lambda}}$.

A challenge in implementing ϵ-DP is the determination of the global sensitivity $S(\mathcal{F})$. It is hard to determine global sensitivity after convolutional layers. Theoretically, the output of convolutional layers can continuously increase or decrease during training epochs. However, too large or too small outputs of the convolutional layers may cause the *gradient exploding problem* or *gradient vanishing problem* Ioffe and Szegedy (2015). Batch normalization (BN) Ioffe and Szegedy (2015) is developed to normalize the output of hidden layers to avoid the problems in supporting neural network training. Using each batch as a unit, BN normalizes the output of specific layers and then forwards it to the next layer. It limits the range of the output, enabling the determination of the global sensitivity $S(\mathcal{F})$. In our approach, we apply standard BN parameters: fixed variance 1 and fixed mean 0. In the following, we explain the method to compute the global sensitivity $S(\mathcal{F})$.

For simplicity, we assume there is only one channel in the CNN and the dimension of the output of the convolutional layers is $L \times W$. Denote the batch size in the convolutional neural network as N, the output of convolutional layers in a position $\langle i,j \rangle$ of element k in the batch as $X_{i,j,k}$. The difference between two adjacent datasets D and D' in our scenario is $X_{i,j,k}$ and $X'_{i,j,k}$, while the other elements are the same. The query request in our scenario is to read each element in the dataset because the coordinator can access all data sent from the participants. Thus, the global sensitivity $S(\mathcal{F})$ is equal to the maximum difference between $X_{i,j,k}$ and $X'_{i,j,k}$, $S(\mathcal{F}) = \max_{\langle i,j,k \rangle \in \langle L,W,N \rangle}\{X_{i,j,k} - X'_{i,j,k}\}$. Due to the constraint imposed by BN, we have $\sum_{k=1}^{N} X_{i,j,k} = 0$ and $\sum_{k=1}^{N} X_{i,j,k}^2 = N$.

To analyze $S(\mathcal{F})$, we now prove for any $\ell \in \{1, 2, \dots, N\}$ that $-\sqrt{N-1} \leq X_{i,j,\ell} \leq \sqrt{N-1}$ and both equal signs are applicable in special cases.

From the Cauchy–Schwarz inequality, we have

$$\left(\sum_{t \in \{1,2,\dots,N\} \setminus \{\ell\}} X_{i,j,t} \right)^2 \leq (N-1) \sum_{t \in \{1,2,\dots,N\} \setminus \{\ell\}} X_{i,j,t}^2. \tag{5.1}$$

Applying $\sum_{t\in\{1,2,...,N\}\setminus\{\ell\}}X_{i,j,t} = -X_{i,j,\ell}$ and $\sum_{t\in\{1,2,...,N\}\setminus\{\ell\}}X_{i,j,t}{}^2 = N - X_{i,j,\ell}{}^2$ to inequality (5.1), we obtain $X_{i,j,\ell}{}^2 \leq (N-1)\cdot(N-X_{i,j,\ell}{}^2)$, which leads to

$$-\sqrt{N-1} \leq X_{i,j,\ell} \leq \sqrt{N-1}. \tag{5.2}$$

The equal sign in the first "$\leq$" of inequality (5.2) is applicable when $X_{i,j,\ell} = -\sqrt{N-1}$ and $X_{i,j,t} = 1/\sqrt{N-1}$ for $t \in \{1,2,...,N\}\setminus\{\ell\}$. Similarly, the equal sign in the second "$\leq$" of inequality (5.2) is taken when $X_{i,j,\ell} = \sqrt{N-1}$ and $X_{i,j,\ell} = -1/\sqrt{N-1}$ for $t \in \{1,2,...,N\}\setminus\{\ell\}$.

From the above analysis, $S(F)$ denoting $\max_{\langle i,j,k\rangle\in\langle L,W,N\rangle}\{X_{i,j,k} - X'_{i,j,k}\}$ is equal to $2\sqrt{N-1}$. By adding a random noise from $\text{Lap}(S(F)/\epsilon)$ Dwork et al. (2006a), we can achieve ϵ-DP to protect original data privacy. It also succeeds when there are multiple channels in CNN. The detailed proof is omitted here due to space constraints.

5.3.2.3 Performance Evaluation

Our evaluation is based on a public dataset MNIST LeCun et al. (2018). MNIST is a handwritten dataset that consists of 60,000 training samples and 10,000 testing samples. Each sample is a 28 × 28 grayscale image showing a handwritten number in the range 0–9. It is widely used in machine learning literature as a basic benchmark dataset to evaluate the learning performance.

In our model, the CNN deployed at the participants has two convolutional layers with 30 and 80 channels, respectively. After each conventional layer, we apply max-pooling layers to reduce the size of the output. In neural networks, the max-pooling layer can accelerate learning with reduced parameter dimension while extracting features of the sub-region in a sample. In dense layers, the ReLU activation layer is used to increase the nonlinearity of the neural network. After the second conventional layer, we use a BN layer to accelerate the learning rate and prevent gradient vanishing problem and gradient exploding problem. Then, the DP technique is applied to perturb the output of the BN layer to preserve data privacy.

After adding the DP noise, the participants send perturbed features to the coordinator as the input for the dense layers. In our model, the dense layers contain four hidden layers for reducing the data dimension gradually and one output layer with a dimension of 10 which is the dimension of labels. Finally, we use softmax layer to predict label and compute loss. The structure of the CNN is shown in Figure 5.12.

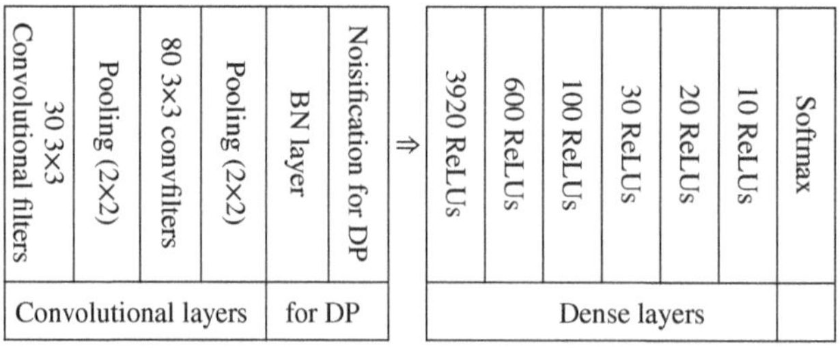

30 3×3 Convolutional filters	Pooling (2×2)	80 3×3 convfilters	Pooling (2×2)	BN layer	Noisification for DP	⇒	3920 ReLUs	600 ReLUs	100 ReLUs	30 ReLUs	20 ReLUs	10 ReLUs	Softmax
Convolutional layers				for DP			Dense layers						

Figure 5.12   CNN structure.

Figure 5.13 Impact of privacy loss level ϵ on the test accuracy of the collaboratively learned model with DP.

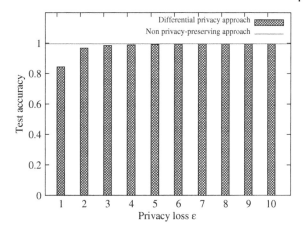

In our experiments, we set the hyper parameters of CNN as follows: the learning rate is equal to 0.01 and the batch size is equal to 64. Thus, the global sensitivity $S(\mathcal{F})$ is equal to $\sqrt{63}$. Therefore, we apply various privacy loss levels ϵ to evaluate the performance of the differentially private collaborative learning based on MNIST.

For comparison, we use the centralized training approach without any privacy consideration as the baseline. The corresponding CNN excludes the noisification layer as shown in Figure 5.12. This centralized non-DP approach achieves 99.58% test accuracy. From Figure 5.13, with our differentially private collaborative learning approach, the test accuracy increases with the privacy loss level ϵ. Note that a large ϵ means less privacy protection. Thus, there exists a trade-off between the test accuracy and the degree of privacy protection. Generally, when the ϵ is chosen to be 5, which is often considered to provide satisfactory privacy protection Abadi et al. (2016), Hamm et al. (2015a), our system can still achieve 99.18% test accuracy. When ϵ is reduced to 1, the test accuracy decreases to 84.33%, because large DP noises start to undermine the performance of the classification system. However, when ϵ is around 2 to 5, the system shows good classification performance. Specifically, only 3% of accuracy reduction is observed when ϵ reduces from 5 to 2.

In the second set of experiments, we investigate the impact of the BN batch size on the classification performance of the collaboratively learned model. For a training CNN, a smaller batch size often results in more accurate estimation of the gradient descent, but longer convergence time of the training process. Moreover, in our approach, the batch size N determines the global sensitivity $S(\mathcal{F})$, i.e. $S(\mathcal{F}) = \sqrt{N-1}$. Thus, the smaller batch size also results in lower noise levels for the same ϵ setting. We set $\epsilon = 2$. Figure 5.14 shows the test accuracy of the CNNs trained by our differentially private collaborative learning approach and the centralized learning approach without privacy preservation, under different batch size settings. When the batch size $N = 32$, the test accuracy is 99.5% and 98.1% for the centralized non-DP learning approach and our DP approach, respectively. When N increases to 128, the accuracy drops to 99.3% for the centralized non-DP approach and 94.0% accuracy for our approach. For our approach, with a larger batch size, both the global sensitivity and noise level become larger, leading to performance drop.

Adding Laplacian noises to the original data to achieve ϵ-DP is an alternative approach. In this section, we also investigate its effectiveness. Under this alternative approach, the

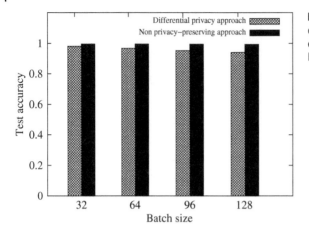

Figure 5.14 Impact of batch size on the test accuracy of the collaboratively learned model with DP ($\epsilon = 2$).

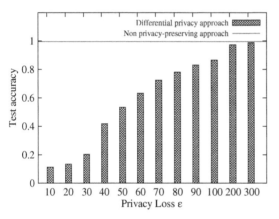

Figure 5.15 Impact of privacy loss level ϵ on the test accuracy of the collaborative learning approach that perturbs the original data for DP.

global sensitivity $S(F)$ of the original data (i.e. the pixel values) is the maximum difference between any two pixels. Since the pixel value in MNIST is within the range (0,255), the global sensitivity is a fixed value of 255. Figure 5.15 shows the test accuracy of the CNN trained by this alternative approach under various ϵ settings. We can see that when $\epsilon = 10$ the test accuracy is 11.35% only, which is close to the performance of random guessing (i.e. 10%). When $\epsilon \geq 100$, although the approach can achieve good test accuracy, the privacy loss is too high to be meaningful in a collaborative learning system. Thus, the results show that adding Laplacian noise to the original data significantly degrades the learning performance. Moreover, by comparing the results obtained with this alternative approach and our approach, we can see that the unsupervised feature learning performed by the convolutional layers is susceptible to the DP noise, whereas the classification boundary learning performed by the dense layers is more robust to the DP noise.

5.3.3 A Lightweight and Unobtrusive Data Obfuscation Scheme for Remote Inference

Different from the above two proposed approaches which are in the training phase, we propose a *lightweight* data obfuscation approach suitable for resource-constrained fog nodes to protect inference data privacy in the remote inference scheme. With the lightweight approach, the fog node spends little time and energy on obfuscating the inference data before transmitting to the backend. Moreover, we aim to achieve another feature of *unobtrusiveness*, in that (i) the inference model at the backend admits both original and obfuscated inference data, and (ii) the fog node does not need to indicate whether obfuscation is applied. The unobtrusiveness feature provides three advantages. First, the system is back-compatible with old fog nodes that cannot be upgraded to perform the data obfuscation. Second, the fog node can easily choose to opt into or out of data obfuscation given its run-time computation and battery lifetime statuses. Third, the exemption of obfuscation indication helps improve privacy protection.

In this section, we present *ObfNet*, an approach to realizing the lightweight and unobtrusive data obfuscation at the IoT edge for remote inference. ObfNet is a small-scale neural network that can run at resource-constrained fog nodes and introduces light compute overhead. ObfNet's sophisticated, many-to-one nonlinear mapping from the input vector to the output vector offers a form of data obfuscation that can well protect the confidentiality of the raw forms of the input data. To achieve unobtrusiveness, we design a training procedure for ObfNet as follows. We assume that the backend has an in-service deep inference model (referred to as *InfNet*). The backend concatenates an untrained ObfNet with the InfNet and then trains the concatenated model using the training dataset that was used to train InfNet. During the training, only the weights of ObfNet are updated by back propagation until convergence. The backend repeats the above procedure to generate sets of distinct ObfNets and transmits a unique set to each of the fog nodes. Then, each fog node chooses an ObfNet randomly and dynamically from the received set and uses it for obfuscating the data for remote inference.

We evaluate the ObfNet approach by three case studies of (1) free spoken digit (FSD) recognition, (2) MNIST handwritten digit recognition, and (3) American sign language (ASL) recognition. The case studies show the effectiveness of ObfNet in protecting the confidentiality of the raw forms of the inference data while preserving the accuracy of the remote inference. Specifically, the obfuscated samples are unrecognizable auditorily by invited volunteers for FSD and visually for MNIST and ASL, while the obfuscation causes inference accuracy drops of generally within 1% from the original inference accuracy of about 99%.

We also benchmark the ObfNet approach on a testbed consisting of (i) a Coral development board equipped with Google's edge tensor processing unit (TPU) that acts as an fog node and (ii) an NVIDIA Jetson AGX Xavier equipped with a Volta graphics processing unit (GPU) that acts as the backend. Measurements on the testbed show the effectiveness of ObfNet and the advantage of remote inference in terms of processing times.

The remainder of this section is organized as follows. Section 5.3.3.1 overviews our approach. Sections 5.3.3.2–5.3.3.4 present performance evaluation via three case studies. Section 5.3.3.5 presents benchmark results on the testbed.

5.3.3.1 Approach Overview

We propose an <u>obfuscation</u> neural <u>network</u> (ObfNet) approach to obfuscating the inference data sample before being transmitted to the backend. In particular, the design of ObfNet aims to provide two properties of *light weight* and *unobtrusiveness*.

ObfNet is a small-scale neural network executed on the fog node to obfuscate the inference data samples. In our proposed approach, the backend generates multiple sets of ObfNets by following an approach detailed in the next paragraph and then transmits a unique set to each of the fog nodes. An fog node that wishes to obfuscate the inference data chooses one ObfNet from the received set and feeds the inference data to the chosen ObfNet. Then, the fog node transmits the output of the ObfNet, i.e. the obfuscated inference data, to the backend for inference. The old fog nodes that cannot be upgraded to perform the data obfuscation and the fog nodes that do not wish to obfuscate the inference data can transmit the original inference data to the backend for inference. The backend executes the InfNet using the received inference data and sends back the inference result to the fog node. Existing cryptographic approaches can be applied to (i) protect the confidentiality and integrity of the data exchanged between the fog nodes and the backend and (ii) the authentication of the fog nodes and the backend. Figure 5.16 illustrates the remote inference system where each fog node can choose to opt into or out of the ObfNet-based privacy preservation.

Now, we present the approach to generating the sets of ObfNets at the backend. Note that the ObfNets in any set are distinct and all sets are also distinct (i.e. any two sets do not share an identical ObfNet). Figure 5.17 illustrates the approach. It has two steps as follows.

- **ObfNet design.** The system designer designs a small-scale and application-specific neural network architecture for ObfNet. The input to ObfNet is the original inference data sample. The output of ObfNet is the obfuscated inference data sample. Note that there

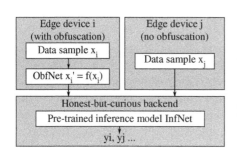

Figure 5.16 ObfNet for remote inference. The fog node *i* desires privacy protection and thus applies ObfNet to obfuscate inference data sample x_i to x_i'. The fog node *j* does not desire privacy protection and thus directly transmits the original inference data sample x_j to the backend. The backend feeds x_i' and x_j to the pre-trained inference model InfNet to generate the results y_i and y_j.

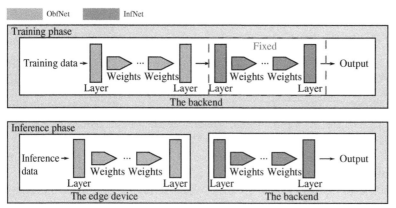

Figure 5.17 The procedure to generate ObfNets.

is no rule of thumb to design ObfNet's architecture; similar to the design of DNNs for specific applications, the design of ObfNet also follows a trial-and-error approach using the validation results of the training process as the feedback (the training of ObfNet will be presented shortly). The designer should try to reduce the scale of ObfNet to make it affordable to resource-constrained fog nodes. Moreover, the ObfNet design should meet the following requirements. First, to be unobtrusive, the dimensions of the input and output should be identical. Second, ObfNet should adopt many-to-one nonlinear mapping activation functions (e.g., ReLU) to prevent the backend from estimating the exact original inference data from the obfuscated one.

- **ObfNet training.** First, the backend initializes the weights of an ObfNet with random numbers. Then, the backend concatenates the ObfNet with the InfNet, forming a concatenated DNN, where the output of ObfNet is used as the input to InfNet. The backend trains the concatenated DNN using the training dataset that was previously used to train InfNet. During the back propagation stage of each training epoch, the loss is back propagated normally. However, only the weights of ObfNet are updated, while the weights of InfNet are fixed. When the training of the concatenated DNN converges, the backend retrieves the trained ObfNet from the concatenated DNN. By repeating the above procedure, the backend generates multiple distinct sets of distinct ObfNets. Note that due to the randomization of ObfNet's initial weights and the randomization techniques (e.g., training data sampling) during the training phase, the trained ObfNets are distinct. The backend can determine the cardinality of each set according to the available storage volume of the corresponding fog node that desires data obfuscation. Finally, the backend transmits the set to the fog node.

We have a few remarks regarding the ObfNet approach. First, since InfNet is not changed during the training of ObfNet, the InfNet can classify both the original and the obfuscated inference data samples. The execution of InfNet does not require any indication of whether the input inference data sample is obfuscated. Thus, the unobtrusive requirement is achieved. Second, as the fog nodes use distinct ObfNets during remote inference, the collusion between any/all other fog nodes with the backend (i.e. the colluding fog nodes let the backend know which ObfNets they use) will not affect the non-colluding fog nodes. Third, as the ObfNet uses many-to-one nonlinear activation functions, it is highly difficult (virtually impossible) for the backend to estimate the exact original inference data sample from the obfuscated one. Moreover, as each non-colluding fog node selects an ObfNet from its received set randomly and dynamically for obfuscation, the difficulty for the backend's inverse attempt is strengthened due to the introduced uncertainty.

In the following section, we present the application of ObfNet to three case studies.

5.3.3.2 Case Study 1: Free Spoken Digit (FSD) Recognition

Our first case study concerns human voice recognition. Recently, voice recognition has been integrated into various edge systems such as smartphones and voice assistants found in households and cars. In many scenarios, voice recordings are privacy sensitive. Thus, it is desirable to obfuscate the voice data for privacy protection, while preserving the performance of voice recognition. In this section, we apply the ObfNet approach to FSD recognition, which can be viewed as a minimal voice recognition task. Using this minimal task as a case study brings the advantage of easy exposition of the results and the associated insights.

We use the FSD dataset Jackson, Jackson et al. that consists of 2000 WAV recordings of spoken digits from 0 to 9 in English. We split the data as 80% for training, 10% for validation, and 10% for testing. We extract the mel-frequency cepstral coefficients (MFCCs) Davis and Mermelstein (1980) as the features to represent a segment of audio signal. According to Davis and Mermelstein (1980), the MFCC is empirically shown to well represent the pertinent aspects of the short-term speech spectrum and form a particularly compact representation compared with other features such as linear frequency cepstrum coefficients (LPCs), reflection coefficients (RCs), and cepstrum coefficients derived from the Linear Prediction Coefficients (LPCCs). As the recordings are of different lengths, we apply constant padding to unify the number of MFCC feature vectors for each recording. As a result, the extracted MFCC feature vectors over time for each recording form a 20×45 two-dimensional image.

Multi-layer perceptron (MLP) and convolutional neural network (CNN) are two types of DNN widely adopted for speech recognition and image classification Abdel-Hamid et al. (2013), LeCun and Bengio (1995), Sainath et al. (2013). An MLP consists of multiple fully connected layers (or dense layers). Specifically, each neuron in any hidden layer is connected to all the neurons in the previous layer. The CNN incorporates the features of shared weights, local receptive fields, and spatial subsampling to ensure shift invariance LeCun and Bengio (1995), LeCun et al. (1998). In this case study, we design an MLP-based InfNet and a CNN-based InfNet, which are denoted by I_M and I_C, respectively. Their details are as follows.

- I_C consists of three convolutional layers, one max-pooling layer, and three dense layers. Zero padding is performed on the input image in the convolutional layers and the max-pooling layer. ReLU activation is applied to the output of every convolutional and dense layer except for the last layer. ReLU rectifies a negative input to zero. The last dense layer has 10 neurons with a softmax activation function corresponding to the 10 classes of FSD. Three dropout layers with dropout rate 0.25, 0.1, and 0.25 are applied after the max-pooling layer and in the first two dense layers. Specifically, 25%, 10%, and 25% of the neurons will be abandoned randomly from the neural network during the training process. Dropout is an approach to regularization in neural networks which helps reduce interdependent learning among the neurons. It is widely leveraged during model training to avoid overfitting Srivastava et al. (2014). Figure 5.18 shows the structure of I_C. The I_C has about 1.1 million parameters in total.
- I_M has five dense layers. ReLU activation is applied to the output of every hidden layer. The last dense layer has 10 neurons with a softmax activation function. To prevent overfitting, four dropout layers are applied after the hidden layers. Figure 5.19 shows the structure of I_M. The I_M has about one million parameters in total.

Similar to InfNets, we design CNN-based and MLP-based ObfNets, which are denoted by O_C and O_M, respectively. Their details are as follows.

- O_C consists of two convolutional layers, one max-pooling layer and one dense layer as the output layer. The first convolutional layer filters the 20×45 input image with three output filters of kernel size 2×4. The second convolutional layer applies five output filters with kernel size 3×6. All convolutional filters use a stride of one pixel. Batch normalization follows both convolutional layers, which is expected to mitigate the problem of internal

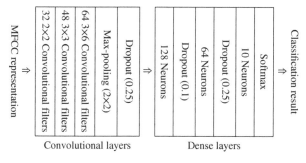

Convolutional layers   Dense layers

Figure 5.18 Structure of I_C for FSD recognition.

Figure 5.19 Structure of I_M for FSD recognition.

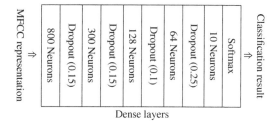

Dense layers

covariate shift to improve model performance. A max-pooling layer with pool size of 2×2 and stride of two is then used to reduce the data dimensionality for computational efficiency. Zero padding is added in each convolutional layer and the max-pooling layer, to ensure that the filtered image has the same dimension as the input image in each layer. The dense layer with 900 neurons is then connected after flattening the output of the max-pooling layer. ReLU activation is applied to the output of every convolutional and dense layer. This introduces many-to-one mapping that is needed in our scheme. Two dropout layers of with dropout rates of 0.25 and 0.15 are applied respectively after the max-pooling layer and in the dense layer. In order to ensure that the output of ObfNet has the same size as the input, a reshape layer is applied in the end to reshape the output size to 20×45. The O_C has about 0.65 million parameters.

- O_M has two dense layers as hidden layers. The first layer has 200 neurons and is fully connected to the second layer of 900 neurons. ReLU activation and batch normalization are applied to the output of both layers. A reshape layer is used as the output layer. The O_M has about 0.37 million parameters.

Following the procedures described above, we train I_C and I_M using the training dataset and then train O_C and O_M in the four concatenations of ObfNet and InfNet (i.e. O_C–I_C, O_M–I_C, O_C–I_M, O_M–I_M). During the training phase, we adopt the AdaDelta optimizer, which introduces minimal computation overhead over stochastic gradient descent (SGD) and adapts the learning rate dynamically Zeiler (2012). Note that during the training phase, only the model achieving the highest validation accuracy is yielded as the training result.

The test accuracy of the trained InfNets I_C and I_M is 99.5%. Thus, the InfNets are well trained. The four ObfNet–InfNet concatenations give distinct test accuracy. For each concatenation, we trained ten different ObfNets. Figure 5.20 shows the inference accuracy of applying ten different ObfNets before the well-trained InfNet. The average test accuracy of

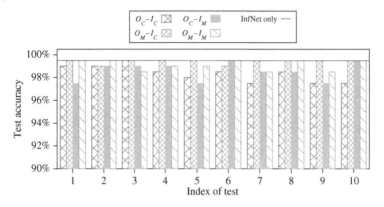

Figure 5.20 Test accuracy of different ObfNet–InfNet concatenations in 10 tests.

applying O_C and O_M before I_C is 98.35% and 99.40%, respectively. The average test accuracy of applying O_C and O_M before I_M is 98.55% and 99.10%, respectively. Compared with the test accuracy of the I_C and I_M on the original data (i.e. 99.5%), the test accuracy drops caused by the obfuscation are merely 1.15%, 0.10%, 0.95%, and 0.40% for different combinations of the ObfNet and the InfNet. Thus, the inference accuracy is well preserved when ObfNet is employed.

To understand the quality of obfuscation, we apply the MFCC inverse using a Python package LibROSA lib to convert the MFCC representations back to WAV audio. The audio converted from the original MFCC representations can be easily recognized by humans, despite some distortions. We also design an experiment to investigate whether humans can interpret the audios inverted from the outputs of ObfNet, i.e. the obfuscated MFCC representations. The details and results of the experiment are as follows.

We invited ten student volunteers (five males and five females) aged from 21 to 23 from Xi'an Jiaotong-Liverpool University. All volunteers have good hearing. In the experiment, we randomly selected ten original MFCC representations from the test dataset (one for each class of the FSD dataset). Then, we applied the MFCC inverse using LibROSA to convert the ten MFCC representations back to audio. The four different ObfNets (two O_Cs and two O_Ms) used in our evaluation were applied to obfuscate the two selected MFCC representations. The obfuscated MFCC representations were inverted using LibROSA to audio. Therefore, in total, there were 50 audio files: ten for the original MFCC representations and 40 for the obfuscated MFCC representations. All volunteers sat in a classroom. The 10 audio files inverted from the original MFCC representations were firstly played in the classroom in a shuffled order. All volunteers can correctly recognize the FSDs. Then, the 40 audio files inverted from the obfuscated MFCC representations were played in a shuffled order. Every volunteer was required to write down the FSD label (from 0 to 9) that they perceived. The volunteers' answers were distributed over all labels without any consensus. This suggests that the volunteers cannot perceive useful information from the audio in recognizing the FSD. The overall accuracy, which is defined as the number of correct answers divided by a total of 100 answers (10 volunteers × 10 audio), is 5%, 7%, 7%, and 4% for the four ObfNets, respectively. Thus, the volunteers' answers seem to be random guesses with an expected accuracy of 10%. Therefore, the ObfNets achieve satisfactory obfuscation quality.

5.3.3.3 Case Study 2: Handwritten Digit (MNIST) Recognition

The MNIST dataset consists of 70,000 handwritten digit images with ten classes corresponding to the digits from 0 to 9, as shown in Figure 5.24(a). Each image has a single channel (i.e. grayscale image). We resize each image to 28×28.

We adopt two InfNets: a CNN-based I_C and an MLP-based I_M. Their details are as follows.

- I_C is similar to LeNet LeCun. It consists of five layers: two convolutional layers, a pooling layer, and two dense layers with ReLU activation. Figure 5.21 shows the architecture. The I_C has about 1.2 million parameters in total.
- I_M has four dense layers, as illustrated in Figure 5.22. It has about 0.93 million parameters in total.

An MLP-based ObfNet O_M and a CNN-based ObfNet O_C are adopted. Details are as follows.

- O_M has two dense layers with ReLU activation. This two-layer design helps reduce the scale of ObfNet. Specifically, to be unobtrusive, ObfNet's output must have the same size as its input. For input size of $28 \times 28 = 784$, a single-layer MLP with bias has $784 \times 784 + 784 = 615440$ parameters. In contrast, a two-layer MLP with 16 neurons within each layer has $784 \times 16 + 16 + 16 \times 784 + 784 = 25888$ parameters only, which is 23.8 times smaller than the single-layer MLP. We configure the number of neurons for the first hidden layer to be 8, 16, 32, 64, or 128. We will investigate the impact of ObfNet's scale on the accuracy of InfNet. The amounts of parameters corresponding to the above configurations are from 0.013 to 0.804 million.
- O_C has a convolutional layer, a pooling layer, a dropout layer, and two dense layers with ReLU activation. The convolutional layer filters the 28×28 input image with 32 output filters of kernel size 3×3 and uses stride of one pixel. The max-pooling layer with pool size of 2×2 and stride of two follows to reduce spatial dimensions. Two dense layers are then connected with ReLU activation.

Figure 5.21 Structure of I_C for MNIST recognition.

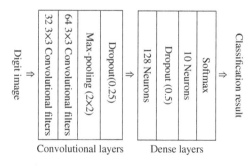

Figure 5.22 Structure of I_M for MNIST recognition.

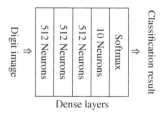

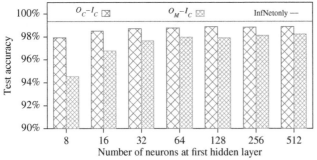

(a) Test accuracy of InfNet I_C and two ObfNet–InfNet concatenations

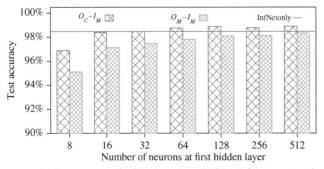

(b) Test accuracy of InfNet I_M and two ObfNet–InfNet concatenations

Figure 5.23 Test accuracy of InfNets and ObfNet–InfNet concatenations for MNIST recognition.

The test accuracy of the trained InfNets I_C and I_M is 99.35% and 98.47%, respectively. This suggests that the InfNets are well trained. We vary the number of neurons of the first hidden layer of the ObfNets and train the ObfNets following the procedure. Figure 5.23 shows the test accuracy of various concatenations of ObfNets and InfNets when the number of neurons in the first hidden layer of the ObfNet varies. From Figure 5.23(a), compared with the test accuracy of I_C, the concatenation O_M–I_C has test accuracy drops ranging from 0.46% to 1.43% over various neuron number settings. When the InfNet I_M is adopted, more neurons in the first hidden layer of ObfNet result in higher test accuracy of the ObfNet–InfNet concatenation, as shown in Figure 5.23(b). In particular, some ObfNet–InfNet concatenations even outperform the corresponding InfNet. This is possible because the ObfNet–InfNet concatenations are deeper neural networks compared with the corresponding InfNet.

Figure 5.24 shows the obfuscation results of O_M when the number of neurons in the first hidden layer varies. From the figure, we cannot interpret the obfuscation results into any digits. When the number of neurons is few (e.g., 8 to 32), the obfuscation results of the digit one are darker than the obfuscation results of other digits. This is because the values of the pixels in the original inference data of digit one are zero, leading to lower pixel values in the obfuscation results. However, when more neurons are used in the first hidden layer of O_M, the overall darkness levels of the obfuscation results of all digits are equalized, suggesting a better obfuscation quality.

(a) Original inference data

(b) Obfuscation results of O_M with 8 neurons in the first hidden layer

(c) Obfuscation results of O_M with 16 neurons in the first hidden layer

(d) Obfuscation results of O_M with 32 neurons in the first hidden layer

(e) Obfuscation results of O_M with 64 neurons in the first hidden layer

(f) Obfuscation results of O_M with 128 neurons in the first hidden layer

(g) Obfuscation results of O_M with 256 neurons in the first hidden layer

(h) Obfuscation results of O_M with 512 neurons in the first hidden layer

Figure 5.24 Obfuscation results of ObfNet O_M on MNIST.

5.3.3.4   Case Study 3: American Sign Language (ASL) Recognition

In this case study, we consider an application of ASL recognition using camera-captured pictures. ASL is a set of 29 hand gestures corresponding to 26 English letters and three other special characters representing the meanings of deletion, nothing, and space delimiter. While ASL is a predominant sign language of the deaf communities in the US, it is also widely learned as a second language, serving as a lingua franca. Therefore, portable ASL recognition systems Fang et al. (2017) are useful for communication between ASL users and those who do not understand ASL. Porting the ASL recognition capability to smart glasses is desirable but also challenging due to smart glasses' limited compute power. Thus, remote inference is a solution for smart glass-based ASL recognition. As the hand gesture images caused by the embedded cameras can contain privacy-sensitive information (e.g., skin color, skin texture, gender, tattoo, location of the shot inferred from the picture background, etc), it is desirable to obfuscate the images. Thus, we apply ObfNet to ASL recognition.

We use an ASL dataset Kaggle consisting of 87,000 static hand gesture RGB images with each sized 200×200 pixels. Figure 5.27(a) shows the samples corresponding to the 29 classes of the ASL alphabet. To reduce the scale of ObfNet, we down-sample the ASL images to 64×64.

As ASL hand gestures have more complex patterns than the MNIST handwritten digits, we adopt a CNN-based InfNet I_C Note that compared with MLP, CNN often better deals with multi-dimensional spatial data. The I_C consists of three convolutional layers with 32, 64, 128 channels, a max-pooling layer, and three dense layers. We adopt dropout after the pooling layer and the second dense layer with drop rates of 0.25 and 0.5. Figure 5.25 shows the architecture of I_C. The I_C has about 111 million parameters in total. We evaluate both the MLP-based ObfNet O_M and the CNN-based ObfNet O_C:

- O_M has two dense layers with ReLU activation. We vary the number of neurons in the first dense layer and evaluate how it affects the inference accuracy. O_M has about 6.3 to 25.2 million parameters, depending on the number of neurons in the first dense layer.
- O_C consists of a convolutional layer, a pooling layer, two dense layers with ReLU activation. The convolutional layer filters the $64 \times 64 \times 3$ input image (i.e. 64×64 RGB image) with 32 output filters of kernel size 3×3 and uses stride of one pixel. A max-pooling layer with pool size of 2×2 and stride of two pixels follows to reduce spatial dimensions. Two dense layers are then connected with ReLU activation. Two dropout layers with dropout rates of 0.25 and 0.4 are applied after the max-pooling layer and the second dense layer to prevent overfitting. O_C has about 22 to 44 million parameters, depending on the number of neurons in the first dense layer.

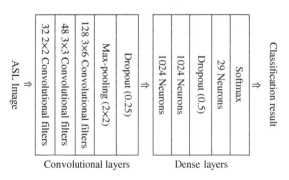

Figure 5.25   Structure of I_C for the ASL dataset.

ASL Image ⇒

| 32 2×2 Convolutional filters | 48 3×3 Convolutional filters | 128 3×6 Convolutional filters | Max-pooling (2×2) | Dropout (0.25) | 1024 Neurons | 1024 Neurons | Dropout (0.5) | 29 Neurons | Softmax |

⇒ Classification result

Convolutional layers   Dense layers

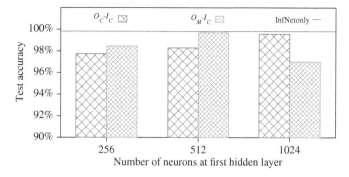

Figure 5.26 Test accuracy of InfNet and ObfNet–InfNet concatenations for ASL recognition.

The test accuracy of the trained I_C is 99.82%. This suggests that the InfNet is well trained. Multiple ObfNets are trained by following the procedure. Figure 5.26 shows the test accuracy of various concatenations of ObfNets and InfNets when the number of neurons in the first hidden layer of the ObfNet varies. From Figure 5.26, compared with the test accuracy of I_C, the concatenation O_M–I_C has test accuracy drops ranging from 0.12% to 2.81% over various neuron number settings. When the ObfNet O_C is adopted, the concatenation O_C–I_C has test accuracy drops ranging from 1.52% to 2.35%. When the number of neurons in the first hidden layer increases from 512 to 1024, the test accuracy of the O_M–I_C drops. This can be caused by overfitting, because compared with the large number of O_M parameters, the number of training samples is not large. Nevertheless, with proper configuration of the ObfNet, the smallest test accuracy drop we can achieve is 0.12%. This shows that the ObfNet introduces little test accuracy drop for ASL recognition.

Figure 5.27 shows the visual effect of the obfuscation on the ASL samples. From Figures 5.27(b) and 5.27(c), we cannot interpret the obfuscation results of O_M and O_C into any hand gestures. Note that the obfuscated samples are still RGB images. Interestingly, the obfuscation results by a certain ObfNet exhibit similar patterns. For instance, each obfuscated sample in Figure 5.27(b) has a dark hole in the center and a greenish circular belt around the dark hole. In fact, as the ObfNet has a large number of parameters (up to tens of millions), the pattern shown in the obfuscation result is mainly determined by the ObfNet, whereas the original inference data sample with a relatively limited amount of information ($64 \times 64 \times 3 = 12288$ pixel values only) can be viewed as a perturbation.

5.3.3.5 Implementation and Benchmark

This section presents the implementation of our ObfNet approach on edge/backend hardware platforms. The benchmark results on the hardware platforms give understanding on the feasibility of ObfNet in practice and interesting observations. For conciseness of presentation, we only present the results of O_M trained for I_C in the three case studies. Our implementation uses the Coral development board (referred to as Coral) and NVIDIA Jetson AGX Xavier (referred to as Jetson) as the fog node and backend hardware platforms, respectively. We implement the ObfNets and InfNets of the three case study applications presented above on Coral and Jetson, respectively.

Coral is a single-board computer equipped with an NXP iMX8M system-on-chip and a Google Edge TPU. Edge TPU is an inference accelerator that cannot perform ML model

(a) Original inference data

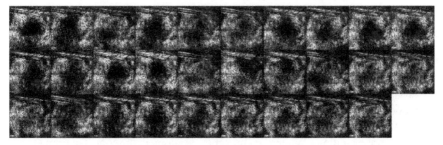

(b) Obfusaction results of O_M

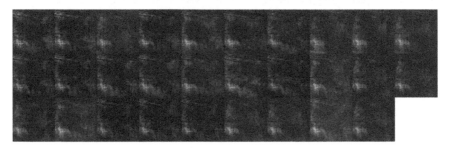

(c) Obfusaction results of O_C

Figure 5.27 Obfuscation results of ObfNet on ASL.

training. Coral size is 8.8×6 cm^2 and weighs about 136 g including a thermal transfer plate and a heat dissipation fan. The power consumption of Coral is no great than 8.5 W. Thus, Coral is a modern fog node platform with hardware-accelerated inference capability. Note that owing to ObfNets' small-scale design, they can also run on fog nodes without hardware acceleration for inference. Coral runs Mendel, a lightweight GNU/Linux distribution. We deploy the ObfNet implemented using the TensorFlow Lite library on Coral.

Jetson is a computing board equipped with a 8-core ARM CPU, 16GB LPDDR4x memory, and a 512-core Volta GPU. The GPU can accelerate DNN training and inference. Jetson size is 10.5×10.5 cm^2 and weighs 280 g including a thermal transfer plate. Jetson's power rating can be configured as 10 W, 15 W, and 30 W. In our experiments, we configure it to run at 30 W to achieve the highest compute power. Jetson can be employed as an embedded backend to serve fog nodes of applications in a locality such as an office building and a

Table 5.2 Per-sample execution time on Coral for O_M trained for I_C.

Case study	Minimum time	Average time	Maximum time
FSD-O_M	2.226 ms	2.312 ms	2.253 ms
MNIST-O_M	0.221 ms	0.221 ms	0.224 ms
ASL-O_M	11.136 ms	11.146 ms	11.170 ms

Table 5.3 Per-sample execution time on Jetson for I_C.

Case study	Minimum time	Average time	Maximum time
FSD-I_C	0.229 ms	0.246 ms	0.289 ms
MNIST-I_C	0.158 ms	0.174 ms	0.212 ms
ASL-I_C	0.201 ms	0.219 ms	0.249 ms

factory floor. To support massive fog nodes, a cloud backend can be used instead. Jetson runs Ubuntu. We deploy the InfNet implemented using TensorFlow on Jetson.

For each case study application, we measure the per-sample execution time for obfuscation on Coral and per-sample inference time on Jetson. To mitigate the uncertainties caused by the operating systems' scheduling, for each tested setting, we run ObfNet or InfNet for 100 times.

Table 5.2 shows Coral's per-sample execution time for the ObfNets designed for the three case studies. We can see that, the ObfNets need little processing time (i.e. a few milliseconds) on Coral. Table 5.3 shows Jetson's per-sample execution time for the InfNets designed for the three case studies. Although the InfNets have larger scales than the ObfNets, the execution times of InfNets are shorter than those of ObfNets due to Jetson's greater compute power. In TensorFlow, batch execution of inferences can improve the efficiency of utilizing the hardware acceleration. Thus, we also evaluate the impact of the batch size on the per-sample execution time of InfNets. Figure 5.28 shows the results. We can see that the per-sample execution time decreases with the batch size and converges. The convergence is caused by the saturation of the hardware acceleration utilization. The above results show that the ObfNets and InfNets introduce little overhead to the fog node and the backend for the considered case study applications.

Inference accelerators such as Edge TPU may enable the execution of deep InfNets on fog nodes (i.e. *local inference*). In contrast, the remote inference scheme considered in this chapter involves the transmissions of the inference data to the backend, which may incur time delays. In this set of benchmark experiments, we put aside the need for protecting the confidentiality of InfNets and compare the local inference and remote inference in terms of total time delay.

Table 5.4 shows the execution time of InfNets on Coral. Compared with the results in Table 5.3, for the FSD and MNIST case study applications, the execution times on Coral are about 50× longer than those on Jetson. For ASL, it is about 480× longer. The data

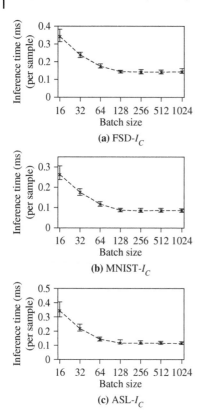

Figure 5.28 InfNet's per-sample execution time on Jetson versus batch size. Error bar represents average, maximum, and minimum over 100 tests.

(a) FSD-I_C

(b) MNIST-I_C

(c) ASL-I_C

Table 5.4 Per-sample execution time of I_C on Coral.

Model	Minimum time	Average time	Maximum time
FSD-I_C	13.484 ms	14.318 ms	15.137 ms
MNIST-I_C	7.606 ms	8.351 ms	9.095 ms
ASL-I_C	100.433 ms	100.467 ms	100.510 ms

transmission delays under the remote inference scheme are often small, because fog nodes often have wideband network connections (e.g., wi-fi and 4G). Based on the average inference data sample sizes of the case study applications (i.e. 10 KB, 13 KB, and 0.6 KB for FSD, ASL, and MNIST, respectively), Figure 5.29 shows the per-sample transmission times versus the network connection data rate. Analysis shows that, compared with the local inference, the remote inference achieves shorter time delays when the connection data rate is higher than 15 Mbps. Note that 4G connections normally provide more than 100 Mbps data rate. Thus, remote inference will be more advantageous in terms of total time delay. The advantage of remote inference can be better exhibited when the scales of the InfNets are larger or the fog nodes are not equipped with inference accelerators.

Figure 5.29 Data sample transmission time versus network connection data rate.

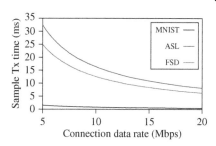

5.4 Conclusion

This chapter reviews the existing privacy-preserving ML approaches that were developed largely in the context of cloud computing and discusses their limitations in the context of IoT. Furthermore, this chapter proposes three different privacy-preserving ML approaches in the context of IoT. In the lightweight privacy-preserving collaborative learning approach the resource-constrained learning participants apply independent Gaussian projections on their training data vectors and the coordinator applies deep learning to train a classifier based on the projected data vectors. This approach protects the confidentiality of the raw forms of the training data against the honest-but-curious coordinator. In the differentially private collaborative learning approach it trains different stages of a deep neural network at the fog nodes and the cloud, respectively. The deep neural network model is constructed based on the training samples contributed by all the participating fog nodes. To protect the privacy contained in the data communicated to the honest-but-curious cloud during the collaborative learning process, Laplacian random noises are added to the communicated data. In the lightweight and unobtrusive data obfuscation approach on remote inference it utilizes a small-scale nonlinear transform in the form of a neural network to obfuscate data samples. In this way, the approach can well protect the confidentiality of the raw form of the inference data samples. Extensive evaluation and implementation show the practicality and efficiency of the three above approaches.

References

Librosa. https://librosa.github.io/librosa/. Accessed: 2019-12-10.

Raspberry Pi 2 Model B, 2018. https://bit.ly/1b75SRj.

PyTorch, 2018. https://pytorch.org/.

M. Abadi, A. Chu, I. Goodfellow, H. McMahan, I. Mironov, K. Talwar, and L. Zhang. Deep learning with differential privacy, In *ACM Conference on Computer and Communications Security (CCS)*, 2016.

Ossama Abdel-Hamid, Li Deng, and Dong Yu. Exploring convolutional neural network structures and optimization techniques for speech recognition. In *Interspeech*, volume 11, pages 73–5, 2013.

Nir Ailon and Bernard Chazelle. The fast johnson–lindenstrauss transform and approximate nearest neighbors. *SIAM Journal on computing*, 39(1):302–322, 2009.

Mauro Barni, Claudio Orlandi, and Alessandro Piva. A privacy-preserving protocol for neural-network-based computation. In *Proceedings of the 8th workshop on Multimedia and security*, pages 146–151. ACM, 2006.

Michel Bierlaire, Ph L Toint, and Daniel Tuyttens. On iterative algorithms for linear least squares problems with bound constraints. *Linear Algebra and its Applications*, 143:111–143, 1991.

Avishek Joey Bose and Parham Aarabi. Adversarial attacks on face detectors using neural net based constrained optimization. In *Proc. Intl. Workshop Multimedia Signal Process.*, 2018.

Stephen Boyd, Neal Parikh, Eric Chu, Borja Peleato, Jonathan Eckstein, et al. Distributed optimization and statistical learning via the alternating direction method of multipliers. *Foundations and Trends® in Machine learning*, 3(1): 1–122, 2011.

Emmanuel J Candès and Michael B Wakin. An introduction to compressive sampling. *IEEE Signal Processing Magazine*, 25(2):21–30, 2008.

Chih-Chung Chang and Chih-Jen Lin. Libsvm – a library for support vector machines, 2018. https://www.csie.ntu.edu.tw/~cjlin/libsvm/.

Zizhong Chen and Jack J Dongarra. Condition numbers of gaussian random matrices. *SIAM Journal on Matrix Analysis and Applications*, 27(3):603–620, 2005.

Steven Davis and Paul Mermelstein. Comparison of parametric representations for monosyllabic word recognition in continuously spoken sentences. *IEEE transactions on acoustics, speech, and signal processing*, 28(4):357–366, 1980.

Jeffrey Dean, Greg Corrado, Rajat Monga, Kai Chen, Matthieu Devin, Mark Mao, Andrew Senior, Paul Tucker, Ke Yang, Quoc V Le, et al. Large scale distributed deep networks. In *Advances in neural information processing systems*, pages 1223–1231, 2012.

Richard Alan DeMillo. Foundations of secure computation. Technical report, Georgia Institute of Technology, 1978.

C. Dwork. Differential privacy. *International Colloquium on Automata, Languages, and Programming (ICALP)*, 2006.

C. Dwork, F. McSherry, K. Nissim, and A. Smith. Calibrating noise to sensitivity in private data analysis. *Conference on Theory of Cryptography*, 2006a.

Cynthia Dwork. Differential privacy. *Encyclopedia of Cryptography and Security*, pages 338–340, 2011.

Cynthia Dwork, Frank McSherry, Kobbi Nissim, and Adam Smith. Calibrating noise to sensitivity in private data analysis. In *Theory of cryptography conference*, pages 265–284. Springer, 2006b.

Biyi Fang, Jillian Co, and Mi Zhang. Deepasl: Enabling ubiquitous and non-intrusive word and sentence-level sign language translation. In *Proceedings of the 15th ACM Conference on Embedded Network Sensor Systems*, page 5. ACM, 2017.

Ran Gilad-Bachrach, Nathan Dowlin, Kim Laine, Kristin Lauter, Michael Naehrig, and John Wernsing. Cryptonets: Applying neural networks to encrypted data with high throughput and accuracy. In *International Conference on Machine Learning*, pages 201–210, 2016.

Oded Goldreich. Secure multi-party computation. *Manuscript. Preliminary version*, 78, 1998.

Ian J. Goodfellow, Jonathon Shlens, and Christian Szegedy. Explaining and harnessing adversarial examples. In *Proc. ICLR*, 2015.

Thore Graepel, Kristin Lauter, and Michael Naehrig. Ml confidential: Machine learning on encrypted data. In *International Conference on Information Security and Cryptology*, pages 1–21. Springer, 2012.

J. Hamm, A. Champion, G. Chen, M. Belkin, and D. Xuan. Crowd-ml: A privacy-preserving learning framework for a crowd of smart devices. In *IEEE International Conference on Distributed Computing Systems (ICDCS)*, 2015a.

Jihun Hamm. Crowd-ml, 2018. https://github.com/jihunhamm/Crowd-ML.

Jihun Hamm, Adam C Champion, Guoxing Chen, Mikhail Belkin, and Dong Xuan. Crowd-ml: A privacy-preserving learning framework for a crowd of smart devices. In *2015 IEEE 35th International Conference on Distributed Computing Systems*, pages 11–20. IEEE, 2015b.

Mark Hopkins, Erik Reeber, George Forman, and Jaap Suermondt. Spambase data set, 2018. https://archive.ics.uci.edu/ml/datasets/spambase.

Sergey Ioffe and Christian Szegedy. Batch normalization: Accelerating deep network training by reducing internal covariate shift. *arXiv preprint arXiv:1502.03167*, 2015.

Zohar Jackson. free-spoken-digit-dataset. https://github.com/Jakobovski/free-spoken-digit-dataset. Accessed:2019-12-10.

Zohar Jackson, César Souza, Jason Flaks, Yuxin Pan, Hereman Nicolas, and Adhish Thite. free-spoken-digit-dataset. https://zenodo.org/record/1342401#.XdlRd-gzY2w. Accessed:2019-12-10.

Linshan Jiang, Rui Tan, Xin Lou, Guosheng Lin. On lightweight privacy-preserving collaborative learning for internet-of-things objects. *In Proceedings of The 4th ACM/IEEE International Conference on Internet of Things Design and Implementation*, 2019a.

Linshan Jiang, Xin Lou, Rui Tan, Jun Zhao. Differentially private collaborative learning for the IoT edge. *In Proceedings of The 2nd International Workshop on Crowd Intelligence for Smart Cities: Technology and Applications (CICS)*, 2019b.

Kaggle. Image data set for alphabets in the american sign language. https://www.kaggle.com/grassknoted/asl-alphabet. Accessed: 2019-12-10.

Yann LeCun. The mnist database of handwritten digits. http://yann.lecun.com/exdb/mnist/. Accessed: 2019-12-10.

Yann LeCun and Yoshua Bengio. Convolutional networks for images, speech, and time series. *The handbook of brain theory and neural networks*, 3361(10): 1995, 1995.

Yann LeCun, Léon Bottou, Yoshua Bengio, and Patrick Haffner. Gradient-based learning applied to document recognition. *Proceedings of the IEEE*, 86(11):2278–2324, 1998.

Yann LeCun, Yoshua Bengio, and Geoffrey Hinton. Deep learning. *Nature*, 521 (7553):436–444, 2015.

Yann LeCun, Corinna Corts, and Christopher J.C. Burges. The mnist database of handwritten digits, 2018. http://yann.lecun.com/exdb/mnist/.

Shancang Li, Li Da Xu, and Xinheng Wang. Compressed sensing signal and data acquisition in wireless sensor networks and internet of things. *IEEE Transactions on Industrial Informatics*, 9(4):2177–2186, 2013.

Bin Liu, Yurong Jiang, Fei Sha, and Ramesh Govindan. Cloud-enabled privacy-preserving collaborative learning for mobile sensing. In *ACM Conference on Embedded Networked Sensor Systems (SenSys)*, 2012.

Kun Liu, Hillol Kargupta, and Jessica Ryan. Random projection-based multiplicative data perturbation for privacy preserving distributed data mining. *IEEE Transactions on knowledge and Data Engineering*, 18(1): 92–106, 2006.

H Brendan McMahan, Eider Moore, Daniel Ramage, Seth Hampson, and Blaise Agüera y Arcas. Communication-efficient learning of deep networks from decentralized data. In *The 20th International Conference on Artificial Intelligence and Statistics (AISTATS)*, 2017.

Frank McSherry and Kunal Talwar. Mechanism design via differential privacy. In *FOCS*, volume 7, pages 94–103, 2007.

Christopher C Paige and Michael A Saunders. Lsqr: An algorithm for sparse linear equations and sparse least squares. *ACM Transactions on Mathematical Software*, 8(1):43–71, 1982.

Yinian Qi and Mikhail J Atallah. Efficient privacy-preserving k-nearest neighbor search. In *2008 The 28th International Conference on Distributed Computing Systems*, pages 311–319. IEEE, 2008.

Yaron Rachlin and Dror Baron. The secrecy of compressed sensing measurements. In *Proc. Allerton*, pages 813–817. IEEE, 2008.

Aaron Roth and Tim Roughgarden. Interactive privacy via the median mechanism. In *Proceedings of the forty-second ACM symposium on Theory of computing*, pages 765–774. ACM, 2010.

Tara N Sainath, Abdel-rahman Mohamed, Brian Kingsbury, and Bhuvana Ramabhadran. Deep convolutional neural networks for lvcsr. In *2013 IEEE international conference on acoustics, speech and signal processing*, pages 8614–8618. IEEE, 2013.

Yiran Shen, Chengwen Luo, Dan Yin, Hongkai Wen, Rus Daniela, and Wen Hu. Privacy-preserving sparse representation classification in cloud-enabled mobile applications. *Computer Networks*, 133:59–72, 2018.

Reza Shokri and Vitaly Shmatikov. Privacy-preserving deep learning. In *Proceedings of the 22nd ACM SIGSAC conference on computer and communications security*, pages 1310–1321. ACM, 2015.

Nitish Srivastava, Geoffrey Hinton, Alex Krizhevsky, Ilya Sutskever, and Ruslan Salakhutdinov. Dropout: a simple way to prevent neural networks from overfitting. *The journal of machine learning research*, 15(1):1929–1958, 2014.

Rui Tan, Sheng-Yuan Chiu, Hoang Hai Nguyen, David KY Yau, and Deokwoo Jung. A joint data compression and encryption approach for wireless energy auditing networks. *ACM Trans. Sensor Networks*, 13(2):9, 2017.

Cong Wang, Bingsheng Zhang, Kui Ren, and Janet M Roveda. Privacy-assured outsourcing of image reconstruction service in cloud. *IEEE Transactions on Emerging Topics in Computing*, 1(1):166–177, 2013.

Dixing Xu, Mengyao Zheng, Linshan Jiang, Chaojie Gu, Rui Tan, Peng Cheng. Lightweight and unobtrusive data obfuscation at IoT edge for remote inference. *IEEE Internet of Things* Journal. Vol. 7, No. 10, October, 2020.

Wanli Xue, Chenwen Luo, Guohao Lan, Rajib Rana, Wen Hu, and Aruna Seneviratne. Kryptein: a compressive-sensing-based encryption scheme for the internet of things. In *Proc. IPSN*, pages 169–180. IEEE, 2017.

Qiang Yang, Yang Liu, Tianjian Chen, and Yongxin Tong. Federated machine learning: Concept and applications. *ACM Transactions on Intelligent Systems and Technology (TIST)*, 10(2):1–19, 2019.

Matthew D Zeiler. Adadelta: an adaptive learning rate method. *arXiv preprint arXiv:1212.5701*, 2012.

Justin Zhijun Zhan, LiWu Chang, and Stan Matwin. Privacy preserving k-nearest neighbor classification. *IJ Network Security*, 1(1):46–51, 2005.

Stephan Zheng, Yang Song, Thomas Leung, and Ian Goodfellow. Improving the robustness of deep neural networks via stability training. In *Proc. CVPR*, pages 4480–4488. IEEE, 2016.

Martin Zinkevich, Markus Weimer, Lihong Li, and Alex J Smola. Parallelized stochastic gradient descent. In *Advances in neural information processing systems*, pages 2595–2603, 2010.

6

Clock Synchronization for Wide-area Applications[1]

6.1 Introduction

IoT devices have been widely deployed for sensing, computing, communication, and actuation capabilities. Data timestamping and clock synchronization are two basic system services for constructing IoT applications. Accurate timestamps are crucial for interpreting data and associating data from different sensors; synchronized clocks enable punctual and well coordinated operations among the nodes. On the other hand, wrong timestamps and malfunctioning clock desynchronization may lead to system chaos and failures.

Atomic clocks, GPS, clock synchronization, and calibration protocols represent principal means to achieve data timestamping and clock synchronization. For massive deployments, chip-scale atomic clocks are still too expensive (e.g., $1500 per unit BusinessWire (2017)). Although GPS receivers can provide global time with μs accuracy, they generally do not work in indoor environments. Existing clock synchronization and calibration approaches can be classified broadly into two categories. The first category consists of clock synchronization protocols (e.g., NTP Foundation (2017a), PTP IEEE (2008), RBS Elson et al. (2002), TPSN Ganeriwal et al. (2003), and FTSP Maróti et al. (2004)) that exchange network or radio messages among nodes to estimate and remove their clock offsets. As these approaches generally involve measuring the delays in transmitting synchronization packets, network outage and dynamic network conditions may degrade their performance significantly Mizrahi (2014). The second category uses time broadcasts from timekeeping radios Chen et al. (2011) and external periodic signals (e.g., periodic data blocks in FM radios Li et al. (2011), wi-fi beacons Hao et al. (2011), and powerline electromagnetic radiation Rowe et al. (2009)) for clock synchronization and calibration. To capture and/or decode the external signals, these approaches need sophisticated hardware peripherals, which hinder their adoption due to the increased node cost and complexity.

Increased heterogeneity and limited resources in both hardware and software platforms are the key characteristics of IoT. For example, unlike the current internet hosts that use a limited number of de facto standard network interfaces (ethernet, wi-fi, and cellular networks), IoT objects additionally use a spectrum of network interfaces (including various 2.4 GHz and sub-GHz technologies). The diverse network interfaces with distinct

---

[1]This chapter is based on the authors' journal papers Li et al. (2018), Yan et al. (2018) and their earlier conference versions Li et al. (2017), Yan et al. (2017a).

Intelligent IoT for the Digital World: Incorporating 5G Communications and Fog/Edge Computing Technologies,
First Edition. Yang Yang, Xu Chen, Rui Tan, and Yong Xiao.
© 2021 John Wiley & Sons Ltd. Published 2021 by John Wiley & Sons Ltd.

physical-layer and link-layer properties introduce challenges for implementing accurate network packet timestamping and clock synchronization. In particular, the existing approaches do not achieve a good trade-off between synchronization accuracy and adaptation to hardware heterogeneity. Although the NTP does not require low-level hardware access to the radio and thus adapts to diverse hardware platforms, it provides sub-second synchronization accuracy only. On the other hand, although PTP and FTSP provides millisecond or even microsecond synchronization accuracy, they require low-level hardware access to the radio and thus lose universality in IoT with diverse hardware platforms.

To address the above challenges, we present a natural timestamping approach and a clock synchronization approach based on a pervasive periodic signal that can be sensed in indoor environments, i.e. electromagnetic radiation from service powerlines. Specifically, we find that the minute fluctuations of the frequency of the powerline electromagnetic radiation contain global time information. Thus, the natural timestamping approach records the powerline electromagnetic radiation trace and uses it as a natural timestamp. In addition, by integrating the periodic signal into the principle of NTP, we avoid requiring low-level access to the radio and achieve the synchrony provided by the power grid frequency. We also exploit the coupling between human body and the electric field generated by the powerlines to apply our approach to wearable IoT devices. Extensive experiments show that our natural timestamping approach for wireless IoT devices achieves sub-second accuracy and our clock synchronization approach for wearable IoT devices achieves millisecond accuracy.

6.2 System Model and Problem Formulation

In this section, we describe the system model and the problems of natural timestamping and clock synchronization in Sections 6.2.1 and 6.2.2, respectively.

6.2.1 Natural Timestamping for Wireless IoT Devices

For collaborative IoT systems, having a Common Notion of Time (CNoT) among nodes is a fundamental and critical requirement. Resilient CNoT against faults such as hardware faults and network outage is important for diverse wireless sensor network (WSN) and IoT systems.

We study *natural timestamping*, a low-cost and novel approach to resilient CNoT. In general, a natural timestamp is a finite-length sampled signal from the environment that yields a unique form during any given sampling time. In other words, the signal form "encodes" when it is sampled. A prior work Viswanathan et al. (2016) shows that the electric network frequency (ENF) of an alternating current (ac) power grid contains natural timestamps that encode time information at sub-millisecond accuracy. Specifically, as the ENF fluctuates around its nominal value (50 Hz or 60 Hz) Wood and Wollenberg (2012) and the fluctuation is almost identical across the whole power grid, an ENF trace collected by a node, say A, can be matched against a longer ENF trace collected by another node, say B. The best matching position indicates a reliable common time when A and B obtained their respective ENF samples. Hence, CNoT is established between A and B.

As power grids pervade almost all civil infrastructures in an urban geographic area, ENF natural timestamping is promising for establishing reliable time for WSN and IoT in these environments. Moreover, based on the natural timestamps, various time-critical services and protocols with much improved resilience can be developed. For instance, the ENF natural timestamps recorded by a node during an extended network outage can be decoded afterwards to recover time information of any critical sensor data and system logs. ENF natural timestamps can also be used for clock synchronization, as shown in Viswanathan et al. (2016). Specifically, by exchanging ENF natural timestamps that are associated with the clocks of A and B, the offset between the two clocks can be identified through natural timestamp matching Viswanathan et al. (2016). Since this approach does not depend on measuring packet transmission delays, it is immune to dynamic delays due to varying network conditions.

However, prior work Viswanathan et al. (2016) collected ENF traces from wall power outlets directly, which limits its applicability to nodes that are plugged into the power grid only. To advance the state of the art, in this section, we systematically investigate the existence and time accuracy of the natural timestamps in wireless electromagnetic radiation (EMR) emitted from ac powerlines. The basis is that, since EMR is mainly excited by powerline ac voltage that oscillates at the same rate as the ENF, it should also carry the ENF natural timestamps. In addition, to foster adoption at low cost, we investigate simple EMR reception devices that can nevertheless capture the ENF natural timestamps effectively. Our results reveal that significant time information is carried by wireless EMR. This understanding extends the applicability of the ENF natural timestamping to nodes that are not connected directly to the grid. These nodes represent a major sector of sensor and IoT deployments.

Unfortunately, compared with strong sinusoidal ac voltage signals sensed directly at wall power outlets, the EMR signal is weak and noisy. First, as EMR has an extremely long wavelength, its signal reception at small sensors is weak. Second, although EMR is mainly excited by the ac voltage, it is nevertheless also affected by changing electric currents in powerlines and appliances. Thus, changes in the appliances' states and operations can create continuous and significant random noises and transient disturbances to the EMR. Third, the EMR can be similarly disturbed by other environment factors such as human movement. These characteristics pose significant challenges to capturing minute ENF fluctuations of less than 0.1 Hz that constitute the natural timestamps.

6.2.2 Clock Synchronization for Wearable IoT Devices

The annual worldwide shipment of consumer wearables (e.g., smart watches, wristbands, eyewear, and clothing) grew by 17% in 2017 Gartner (2017). This rapid growth is expected to continue, projected to be 504 million units shipped in 2021 Gartner (2017). Along with the proliferation of consumer wearables, specialized domains such as clinical/home healthcare Chan et al. (2012) and exercise/sport analysis Mokaya et al. (2016) are increasingly adopting smart wearable apparatus. In the body-area networks formed by these wearables, a variety of system functions and applications depend on tight clock synchronization among the nodes. For instance, motion analysis Lorincz et al. (2009) and muscle activity monitoring

Mokaya et al. (2015, 2016) require sensory data from multiple tightly synchronized nodes.

While current wearable systems adopt customized, proprietary clock synchronization approaches Apple Inc. (2015), we envisage a wide spectrum of interoperable wearables that can synchronize with each other to enable more novel applications. In the envisaged scheme, an application developer can readily synchronize any two communicating wearables using high-level and standard system calls provided by their operating systems (OSs), such as reading system clocks, and transmitting and receiving network messages. However, the design of clock synchronization approaches faces a fundamental trade-off between the synchronization accuracy and the universality for heterogeneous platforms. This is because a high synchronization accuracy generally requires low-level timestamping for the synchronization packets, which may be unavailable on the hardware platforms or inaccessible to the application developer.

We illustrate this accuracy-universality trade-off using the Network Time Protocol (NTP) Foundation (2017a) and the Precision Time Protocol (PTP) IEEE (2008). NTP synchronizes a slave node and a master node by recording their clock values when a UDP synchronization packet is passed to and received from the sender's and receiver's OSs, respectively. Thus, NTP is universal in that it can be deployed on any host through UDP. However, as the application-layer timestamping cannot capture the details of the non-deterministic OS and network delays, NTP may yield significant synchronization errors up to hundreds of milliseconds (ms) in a highly asymmetric network. To solve this issue, PTP uses hardware-level timestamping provided by PTP-compatible ethernet cards at the end hosts and all the switches on the network path to achieve microsecond (μs) accuracy. However, the need for special hardware inevitably negates its universality and restricts PTP's adoption to time-critical local-area networks only, e.g., those found in industrial systems.

In wireless networks, due to the more uncertain communication delays caused by media access control (MAC), NTP performs worse. Thus, similar to PTP, most existing clock synchronization approaches for wireless sensor networks (WSNs) (e.g., reference-broadcast synchronization (RBS) Elson et al. (2002), a timing-sync protocol for sensor networks (TPSN) Ganeriwal et al. (2003), and a flooding time synchronization protocol (FTSP) Maróti et al. (2004)) have resorted to MAC-layer timestamping provided by the nodes' radio chips to pursue synchronization accuracy. A study Casas et al. (2005) uses MAC-layer access of Bluetooth called the host controller interface (HCI) to synchronize Bluetooth devices. The classic Bluetooth devices can use the Bluetooth clock synchronization protocol (CSP) included in the Bluetooth health device profile (HDP). However, Bluetooth low energy (BLE) does not support HDP and CSP. The need for MAC-layer timestamping or specific radio links presents a barrier for the wide adoption of these approaches to the broader IoT domain, where the IoT platforms use diverse radios and OSs, and in general they do not provide an interface for the MAC-layer timestamping.

A new clock synchronization approach for wearables is desirable. The approach can establish an accuracy–universality trade-off point between the two extremes represented by NTP and PTP to well address the momentum of IoT platform heterogenization. In particular, the approach needs to comply with the sensor nature of wearables to explore ambient signals that can assist clock synchronization.

6.3 Natural Timestamps in Powerline Electromagnetic Radiation

In this section, we introduce *natural timestamping* to address the challenges described in Section 6.2.1. Section 6.3.1 introduces the background knowledge about ENF fluctuations and powerline EMR. Section 6.3.2 presents key hardware and algorithm designs for EMR natural timestamping. Section 6.3.3 presents a low-cost implementation of the proposed technique, and discusses benchmark results that guide the settings of several important parameters of the natural timestamping algorithm. Section 6.3.4 presents the results of extensive experiments conducted at five sites of office and residential environments. Section 6.3.5 presents the results of experiments in factory environment. Section 6.3.6 presents the applications of EMR natural timestamping.

6.3.1 Electrical Network Frequency Fluctuations and Powerline Electromagnetic Radiation

In an ac power grid, the ENF is regulated at a nominal value (50 or 60 Hz) by a grid-wide centralized control system Wood and Wollenberg (2012). ENF has the following two properties from power engineering Wood and Wollenberg (2012). First, at any given time instant, the ENF is almost identical across all locations within a geographical region (e.g., building, factory, or city). Second, as the load of a power grid is a random process, the ENF regulated by the centralized control system is also a random process. We note that distributed generation is increasingly connected to the power grid. The ac electricity locally generated must be synchronized precisely both in frequency and phase with the power grid. Thus, the local power injection back to the power grid will not affect the above two properties. Figure 6.1 shows the per-second ENF measurements during one hour collected by two high-precision voltage sensors connected to the wall power outlets at two locations 12 km apart in a city. We can see that (i) the ENF fluctuates by less than 0.1 Hz, and (ii) the fluctuations are identical at the two locations. Thus, it is possible to match a short segment of ENF trace collected at a location against a long ENF trace collected at the other location. The best match point indicates a common time instant when the two nodes obtain their respective ENF samples.

Ac powerlines and electrical appliances can emit EMR that oscillates at the same rate as the ENF. Figure 6.2(a) shows the ac voltage signal collected by a high-precision voltage

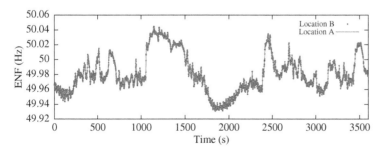

Figure 6.1 Per-second ENF measured at two locations 12 km apart in a city. Mean error: 0.1 mHz; RMSE: 0.2 mHz.

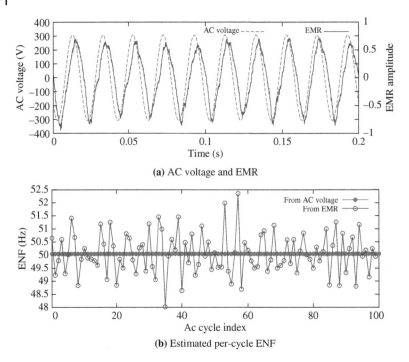

(a) AC voltage and EMR

(b) Estimated per-cycle ENF

Figure 6.2 Ac voltage and EMR data.

sensor from a wall power outlet and the EMR signal recorded by another sensor. The two sensors are located in the same building. We can see that the EMR signal is not an ideal sinusoid. Moreover, the EMR signal has a phase lag compared with the ac voltage signal collected from the power grid directly. Figure 6.2(b) shows the ENF traces estimated from the cycle lengths of the ac voltage and the EMR signal, respectively. The EMR-based ENF fluctuates around the actual ENF by up to 2 Hz, significantly exceeding the peak fluctuation of the actual ENF on the order of 0.1 Hz, as shown in Figure 6.1. The key question we ask in this section is whether the noisy EMR data contains ENF natural timestamps at sufficient time accuracy.

6.3.2 Electromagnetic Radiation-based Natural Timestamping

This subsection presents the hardware and algorithms for validating the EMR natural timestamping.

6.3.2.1 Hardware

We use three types of natural timestamping hardware devices in this subsection, i.e. Raspberry Pi (RPi) Foundation (2017b) based EMR sensors, Zolertia Z1 motes Zolertia (2017), and an ENF database node. The RPi-based node has a dedicated EMR sensor and samples the EMR signal at up to 44.1 kHz. The Z1 mote uses a normal conductor wire as the EMR antenna and samples at sub-kiloHertz such that the EMR data can be processed in real time given the Z1's limited processing capability. Thus, the RPi-based sensor and Z1 mote

Figure 6.3 Illustration of an EMR natural timestamping system.

represent high-end and low-end sensors, respectively. The results obtained on the two plat-forms offer important insights into how hardware and processing capability may impact the effectiveness of natural timestamping. The ENF database node, synchronized to global time, uses a high-precision ENF sensor to record ground-truth ENF data directly from a wall power outlet. As we will discuss later (Section 6.3.2.3), the ground-truth ENF data is used to decode the EMR natural timestamps captured by the RPi-based sensors and Z1 motes into global time. Figure 6.3 illustrates a simple EMR natural timestamping system, in which an EMR sensor transmits a natural timestamp to the database node, and the database node decodes it and sends back the global time at which the natural timestamp was captured. The database node can serve multiple EMR sensors.

RPi-based EMR sensor

Figure 6.4(a) shows the schematics of the proposed EMR sensor. It shares a similar design as the one in Rowe et al. (2009). Specifically, it includes an Analog Front-end (AFE) and an LC antenna consisting of a 470 mH coil inductor and a 22 μF capacitor. The inductor generates currents in the EMR's changing magnetic field. Through impedance matching by the capacitor, the LC circuit's theoretic resonant frequency is 49.49 Hz, which is very close to the nominal ENF of 50 Hz in our region. Thus, we expect good EMR reception. We also tested other combinations of candidate inductors and capacitors that produce the-oretic resonant frequencies close to 50 Hz. However, compared with the 470 mH and 22 μF combination, the other combinations gave inferior EMR reception, potentially due to their larger resistance which leads to actual resonant frequencies different from 50 Hz.

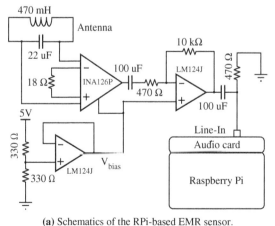

(a) Schematics of the RPi-based EMR sensor.

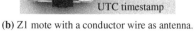

(b) Z1 mote with a conductor wire as antenna.

Figure 6.4 EMR capture devices.

The output of the LC circuit is processed by the AFE to adapt to a sampling board. Specifically, the signal is amplified by a two-stage amplification circuit. The first stage is by an instrumentation amplifier INA126P that gives an open-loop gain of 4450. The second stage, based on an operational amplifier LM124J, gives a gain of about 21. Two high-pass filters are applied after the two amplifiers respectively to remove the direct current (dc) component. The output of the AFE swings within [−1 V, 1 V].

We use a Cirrus Logic audio card mounted on the RPi to sample the output of the AFE. This audio card provides a 24-bit resolution and up to a 192 kHz sampling rate, both of which are much higher than that of other off-the-shelf Analog-to-digital Converter (ADC) boards for RPi. The output of the AFE is wired to the line-in port of the audio card. The RPi runs Raspbian OS. In our experiments, we configure the audio card to operate at 16-bit resolution, and 44.1 kHz or 8 kHz sampling rate. As the audio card sampling introduces a 4% CPU utilization only, we do not observe significant jitters in the sampling. The RPi uses NTP Foundation (2017a) to synchronize its clock with a local stratum-1 time server equipped with a GPS receiver. From NTP's status report, the RPi's synchronization error is around one millisecond. With the synchronized clock, the RPi adds Coordinated Universal Time (UTC) timestamps to the EMR measurements. The UTC timestamps are used as ground truth to evaluate the error of the time information derived from the ENF fluctuations. The whole setup is powered by a high-capacity dc power bank to make sure that the node senses EMR only, rather than ac voltage leaked from an ac/dc power supply unit.

Z1 mote

The Z1 mote Zolertia (2017) is equipped with an MSP430 microcontroller (MCU). It has two Phidgets sensor ports connected to several ADC pins of the MCU. The Z1 mote runs TinyOS 2.1.2. As shown in Figure 6.4(b), we connect a normal conductor wire to a pin in a Phidgets port and use the MCU's ADC to sample the electric potential generated by the EMR on the wire. Such a configuration minimizes the use of extra hardware. In addition, the power consumption of capturing EMR is merely the MCU's power in sampling and processing the signal. In applications of EMR natural timestamping, the EMR data is typically collected for a short duration infrequently or on an on-demand basis. These applications will introduce little power consumption. The simplicity of the EMR reception device and the low power footprint can foster the adoption of EMR natural timestamping in WSN and IoT systems.

To facilitate evaluation, we connect the Z1 mote to an RPi via a USB cable for power supply and data storage. The RPi is powered by a dc power bank and synchronized with the aforementioned local stratum-1 time server. Once an ENF measurement is available, the Z1 mote sends it over the USB cable to the RPi for ground-truth UTC timestamping. Note that in real applications, the RPi is not needed.

ENF database node

The ENF database node consists of an ENF capture device connected to a wall power outlet and an RPi equipped with an Adafruit GPS receiver for UTC synchronization. The ENF capture device uses an ac/ac transformer to step down the 220 V ac voltage and a signal processing circuit to condition the signal and generate interrupts upon zero crossings of the ac voltage to an MCU. The MCU uses a hardware timer running at 8.4 MHz to measure the time periods between consecutive interrupts, i.e. the cycle lengths of the ac voltage.

The MCU streams the measurements to the RPi for UTC timestamping. The timestamped ac cycle length measurements are further streamed to a server for long-term storage. The ac cycle length measurements can be easily translated to ENF measurements. Our ENF capture device provides a resolution of 0.3 mHz for measuring the ENF.

From our measurements using multiple database nodes distributed in our city, the recorded ENFs across the city are almost identical with a mean error of 0.1 mHz. Thus, for our experiments conducted in the city, a single ENF database node suffices. We leave the problem of how to deploy ENF database nodes for an IoT system distributed in a large-scale (e.g., state-scale) power grid to future work.

6.3.2.2 ENF Extraction

This section presents the algorithm for extracting the ENF trace from the EMR measurements. This algorithm will be used by the RPi-based EMR sensor and the Z1 mote. We adopt a time-domain approach to detect the zero crossings of the near-sinusoidal EMR signal. From our result, this approach can achieve a precision of a few mHz in estimating an ENF value based on EMR measurements over one second. Alternatively, a frequency-domain approach can apply the Fast Fourier Transform (FFT) to the EMR measurements and then find the principal frequency as the ENF. However, as the FFT needs x s of data to generate an ENF estimate with a resolution of $1/x$ Hz, it will need more than 16 min of EMR data to achieve a 1 mHz resolution. Thus, the FFT approach cannot track the ENF fluctuations at the sub-second time resolutions that are critical for natural timestamping.

With the detected zero crossings, we use a sliding window approach to compute the ENF trace. Specifically, for each window consisting of K ac cycles, an ENF measurement is computed as $f = \frac{K}{\sum_{i=1}^{K} t_i}$, where t_i is the time duration of the ith ac cycle that is timed by the RPi or Z1 local clock. The window slides by one ac cycle each time. A larger setting of K reduces the noise of f due to the averaging, but also reduces the sensitivity to ENF changes. In Section 6.3.3, we will evaluate its setting to achieve the best overall accuracy of the natural timestamping. Robust detection of zero crossings is a key to the success of the time-domain approach for extracting ENF. However, this task is non-trivial, due to primarily considerable noises in the EMR signal. In the rest of this section, we firstly discuss the challenges of this task and then propose a band-pass filtering approach as solution.

Challenges of zero crossing detection

The EMR signal often contains both high- and low-frequency noises. A hardware band-pass filter with a steep passband boundary is often complicated. Moreover, the Z1 mote has no circuits to condition the EMR signal. Thus, it is important to remove the noises in software.

We use real signals captured by an RPi-based EMR sensor to illustrate the challenges. The solid curve in Figure 6.5(a) shows the EMR signal obtained by the EMR sensor shown in Figure 6.4(a). We can see that the near-sinusoidal signal has many high-frequency spurs that can lead to false zero crossing detection. For instance, Figure 6.5(b) shows a zoomed-in view of the time duration within the two vertical lines in Figure 6.5(a). The signal staggers around zero, leading to repeated zero crossing detection if no special signal processing step is used to deal with this issue. A *dead-zone* approach could suppress the repeated detection. Specifically, it will not report successive zero crossings within a dead zone of several milliseconds after a detection. However, this approach will lead to missed detection

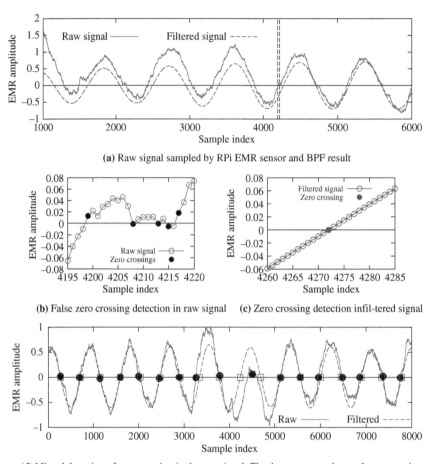

(a) Raw signal sampled by RPi EMR sensor and BPF result

(b) False zero crossing detection in raw signal   **(c)** Zero crossing detection in filtered signal

(d) Missed detection of zero crossing in the raw signal; The dots represent detected zero crossings

Figure 6.5 Challenges of zero crossing detection in a noisy EMR signal, and effectiveness of BPF.

in the presence of low-frequency noises, such as those caused by nearby human movement. In Figure 6.5(d), from the 3000th to the 6000th samples, the EMR sensor experiences a transient low-frequency disturbance and the signal barely crosses the zero line. Because of the dead zone, the detector misses a zero crossing.

Figs. 6.6(a) and 6.6(b) show the results of the dead-zone approach. When the dead zone is set to be 3 ms, the approach generates repeated zero crossings due to high-frequency noises. Thus, the estimated ENF can be a few Hz higher than the ground-truth ENF. When the dead zone is set to be 5 ms, although the approach generates fewer repeated zero crossings, it will miss more actual zero crossings. As a result, the approach yields many wrong ENF measurements that are several Hz lower than the ground truth. The above results illustrate that the dead-zone approach cannot effectively eliminate false positives and negatives simultaneously. Although the dead zone can be fine tuned and more heuristic rules can be added to reduce false positives/negatives, the performance of the tuning and piecemeal heuristics tend to be data dependent. A sharp approach that directly addresses the root cause of the high-/low-frequency noises is desirable.

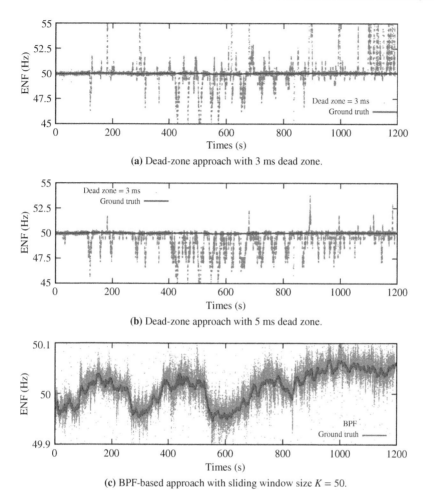

(a) Dead-zone approach with 3 ms dead zone.

(b) Dead-zone approach with 5 ms dead zone.

(c) BPF-based approach with sliding window size $K = 50$.

Figure 6.6 Comparison between the dead-zone approach and our BPF-based approach.

In Rowe et al. (2009), a phase-locked loop (PLL) is used to deal with the zero crossing jitters and misses due to low-frequency disturbances. However, the objective of the PLL is to generate an extremely stable periodic signal driven by EMR to calibrate the mote's clock, which is different from our objective of tracking ENF fluctuations at fine time resolutions. Although the parameters of the PLL can be tuned to reduce its convergence time and improve tracking performance, the stability of the PLL will be worsened due to its closed-loop nature.

Band-pass filtering

The results in Section 6.3.2.2 clearly suggest the necessity of conditioning the EMR signal to remove the high-/low-frequency noise simultaneously. To this end, we apply a fifth-order/six-tap Infinite Impulse Response (IIR) Band-pass Filter (BPF) that gives steep boundaries of the passband. We use Scipy's IIR design tool to choose the filter parameters to give a passband of [45 Hz, 55 Hz]. Then, we implement the filter in C and nesC for the RPi-based EMR sensor and the Z1 mote, respectively. As shown in Figures 6.5(a) and 6.5(c), the filtered signals represented by the dashed curves are smooth and have clear zero crossings. In Figure 6.5(d), during the transient time period of the low-frequency distur-bance, the EMR signal is restored by the BPF. Figure 6.6(c) shows the ENF measurements by our BPF-based approach. We can see that our approach can well track the ground-truth ENF measured by the database node. The mean error is 0.7 mHz and the Root Mean Square Error (RMSE) is 44 mHz.

The above results show the effectiveness of BPF in achieving robust zero crossing detec-tion. However, the implementation of the IIR filter on resource-constrained platforms such as Z1 mote is difficult, due to its compute complexity. The implementation challenges will be discussed in Section 6.3.3.2.

6.3.2.3 Natural Timestamp and Decoding

From Figure 6.6(c), the EMR sensors can track the fluctuations of ENF. This implies that it is possible to *time-align* an ENF trace captured by an EMR sensor against an ENF trace cap-tured by the database node. In other words, the database node can "decode" the time during which the ENF trace was captured by the EMR sensor. Thus, we call a finite-length ENF trace captured by an EMR sensor a *natural timestamp*. The process of identifying the UTC time when the natural timestamp was captured is called *decoding*. This section presents the decoding algorithm. Extensive evaluation of the accuracy will be presented in Section 6.3.3 and Section 6.3.4.

A natural timestamp, denoted by $\mathbf{f}$, is a vector of n consecutive ENF measurements by an EMR node. Let a vector $\mathbf{g}$ denote a trace of m consecutive ground-truth ENF measurements by the database node. The measurements in $\mathbf{g}$ are timestamped with UTC. Moreover, the following condition needs to be satisfied for natural timestamp decoding:

Definition 6.1 *(Decoding condition C1)* The time duration of measuring $\mathbf{f}$ is within the time duration of measuring $\mathbf{g}$.

The above condition can be verified using an approach presented in Section 6.3.6.2. The decoding is to match $\mathbf{f}$ with a ground-truth natural timestamp within a window of size n in $\mathbf{g}$ using a dissimilarity metric, e.g., RMSE. Specifically, by sliding the window within $\mathbf{g}$, the UTC timestamp of the first element of the window that yields the minimum dissimilarity is identified as the UTC time for the first element of $\mathbf{f}$. Formally, the index of the window that yields the minimum dissimilarity is given by

$$i^* = \underset{i \in [1, m-n+1]}{\arg \min} \; \text{dissimilarity}(\mathbf{f}, \mathbf{g}[i : i + n - 1]), \tag{6.1}$$

where $\mathbf{g}[i : i + n - 1]$ represents a vector consisting of the ith to $(i + n - 1)$th elements of $\mathbf{g}$. Then, the decoding algorithm outputs the UTC timestamp of $\mathbf{g}[i^*]$ for $\mathbf{f}[0]$. Thus, the time when $\mathbf{f}$ is captured is known.

As discussed in Section 6.3.2.1, both the RPi-based EMR sensor and the Z1 mote add UTC timestamps to the captured EMR measurements for evaluation purpose. For a natural timestamp, the difference between the decoded UTC time and the UTC time recorded by the sensor is the *decoding error*. To understand the overall accuracy, we decode many natural timestamps from a long ENF trace captured by an EMR sensor and present various statistics of the decoding errors.

6.3.3 Implementation and Benchmark

This section presents a set of benchmark experiments to guide the settings of several important parameters of the natural timestamping. Extensive performance evaluation of the natural timestamping will be presented in Section 6.3.4.

6.3.3.1 Timestamping Parameter Settings

Our natural timestamping approach presented in Section 6.3.2.2 has three key parameters: the EMR signal sampling rate, the window length K for estimating an ENF value, and the natural timestamp length n. In this section, we extensively evaluate the impact of their settings on the natural timestamping performance. We use the RPi-based EMR sensor to collect a one-hour EMR data trace at a sampling rate of 44.1 kHz. The natural timestamps generated from this data trace are decoded using a one-hour ground-truth ENF trace collected by the database node. The RPi-based EMR sensor is deployed in an office pantry with several electrical appliances; the ENF database node is deployed in a workspace in the same office. The decoding results under different parameter settings are presented below.

First, we jointly evaluate the settings of K and n. Figure 6.7 shows the decoding error versus the natural timestamp length n under different settings of K. To facilitate comparison, we translate the natural timestamp length n to the time duration of the natural timestamp and use it as the X-axis of Figure 6.7. Under the setting $K = 50$ (i.e. each ENF value is computed from the EMR measurements in one second), the average decoding errors are generally lower than or comparable to those under other K settings. Thus, in the rest of this section, we adopt $K = 50$ unless otherwise specified. Under this setting, when the natural timestamp length is 300 s, the average decoding error is 37 ms only.

Second, we jointly evaluate the settings of sampling rate and natural timestamp length. To simulate different sampling rates, we downsample the original EMR trace to 7.4, 3.7, and 0.7 kHz. Note that 0.7 kHz and 3.7 kHz are the limits for the Z1 to perform real-time and offline ENF extraction, respectively (cf. Section 6.3.3.2). Figure 6.8 shows box plots of the decoding errors versus the timestamp length under various sampling rates. The sampling rates of 7.4 kHz and 3.7 kHz give similar decoding errors. Compared with the results in Figure 6.7, the decoding errors in Figure 6.8 are generally higher due primarily to the reduced sampling rates, which is consistent with intuition. Nevertheless, with a sampling rate of 3.7 kHz and a timestamp length of 300 s, the average decoding error is 44 ms, which

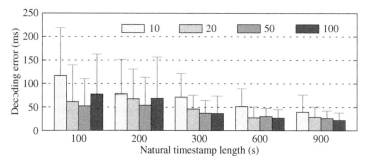

Figure 6.7 Mean decoding error versus natural timestamp length under different K. Error bar represents standard deviation.

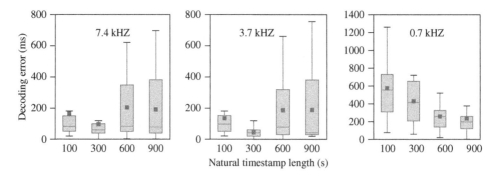

Figure 6.8 Impact of sampling rate and timestamp length on the RPi-based EMR sensor. Gray line represents median; square dot represents mean; box represents the (25%, 75%) range; upper/lower bar represents maximum/minimum.

is similar to the result in Figure 6.7. If the sampling rate is 0.7 kHz, the average decoding error is below 500 ms when the timestamp length is larger than 300 s. Thus, we can see that the node's processing capability may limit the performance of natural timestamping due to limited achievable sampling rate.

In the first two plots of Figure 6.8, when the natural timestamp length is larger than 300 s, the decoding error increases with the natural timestamp length. In the third plot of Figure 6.8, the decoding error decreases with the natural timestamp length. This is because the EMR sampling rate and timestamp length jointly affect the decoding errors. When the EMR sampling rate is high, the sensor can capture many random noises in EMR. Thus, a longer timestamp will likely include more noise and potentially increase the decoding errors. This understanding is consistent with the observation from the first two plots, where the sampling rate is relatively high (7.4 kHz and 3.7 kHz). When the sampling rate is relatively low (e.g., 0.7 kHz in the third plot), a longer timestamp generally results in smaller decoding errors.

6.3.3.2 Z1 Implementation and Benchmark

This section presents preliminary benchmark experiments conducted on a Z1 mote deployed in an office pantry to understand its natural timestamping capability. The first set of experiments specifically evaluates the fidelity of the EMR signal captured by the Z1's simple EMR reception device. The Z1 mote transmits the EMR measurements directly to the attached RPi through the USB cable. The Z1 mote adopts the maximum sampling rate of 3.7 kHz that can be supported by the UART interface. Figure 6.9 shows a one-hour ENF trace computed from the EMR measurements captured by the mote. We can see that the ENF trace consists of discrete values, with a resolution of about 13 mHz. This is because of the limited sampling rate.[2] Nevertheless, the Z1 mote's ENF trace tracks well the ground-truth ENF trace. Figure 6.10(a) shows a box plot of the decoding errors under various natural timestamp lengths. By comparing Figure 6.10(a) and Figure 6.8, the Z1 mote generally yields larger decoding errors. This is expected, since the Z1 mote uses

---

2 With a sampling rate of 3.7 kHz, a zero-crossing detection error of one EMR sample leads to an ENF estimation error of 660 mHz. The averaging in each sliding window reduces the error to 13 mHz.

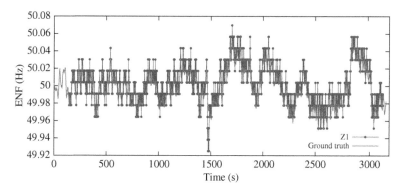

Figure 6.9 ENF trace computed based on EMR measurements captured by the Z1 mote. Sampling rate is 3.7 kHz.

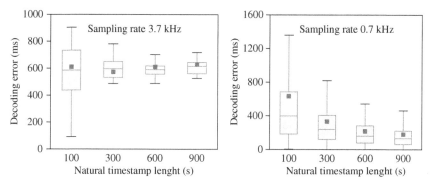

Figure 6.10 Impact of sampling rate and timestamp length on the timestamping performance of the Z1 mote.

a normal conductor wire for capturing the EMR. Nevertheless, the Z1 mote still gives sub-second average decoding errors.

The second set of experiments evaluates the impact of the Z1's limited processing capability on the natural timestamping performance. We have implemented the IIR BPF algorithm using nesC on the Z1. By extracting the ENF trace in real time, the Z1 does not need to store the massive EMR data. However, without special care in implementing the IIR filter, its high compute complexity may greatly limit the mote's ability to achieve a satisfactory sampling rate. The filter needs many floating-point multiplications that consume a significant number of MCU cycles. Our test shows that, without code optimization, the Z1 mote can sustain a sampling rate of 90 Hz only. To address this, we apply the Horner algorithm Instruments (2017), which implements floating-point multiplication with integer operations. As a result, the Z1 mote can sustain real-time ENF extraction at a sampling rate of 0.7 kHz. Figure 6.11 shows a one-hour ENF trace estimated by the Z1 mote based on the EMR signals sampled at 0.7 kHz. We can see that the ENF estimates are at discrete levels with a low resolution, which is caused by the limited sampling rate, as discussed previously. It is interesting to understand whether the low-resolution natural timestamps contain global time information of satisfactory accuracy. Figure 6.10(b) shows a box plot of the decoding errors. The decoding errors in Figure 6.10(a) and Figure 6.10(b) are comparable. For

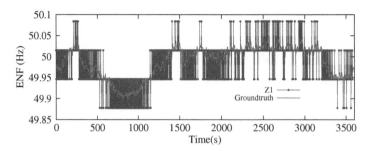

Figure 6.11 ENF trace estimated by the Z1 mote based on EMR signals sampled at 0.7 kHz.

some natural timestamp length settings (e.g., 900 s), the decoding error in Figure 6.10(b) is even lower than that in Figure 6.10(a). As the Z1 mote uses a simple conductor wire as the EMR antenna, its measurements contain larger noise. Thus, as discussed in Section 6.3.3.1, a higher sampling rate in Figure 6.10(a) may result in the inclusion of more noise and increase the decoding errors.

The above benchmark results show that the Z1 mote can achieve satisfactory natural timestamping performance.

6.3.4 Evaluation in Office and Residential Environments

This section presents an extensive set of experiments at five office and residential deployment sites conducted over long periods of time (up to one month) to validate the EMR natural timestamping.

6.3.4.1 Deployment and Settings

We chose five sites, as summarized in Table 6.1, to evaluate the natural timestamping performance. For each site, we deploy a RPi-based EMR sensor and a Z1 mote. Note that we deploy a database node at Site A.0. The distances given in Table 6.1 are roughly estimated Euclidean distances between each deployment site and the database node. The sites A.1 and A.2 are two spaces in site A. At site A.1, the nodes are placed on an occupied working

Table 6.1 Various sites for performance evaluation.

Site		Description	Distance from A.0
A		Office building	
	A.0	Database node	
	A.1	Work space	9 m
	A.2	Pantry	14 m
B		Living room of Apartment I	10 km
C		Study room of Apartment II	12 km
D		Office at Campus I	19 km
E		Office at Campus II	24 km

desk with a desktop computer, a laptop, and various chargers. Site A.2 is a pantry with various appliances including a fridge and a microwave. The nodes are placed in a corner of the pantry. Site B is the living room of an apartment. The nodes are placed on a TV stand. Site C is the study room of a different apartment. The nodes are placed on a computer desk. Site D is an office on a campus. Site E is an open office space on a different campus. For evaluation purpose, the nodes at sites A.1, A.2, and E are synchronized with local stratum-1 time servers equipped with GPS receivers; at the other sites, due to difficulty of deploying GPS receivers, the nodes are synchronized with internet time servers in our city.

For the RPi-based sensor and the Z1 mote, the sampling rates are 8 kHz and 0.7 kHz, and the natural timestamp lengths are 300 s and 600 s, respectively. A one-hour ENF trace that is recorded by the database node and satisfies the decoding condition C1 is used for natural timestamp decoding.

6.3.4.2 Evaluation Results
Performance evaluation at various sites

Figure 6.12 shows the distribution of the decoding errors of the RPi-based EMR sensor at sites A.2 and B over eight days. As discussed in Section 6.3.2.2, we slide the window by an ac cycle each time to compute an ENF value. Thus, ideally, the decoding error should also be a multiple of the ac cycle. From Figure 6.12, the decoding error clusters are indeed centered at multiples of the ac cycle 20 ms. As the EMR signal is also affected by the environment, we can see that the decoding errors are also distributed around the multiples of 20 ms. At site A.2, a majority of the decoding errors are below 2 ms and all the decoding errors are below 298 ms. Site B gives a different decoding error distribution, with the highest bar at around 60 ms. This is because the deployment position at site B gives a weaker EMR signal strength.

Table 6.2 summarizes the medians and 95th percentiles of the RPi-based EMR sensor and the Z1 mote at the different deployment sites. For the RPi-based EMR sensor, the median decoding error ranges from 37 ms to 82 ms and the 95th percentile is below 357 ms. Due to the simple EMR antenna and limited sampling rate, the Z1 mote yields

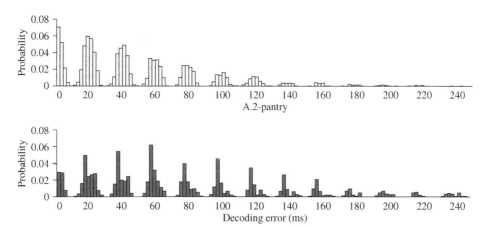

Figure 6.12 Distribution of decoding errors of RPi-based EMR sensor at sites A.2 and B. Histogram bin width is 2 ms.

Table 6.2 Decoding error statistics at various sites.

Site	RPi-based EMR sensor		Z1 mote	
	Median (ms)	95th percentile (ms)	Median (ms)	95th percentile (ms)
A.1	59	204	150	786
A.2	40	128	146	499
B	82	357	165	580
C	56	138	142	523
D	37	136	160	542
E	59	240	220	780

higher decoding errors. The median is from 142 ms to 220 ms and the 95th percentile is below 786 ms. The results show that the EMR natural timestamping works well in various environments, when the distance between an EMR sensor and the database node is up to 24 km. This set of experiments validates the applicability of the EMR natural timestamping for city-scale WSN and IoT systems.

One-month evaluation
We deploy two RPi-based EMR sensors at sites A.2 and B for one month. Figure 6.13 shows the error bars of the decoding errors in different time slots at the two deployment sites. At the two sites, the average decoding errors are around 50 ms and 150 ms, respectively. Larger decoding errors are observed when human activity and electrical appliance usage increase. The one-month deployments at the two sites show that the timestamping approach can work reliably over long periods of time.

Impact of deployment position
The EMR signal attenuates with physical distance and it can be affected by the operation of nearby electrical appliances. In this set of experiments, we deploy the RPi-based EMR sensor at different spots at site B, which is a real home environment. The triangles

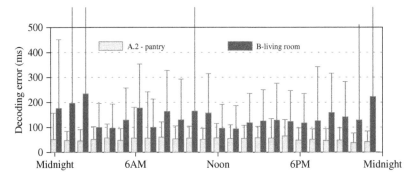

Figure 6.13 Decoding errors in one month. Each error bar is computed based on the decoding errors in the same time slot of the 30 days. The error bars represent standard deviation.

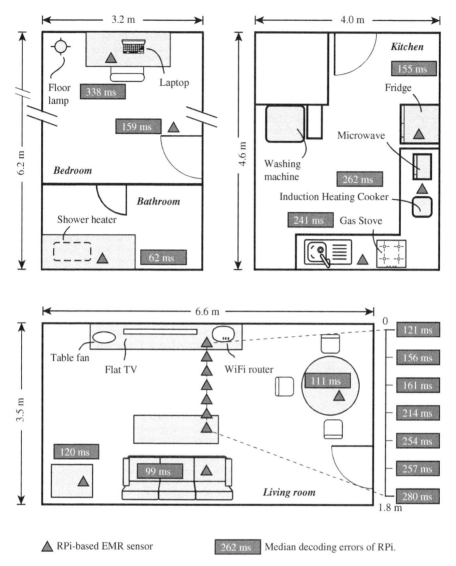

▲ RPi-based EMR sensor   262 ms   Median decoding errors of RPi.

Figure 6.14 Deployment spots at site B. An RPi-based EMR sensor is deployed at each spot for one hour.

in Figure 6.14 represent the deployment spots. At each spot, the sensor collects the EMR trace for one hour. The rectangle labels in Figure 6.14 show the median decoding errors at different deployment spots. We can see that the decoding errors vary with the deployment spots, from 62 to 338 ms. From more extensive measurements in the apartment, the RPi-based sensor can always receive a salient EMR signal for natural timestamping.

We also conduct a set of experiments to evaluate the impact of distance from powerlines on the decoding errors. As shown in Figure 6.14, we increase the distance of the EMR sensor from a TV stand in the living room up to 1.8 m. We choose the TV stand because it contains a collection of power cords that supply a TV, a table fan, and a wi-fi router. The rectangular labels at the right bottom of Figure 6.14 show the median decoding errors at different distances from the TV stand. As the EMR signal strength attenuates with distance, the zero-crossing detection will be more likely affected by environmental disturbances. Thus, from Figure 6.14, we can see that the decoding error increases with the distance. In the center of the living room, which is likely the farthest position from the powerlines in the home, the median decoding error is 280 ms.

We also evaluate the performance of the Z1 mote in the apartment. Figure 6.15 shows the median decoding error versus the strength of the EMR signal received by the Z1 mote, where the signal strength is measured by the ratio of the root mean square of the raw EMR signal to the full scale of the Z1 ADC. When the signal strength is 3% only, the median decoding error remains at about 150 ms. When the signal strength reduces to 1.2% and below, the decoding gives large errors. From these results, the wireless sensor platforms with very simple EMR reception devices, like the conductor wire used by the Z1 mote in this section, are suitable for deployments close to powerlines and industrial systems that often produce considerably strong EMR.

Storage volume constraint

As discussed in Section 6.3.2.2, we adopt a sliding window approach to compute the ENF trace. Thus, an ENF value is generated every 20 ms. As a result, a natural timestamp of 600 s will need 60 kByte storage, which may impose too much storage overhead on resource-constrained nodes. Alternatively, we can adopt a neighboring window approach to compute the ENF trace. When the window size is set to be one second, a natural timestamp of 600 s uses 1.2 kByte storage only. However, the time resolution of the decoding will increase from 20 ms to 1 s. Figure 6.16 shows the decoding errors for the natural timestamps

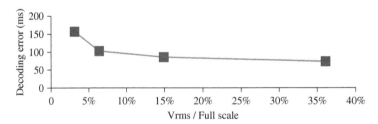

Figure 6.15 Z1 decoding error versus EMR signal strength.

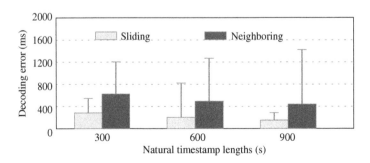

Figure 6.16 Decoding errors of Z1 under the sliding and neighboring approaches. The error bars represent standard deviation.

generated by the sliding and neighboring window approaches in a one-day experiment. The neighboring window approach yields doubled decoding errors, compared with the sliding window approach. This experiment shows the trade-off between the storage overhead and the natural timestamping accuracy.

System overhead

In various applications, the natural timestamp capture is performed by the sensor node periodically or in an on-demand fashion. For instance, in the application of time recovery that will be presented in Section 6.3.6.1, a natural timestamp is collected over five minutes every hour to maintain a node's clock drift below 200 ms. In the applications of run-time clock verification and secure clock synchronization, which will be presented in Sections 6.3.6.2 and 6.3.6.3, a natural timestamp is collected only when needed.

We measure the computation overhead in sampling and processing the EMR signal on the Z1 platform. We enable the direct memory access mode to free the MCU from the sampling process. The EMR signal is sampled block by block. Once a block is ready, it is processed by a signal processing pipeline consisting of band-pass filtering, zero crossing detection, and ENF extraction as described in Section 6.3.2.2. For each block consisting of 118 samples collected during 162.5 ms, the band-pass filtering consumes 135.8 ms and the remaining two signal processing steps consume 1.5 ms. As a result, the MCU is idle for 25.2 ms in the time period of processing a block. Thus, the utilization of the MCU in processing the EMR signal is 84.5%.

We also measure the power consumption of the EMR sampling and processing. When the Z1 is idle, its power consumption is 91.2 mW. When the Z1 is sampling and processing the EMR signal, its power consumption is 106.3 mW. Thus, a net power of 15.1 mW is consumed by the EMR sampling and processing.

6.3.5 Evaluation in a Factory Environment

To evaluate the performance of our natural timestamping approach in an industrial setting, we deploy an RPi-based EMR sensor and a Z1 mote on a printed circuit board (PCB) manufacturer's factory floor housing two production lines. The factory is about 10 km away from the database node at site A.0. The dimension of the factory floor is 30 m × 15 m. As shown in Figure 6.17, the two production lines are powered by 440 V line-to-line voltage and the electrical appliances in an operation area of the factory floor are powered by 220 V line-to-neutral voltage. The RPi-based EMR sensor and the Z1 mote are placed at two ends of the production lines, respectively. A local stratum-1 time server equipped with a GPS receiver is also set up on the factory floor to provide ground truth time to the two EMR sensors. The two EMR sensors continuously record natural timestamps for 17 h, spanning the time periods when the production lines are on and off.

Figures 6.18(a) and 6.18(b) show one-hour ENF traces collected by the RPi-based EMR sensor, the Z1 mote, and the site A.0 database node, when the production lines are on during the daytime. From Figure 6.18(a), the ENF trace sensed by the RPi-based EMR sensor tracks well the ground truth ENF sensed by the database node. From Figure 6.18(b), the ENF trace sensed by the Z1 mote is noisier. In particular, it contains several spikes that can be caused by transient EMR interference from the production lines when they change

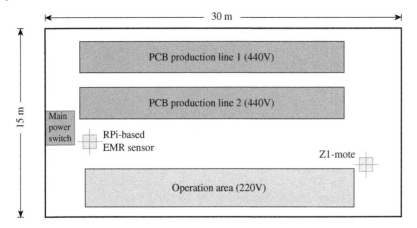

Figure 6.17 Factory floor plan and EMR sensors' locations.

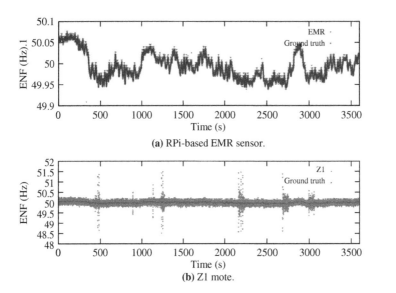

Figure 6.18 ENFs sensed by the RPi-based EMR sensor and the Z1 on the factory floor.

operation states. We note that the RPi-based EMR sensor is insensitive to the interference because its EMR reception circuit is tuned to capture the 50 Hz EMR only, while the Z1 mote can capture EMR in other frequencies.

Figure 6.19 shows the mean values and the standard deviations of the decoding errors within one-hour time slots during the 17-hour deployment. We can see that the RPi-based sensor yields smaller decoding errors than the Z1 mote. We note that the production lines are turned off at around 7 PM and turned on at around 9 AM. From the figure, the impact of the production lines on the natural timestamping accuracy is insignificant. Note that, as shown in Figure 6.18(b), the Z1 mote can be affected by transient EMR interference that lasts for seconds. Since the natural timestamp length is 300 s, the interference has little impact on the natural timestamp decoding accuracy.

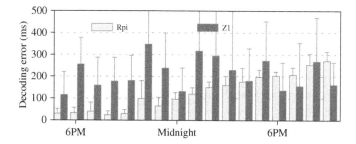

Figure 6.19 Natural timestamp decoding errors in the factory environment. The error bars represent the mean value and the standard deviation of the decoding errors in one hour.

Table 6.3 Natural timestamp decoding error statistics in the factory environment.

Site	RPi-based EMR sensor		Z1 mote	
	Median (ms)	95th percentile (ms)	Median (ms)	95th percentile (ms)
Factory	116	290	157	525

Table 6.3 shows the natural timestamp decoding error statistics during the 17-hour deployment. The RPi-based sensor achieves better natural timestamping accuracy due to its higher sampling rate and dedicated EMR antenna. The 95th percentile decoding errors of the RPi-based EMR sensor and the Z1 mote are 290 and 525 ms, respectively, which are comparable with the results obtained in the office and residential environments (cf. Table 6.2). Hence, our EMR natural timestamping approach achieves similar time accuracy across the office, residential, and factory environments.

6.3.6 Applications

Leveraging the time accuracy of EMR natural timestamps, this section discusses three applications of *time recovery*, *run-time clock verification*, and *secure clock synchronization*.

6.3.6.1 Time Recovery
In the following two scenarios, EMR natural timestamping is a useful method for recovering the global time of critical sensor data and system logs.

Network-free data loggers. In various one-off deployments of data loggers found in diagnostic applications, it can be logistically tedious to set up a supporting communication network. With EMR natural timestamping, the sensor data timestamps based on its local clock can be rectified using EMR natural timestamps in offline data analysis. Specifically, at run time, periodically the logger captures an EMR natural timestamp and adds to it a corresponding local clock timestamp. In the offline data analysis, the offset of the local clock can be identified by decoding the EMR natural timestamps. The sensor data timestamps can thus be rectified given the identified clock offsets. The period of capturing an EMR natural timestamp can be set to make the local clock drift over the period to be commensurate with the inaccuracy of the EMR natural timestamping. Typical crystal oscillators found in MCUs and personal computers have drift rates of 30–50 ppm Hao et al. (2011). Thus, the period can be set to one hour such that the clock drift is below 180 ms within this period, commensurate with the decoding errors in Table 6.2. As a natural timestamp takes five minutes of sampling time, it introduces a small overhead only.

System outages. Various system outages (e.g., loss of communication network and failure of centralized time servers) may significantly degrade the synchronism between nodes in a system. Using the same approach described for the network-free data loggers, the timestamps of the critical system logs can be rectified in a post-outage analysis.

We note that network unavailability presents a key challenge to accurate time keeping and timestamping in the above two scenarios. The nodes' local clocks based on crystal keep drifting from the global time. Thus, an on-chip logging approach cannot guarantee bounded timestamping errors. Other state-of-the-art approaches include Lukac et al. (2009), Su et al. (2014), Grigoras (2005), Fechner and Kirchner (2014), Garg et al. (2011). However, the approach in Lukac et al. (2009) is developed specifically for seismic sensors. The forensic approaches in Su et al. (2014), Grigoras (2005), Fechner and Kirchner (2014), Garg et al. (2011) are for multimedia recorders, which are generally unsuitable for resource-constrained nodes. Moreover, they can achieve accuracies of seconds only. In contrast, EMR is widely available in civil infrastructure and it can be easily captured using simple devices. Moreover, our EMR-based approach provides sub-second accuracy.

6.3.6.2 Run-time Clock Verification

A network node may lose clock synchronization because of various unexpected reasons, e.g., hardware clock faults, transient power failures, operating system faults, blocked NTP traffic due to firewall misconfiguration, etc. The desynchronization has been a practical concern for real systems. For instance, in a two-year deployment of 100 seismic sensors, each equipped with a GPS receiver for clock synchronization, up to 7% of the collected data has time offsets ranging from tens of seconds to 3000 s Lukac et al. (2009). EMR natural timestamping provides a tool for verifying the integrity of network nodes' clocks at run time. Specifically, a node can reply to a clock verification request with a natural timestamp together with its clock value when the natural timestamp was sampled. Once the natural timestamp is decoded, the clock value can be verified. Our approach gives high confidence in the verification, because it is not affected by network transmission delays.

As discussed in Section 6.3.2.3, we use a finite-length ENF trace collected by the database node to decode a natural timestamp. A practical issue is how to determine whether the decoding condition C1 in Section 6.3.2.3 is satisfied or not, given that the clock value from the node in question may be wrong. We now present a method to verify this condition. Specifically, from the ENF database, we choose an ENF trace of sufficient length and centered at the claimed clock value of the natural timestamp. We slide a window within the natural timestamp to generate a total of J sub-natural timestamps, as illustrated in Figure 6.20. Then, we decode each sub-natural timestamp against the chosen ENF database trace. Let i_j^* denote the decoding result for the jth sub-natural timestamp using the algorithm in Equation (6.1). For the jth sub-natural timestamp, where $j \in [1, J]$, we define $(i_j^* - j)$ as its *decoding offset*. If the decoding condition C1 is true and there are no errors in decoding each

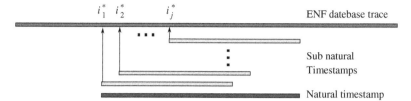

Figure 6.20 Dividing a natural timestamp to sub-natural timestamps to verify the decoding condition C1.

sub-natural timestamp, the decoding offsets of all the sub-natural timestamps should be a constant. If the decoding condition C1 is false due to, say, wrong clock value of the node in question, the decoding offsets will be likely spread out. To address errors in decoding the sub-natural timestamps, we construct a discrete probability density function (PDF) of the decoding offsets. If a significant peak can be found in the PDF, the decoding condition C1 is considered true. For instance, Figure 6.21 shows the discrete PDFs when the decoding condition C1 is respectively false and true. In our experiments, if the highest peak in the PDF is at least three times higher than the second highest peak, the decoding condition C1 is considered true.

We evaluate the true/false positive rates, accuracy, precision, and recall of the above approach in determining the decoding condition C1. Specifically, in each test for evaluating the true/false positive rate, we randomly choose a natural timestamp captured by an EMR sensor and a 660 s ENF database trace that satisfies or violates the decoding condition C1 from a one-week dataset. A true positive is a positive test result when the decoding condition C1 is true; a false positive is a positive test result when the decoding condition C1 is false. We can define true negative and false negative similarly. We follow convention Fawcett (2006) in calculating the true/false positive rates, accuracy, precision, and recall. The results are shown in Tables 6.4 and 6.5. From Table 6.4, for the RPi-based EMR sensor, when the natural timestamp length is 200 s or longer, our approach is extremely accurate in determining the decoding condition C1. From Table 6.5, for the Z1 mote, our approach can also achieve high accuracy.

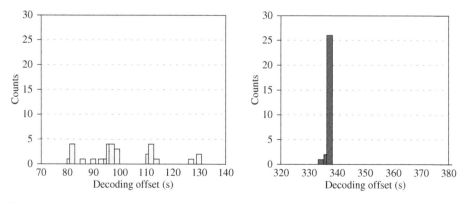

Figure 6.21 PDF of decoding offsets when the decoding condition is respectively false (left figure) and true (right figure).

Table 6.4 Measured true positive rate, false positive rate, accuracy, precision, and recall in verifying the decoding condition C1 on an RPi-based EMR sensor.

Natural timestamp length (s)	RPi-based EMR sensor				
	True positive	False positive	Accuracy	Precision	Recall
60	83.3%	1.0%	91.1%	98.8%	83.3%
100	97.1%	0.0%	98.6%	100.0%	97.1%
200	100.0%	0.0%	100.0%	100.0%	100.0%
300	100.0%	0.0%	100.0%	100.0%	100.0%
600	100.0%	0.0%	100.0%	100.0%	100.0%

Table 6.5 Measured true positive rate, false positive rate, accuracy, precision, and recall in verifying the decoding condition C1 on a Z1 mote.

Natural timestamp length (s)	Z1 mote				
	True positive	False positive	Accuracy	Precision	Recall
60	17.4%	0.6%	58.4%	96.6%	17.4%
100	37.1%	0.9%	68.1%	97.6%	37.1%
200	69.3%	0.0%	84.7%	100.0%	69.3%
300	85.7%	0.6%	92.6%	99.3%	85.7%
600	97.3%	0.3%	98.5%	99.7%	97.3%

If C1 is verified to be false, the node in question does not pass the clock integrity verification; otherwise, we can decode the natural timestamp to obtain the clock offset of the node and determine whether its clock is wrong or not.

6.3.6.3 Secure Clock Synchronization

Many clock synchronization protocols (e.g., NTP Foundation (2017a), PTP IEEE (2008), RBS Elson et al. (2002), TPSN Ganeriwal et al. (2003), and FTSP Maróti et al. (2004)) measure the time delays in transmitting network packets to estimate the offset between the clocks of two nodes. However, these protocols are generally vulnerable to a basic *packet delay attack* that maliciously delays the transmissions of the synchronization packets Mizrahi (2014), Ullmann and Vogeler (2009). This section investigates exploiting the natural timestamps to design a clock synchronization approach that is secure against the attack. In Section 6.3.6.3, we investigate the susceptibility of our natural timestamping approach to malicious EMR spoofing attacks. In Section 6.3.6.3, we formally describe the packet delay attack model and the secure clock synchronization design based on natural timestamping.

Susceptibility of natural timestamping to EMR spoofing

A fake EMR signal emitted by a malicious attacker may mislead the natural timestamping into generating wrong timestamps. this subsection investigates the susceptibility of our approach to EMR spoofing.

First, we discuss the feasibility of launching a long-range EMR spoofing attack that aims at affecting a large geographic area. As the EMR is an extremely low frequency (ELF) signal with a wavelength of about 6000 km, a large-size antenna will be needed to effectively generate fake EMR that can propagate over long distances. For instance, ELF communication systems, primarily for military applications U.S. Navy (2003), need to insert extremely long leads into the ground to use the earth as an antenna, and they consume a considerable amount of power. These requirements represent insurmountable economical and logistical barriers for long-range EMR spoofing attacks that target many EMR sensors distributed within an area (e.g., a city, factory, or building).

Thus, in this section, we focus on investigating short-range EMR spoofing attacks. To understand the practical effective range of EMR spoofing, we build an EMR spoofer and conduct a set of experiments. Figure 6.22 shows a schematic of the EMR spoofer. It is based on a 320 mH coil inductor L. When an oscillating current is applied to the inductor, the inductor will induce EMR of the same frequency due to electromagnetic induction. The attacker can specify an *attack profile* to drive a voltage source V_c to generate an ac voltage signal. The attack profile includes the amplitude, frequency, and phase of the ac voltage signal. To increase the effective range of the EMR spoofer, the ac voltage is amplified by a *power amplifier* (PyleHome Model PTAU45) that weighs 2.6 kg and is applied to the inductor. A smooth capacitor C_s of 1 μF is used to remove unwanted noises during the electromagnetic induction generation. The left part of Figure 6.23 shows the finished EMR spoofer.

The experiment settings are as follows. The EMR spoofer is configured to generate a 50 Hz singular frequency EMR signal. We configure the power amplifier to use its maximum gain. From the measurements by an oscilloscope, the peak-to-peak output voltage of the power amplifier is 44.8 V. A victim RPi-based EMR sensor is used in our experiments. The EMR signal sensed by the victim sensor is used to extract the ENF trace, which is decoded against the ground truth ENF trace recorded directly from a wall power outlet. The timestamp decoding error is the main evaluation metric. We conduct a preliminary experiment over 48 h. In the first 24 h, the EMR spoofer is not activated. The median decoding error is about

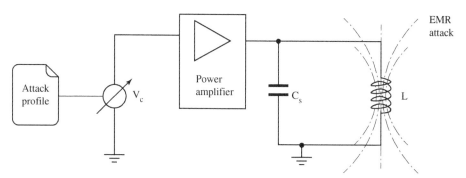

Figure 6.22 A schematic of the EMR spoofer.

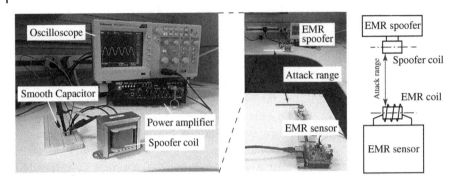

Figure 6.23 EMR spoofer (left) and spoofing experiment (middle and right).

77 ms. In the next 24 h, the EMR spoofer is activated and placed 1 cm from the victim EMR sensor. The EMR sensor yields large decoding errors.

We conduct a set of experiments to evaluate the impact of the distance between the EMR spoofer and the victim EMR sensor on the timestamp decoding errors. The said distance is referred to as the *attack range*. For each attack range setting, the EMR sensor continuously records the powerline ENF data for one hour. In the first 30 min, we turn on the EMR spoofer. Then, we turn the EMR spoofer off and continue the experiment for another 30 min. Figure 6.24 shows the ENF sensed by the victim EMR sensor in each experiment with the attack range varied from 10 to 160 cm. Table 6.6 shows the decoding errors under different attack ranges and natural timestamp lengths.

As shown in Figure 6.24(a), when the EMR spoofer is 10 cm from the EMR sensor, the ENF sensed by the sensor fluctuates around the singular frequency generated by the EMR spoofer. We note that, due to a calibration issue, the frequency of the fake EMR signal is not exactly the specified 50 Hz. As the EMR sensor cannot track the powerline ENF, it cannot correctly decode the natural timestamps. Once the EMR spoofer is turned off, the EMR sensor can track the powerline ENF well and correctly decode the natural timestamps. As shown in Figure 6.24(b), when the attack range increases to 20 cm, the ENF sensed by the EMR sensor shows a complex pattern. This may be because that the fake EMR and the powerline EMR have similar signal strengths. As a result, the superimposition of the two signals is a complex signal that may lead to false zero crossing detection. As shown in Figure 6.24(c), when the attack range increases to 40 cm, the EMR sensor can roughly track the powerline ENF during the attack period. As shown in Figures 6.24(d) and 6.24(e), when the attack range increases to 80 and 160 cm, the effects of the EMR spoofing become insignificant or even invisible. From Table 6.6, when the attack range is 160 cm, the EMR spoofing attack generates little impact on the time accuracy.

The above results show that the effective range of our EMR spoofer is within 160 cm. Although the effective range can be increased by further optimizing the design of the spoofer and using a bulkier power amplifier with higher gains, our results suggest that it is non-trivial to build EMR spoofers that can affect many EMR sensors in a large area. Without high-power and bulky amplifiers, any successful EMR spoofing attacks will have to be in close proximity to the victim nodes. These attacks will be similar to direct physical attacks on the node. Thus, by protecting a certain proximity region of the EMR sensors, we can effectively defend against both the physical attacks and near-field EMR spoofing.

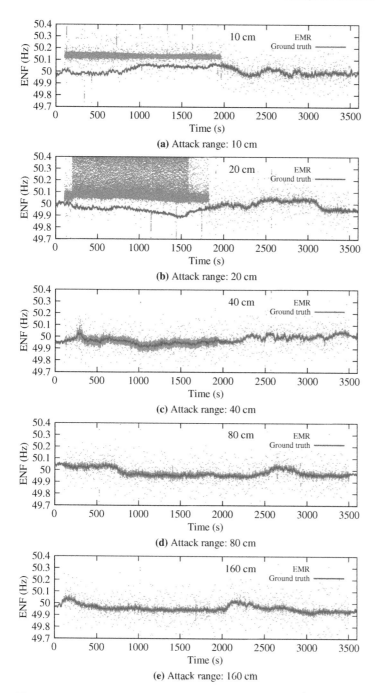

Figure 6.24 The ENF sensed by the victim EMR sensor when the EMR spoofer is on and off in the first and the last 30 min, respectively.

Table 6.6 Decoding errors under EMR spoofing attacks.

Attack range (cm)	Natural timestamp length (s)			
	300		600	
	Median (ms)	95th percentile (ms)	Median (ms)	95th percentile (ms)
40	754	2366	502	1578
80	103	300	73	223
160	78	201	69	140
No EMR spoofing	77	181	60	159

Attack model and secure clock synchronization
Compared with the EMR spoofing attacks that will likely target individual victim nodes due to the requirement of physical proximity, cyberspace attacks that may spread wide and far can be more attractive to attackers. In this section, we consider a cyber-attack in the form of the aforementioned packet delay attack, which is formally described as follows.

Definition 6.2 *(Packet delay attack)* The endpoints of a clock synchronization protocol are trustworthy. The protocol's packets cannot be tampered with because of cryptographic protection. However, one or more attackers on a network path of the protocol's packets may delay the transmission of these packets.

For instance, the packet delay attack can be launched against wireless communications of the EMR sensors by a mix of wireless jamming and replay attacks. When it is launched at compromised nodes in backbone wireline networks, it will generate widespread impact on many nodes. In this section, we propose to use natural timestamps to design a clock synchronization that is secure against packet delay attack. Ganeriwal et al. (2008) has analyzed the impact of the packet delay attack against message passing based WSN clock synchronization protocols (e.g., TPSN Ganeriwal et al. (2003) and FTSP Maróti et al. (2004)). Their approach compares the round-trip time (RTT) between a pair of nodes with a predefined threshold to detect the attack. To reduce the uncertainty of RTT measurements due to MAC layer processing, they need accurate packet timestamping in the PHY layer, but this layer is often inaccessible. Moreover, to address multi-hop networks, the approach Ganeriwal et al. (2008) needs to know the communication path between the two synchronizing nodes to determine the detection threshold. In contrast, based on EMR natural timestamping, our secure clock synchronization is independent of the communication topology and applicable to different kinds of networks beyond WSNs.

In previous work Viswanathan et al. (2016), we have developed a clock synchronization protocol using ENF-based natural timestamps that are collected directly from wall power outlets to solve packet delay attacks. The protocol can likewise use the EMR natural timestamps to solve these attacks. Specifically, as illustrated in Figure 6.25, in a synchronization session, a client node captures an EMR natural timestamp, adds a local clock timestamp, and transmits the pair to the database node acting as a time server. The time server decodes the received natural timestamp and replies to the client node with the difference between

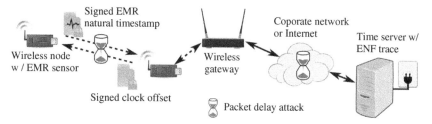

Figure 6.25 Packet delay attack and secure clock synchronization.

the decoding output and the client's local clock timestamp. Finally, the client node sets or calibrates its clock using the received offset. For a proof of this protocol's security against the packet delay attack, refer to Viswanathan et al. (2016).

6.4 Wearables Clock Synchronization Using Skin Electric Potentials

In this section, we present TouchSync, an application-layer clock synchronization approach for wearables. Section 6.4.1 introduces the traditional network clock synchronization, i.e. NTP. Section 6.4.2 presents a measurement study. Section 6.4.3 designs TouchSync. Section 6.4.4 extends the design to address the situations where the iSEPs are too weak. Section 6.4.5 and Section 6.4.6 implement and evaluate TouchSync, respectively.

6.4.1 Motivation

Many clock synchronization approaches adopt the principle of NTP, which is illustrated in Figure 6.26(a). A *synchronization session* consists of the transmissions of a request packet and a reply packet. The t_1 and t_4 are the slave's clock values when the request and reply packets are transmitted and received by the slave node, respectively. The t_2 and t_3 are the master's clock values when the request and reply packets are received and transmitted by the master node, respectively. Thus, the round-trip time (RTT) is RTT $= (t_4 - t_1) - (t_3 - t_2)$. Based on a *symmetric link assumption* that assumes identical times for transmitting the two packets, the offset between the slave's and the master's clocks, denoted by δ_{NTP}, is estimated

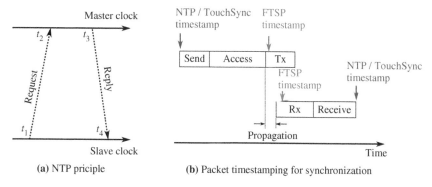

(a) NTP principle

(b) Packet timestamping for synchronization

Figure 6.26 NTP principle and packet timestamping.

as $\delta_{NTP} = t_4 - \left(t_3 + \frac{RTT}{2}\right)$. Then, this offset is used to adjust the slave's clock to achieve clock synchronization. Under the above principle, non-identical times for transmitting the two packets will result in an error in estimating the clock offset. The estimation error is half of the difference between the times for transmitting the two packets.

We use Figure 6.26(b) and the terminology in Maróti et al. (2004) to explain how the timestamps (e.g., t_3 and t_4) are obtained in NTP and existing WSN clock synchronization approaches. The *send time* and the *receive time* are the times used by the OS to pass a packet between the synchronization program and the MAC layer at the sender and receiver, respectively. They depend on OS overhead. The *access time* is the time for the sender's MAC layer to wait for a prescribed time slot in time-division multiple access (TDMA) or a clear channel in carrier-sense multiple access with collision avoidance (CSMA/CA). It often bears the highest uncertainty and can be up to 500 ms Maróti et al. (2004). The transmission (Tx) and reception (Rx) times are the physical layer processing delays at the sender and receiver, respectively. The *propagation time* equals the distance between the two nodes divided by the speed of light, which is generally below 1 μs.

As illustrated in Figure 6.26(b), NTP timestamps the packet when the packet is passed to or received from the OS. Thus, the packet transmission time used by NTP is subjected to the uncertain OS overhead and MAC. Therefore, as measured in Section 6.4.1, NTP over a Bluetooth connection can yield nearly 200 ms clock offset estimation errors. To remove these uncertainties, FTSP uses MAC-layer access to obtain the times when the beginning of the packet is transmitted/received by the radio chip. As the propagation time is generally below 1 μs, FTSP simply estimates the clock offset as the difference between the two hardware-level timestamps. Thus, the two-way scheme in Figure 6.26(a) becomes non-essential for FTSP.

As our objective is to devise a new clock synchronization approach that uses application-layer timestamping as NTP does, this section measures the performance of NTP to provide a baseline. We implement the NTP principle described in Section 6.4.1 on Flora Adafruit (2018), a wearable platform. Our setup includes a Flora node and a Raspberry Pi (RPi) 3 Model B+ single-board computer Foundation (2017c) that perform as the NTP slave and master, respectively. They are connected via Bluetooth Low Energy (BLE). More details of the Flora setup can be found in Section 6.4.2.1. Figure 6.27(a) shows the distributions of the one-way application-layer communication delays over 110 NTP sessions. The

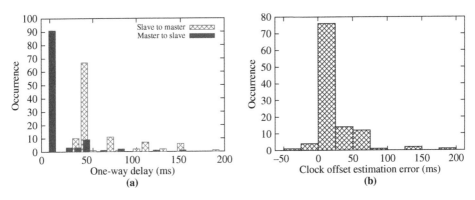

Figure 6.27 Performance of NTP over a BLE connection.

slave-to-master delays are mostly within [40, 50] ms, with a median of 42 ms and a maximum of 376 ms (not shown in the figure). As specified by the BLE standard, the master device pulls data from a slave periodically. The period, called *connection interval*, is determined by the master. The slave needs to wait for a pull request to transmit a packet to the master. In the RPi's BLE driver (BlueZ), the connection interval is set to 67.5 ms by default. As the arrival time of a packet from the slave's OS is uniformly distributed over the connection interval, the expected access time is $67.5/2 = 33.75$ ms. This is consistent with our measured median delay of 42 ms, which is about 8 ms longer because of other delays (e.g., send and receive times). The exceptionally long delays (e.g., 376 ms) observed in our measurements could be caused by transient wireless interference and OS delays. For the master-to-slave link, the delays are mostly within [0, 10] ms, with a median of 8 ms and a maximum of 153 ms. A BLE slave can skip a number of pull requests, which is specified by the *slave latency* parameter, and sleep to save energy. Under BlueZ's default setting of zero for slave latency, the slave keeps awake and listening, yielding short master-to-slave delays.

The asymmetric slave-to-master and master-to-slave delays cause significant errors in the NTP's clock offset estimation. At the end of each synchronization session, the RPi computes this error as $\delta_{\text{NTP}} - \delta_{\text{GT}}$, where δ_{NTP} and δ_{GT} are NTP's estimate and the ground-truth offset, respectively. Figure 6.27(b) shows the distribution of the errors. We observe that 28% of the errors are larger than 25 ms. The largest error in the 110 NTP sessions is 183 ms. Such an error profile does not well meet the millisecond accuracy requirements of many applications Dinescu et al. (2015), Lorincz et al. (2009), Mokaya et al. (2016). Though it is possible to calibrate the average error to zero by using prior information (e.g., the settings of the connection interval and slave latency), the calibration is tedious, non-universal, and incapable of reducing noise variance.

6.4.2 Measurement Study

In this section, we conduct measurements to gain insights for guiding the design of TouchSync.

6.4.2.1 Measurement Setup

Our measurement study uses two Flora nodes and an RPi 3 Model B single-board computer. The Flora is an Arduino-based wearable platform. Each Flora node, as shown in Figure 6.28,

Figure 6.28 Flora.

consists of a main board with an ATmega32u4 MCU (8 MHz, 2.5 KB RAM), a BLE 4.1 module, and a 150 mAh lithium-ion polymer battery. The RPi has a built-in BLE 4.1 module and runs Ubuntu MATE 16.04 with BlueZ 5.37 as the BLE driver. Each node is powered by an independent battery and has no electrical connection to any ground or grounded appliance. We use Adafruit's nRF51 Arduino library and BluefruitLE Python library on the Floras and the RPi, respectively, to send and receive data over BLE in the UART mode. The Floras and RPi operate as BLE peripheral (slave) and central (master), respectively. To obtain the ground truth clocks in each experiment, we synchronize the Floras with the RPi as follows. At the beginning of the experiment, we wire a general-purpose input/output (GPIO) pin of the RPi with a digital input pin of each Flora. Then, the RPi issues a rising edge through the GPIO pin and records its clock value t_{master}. Upon detecting the rising edge, a Flora records its clock value t_{slave} and sends it to the RPi. The RPi computes the ground-truth offset between the Flora's and the RPi's clocks as $\delta_{\text{GT}} = t_{\text{slave}} - t_{\text{master}}$. Then, we remove the wiring and conduct experiments.

6.4.2.2 iSEP Sensing under Various Settings

In this set of measurements, we explore (i) whether a human body is an effective antenna for receiving the powerline radiation and (ii) whether the iSEP signals induced by the radiation on the same wearer or different wearers are synchronous. The Flora's microcontroller (MCU) has a 10-bit ADC that supports a sampling rate of up to 15 kHz. To facilitate experiments, we have made two Flora-based prototypes, as shown in Figure 6.29. We place the Flora into a 3D-printed insulating wristband and use a thin stainless steel conductive thread to create a connection between Flora's ADC pin and the wearer's skin. The Flora samples the ADC at 333 Hz continuously for two minutes and streams the timestamped raw data to the RPi for offline analysis. The sampling rate of 333 Hz is sufficient to capture the powerline radiation or iSEP with a frequency of 50 Hz in our region. All samples are normalized using the reference voltage of the ADC. As only the ADC pin is connected to the researcher, the grounding of the Flora may affect the sampling result. To understand the impact of grounding, we conduct two sets of comparative experiments, where the two Floras have shared and independent grounds, respectively. During the experiment, neither Flora nor human body has contact with grounded appliance. In each experiment set, there are two scenarios: still and moving. For the moving scenario, the researcher keeps changing the body orientation, movement, and location. The experiments are conducted in a computer science laboratory with various appliances such as lights, computers, and printers.

Conductive
thread

Figure 6.29 TouchSync prototypes.

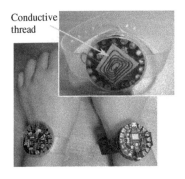

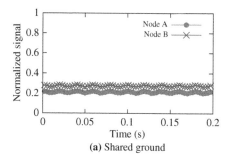 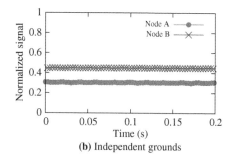

(a) Shared ground

(b) Independent grounds

Figure 6.30 No human body contact.

Shared ground

We wire the ground pins of the two Floras, such that they have a shared ground. We conduct three tests.

First, Figure 6.30(a) shows the signals captured by the two Floras when they have no physical contact with any human body. The signals have small fluctuations with a normalized peak-to-peak amplitude of 0.024. The signals fluctuate at a frequency of 50 Hz. This suggests that the Floras can pick up the powerline radiation. However, the signals are weak.

Second, a researcher touches the ADC pins of the two Floras with his two hands, respectively. Figures 6.31(a) and 6.31(b) show the signals captured by the two nodes during the same time duration, when the researcher stands still and walks, respectively. Under the two scenarios (still and moving), the two nodes yield salient and almost identical signals. The peak-to-peak amplitudes in the two figures are around 0.4 and 0.8, which are 17 and 33 times larger than that of the signal shown in Figure 6.30(a). This suggests that the human body can effectively receive the powerline radiation.

Third, two researchers touch the ADC pins of the two Floras separately. Figures 6.32(a) and 6.32(b) show the signals captured by the two nodes when the two researchers stand still and walk, respectively. The two nodes yield salient signals with different amplitudes. We note that several factors may affect the reception of powerline radiation, e.g., human body size, position and facing of the body in the electromagnetic field generated by the powerlines.

We evaluate the synchrony between the signals captured by the two Floras shown in Figures 6.31 and 6.32. We condition the signals by first applying a band-pass filter (BPF) to remove the direct current (dc) component that may fluctuate as seen in Figure 6.31(b)

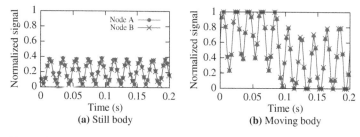

(a) Still body

(b) Moving body

Figure 6.31 iSEPs on the same wearer (shared ground).

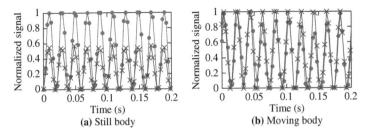

Figure 6.32 iSEPs on different wearers (shared ground).

and then detect the zero crossings (ZCs) of the filtered signals. More details of the BPF and ZC detection are presented in Section 6.4.3.2. We use the *time displacement* between the two signals' ZCs as the metric to evaluate their synchrony. Specifically, the time displacement, denoted by ϵ, is given by $\epsilon = t_A^{ZC} - t_B^{ZC}$, where t_A^{ZC} and t_B^{ZC} represent the ground-truth times of node A's ZC and the corresponding ZC at node B, respectively. Figure 6.33(a) shows the error bars for $|\epsilon|$, which correspond to the scenarios in Figures 6.31(a), 6.31(b), 6.32(a), and 6.32(b), respectively. In Figure 6.33(a), "same" and "diff" mean the same wearer and different wearers, respectively; "still" and "move" mean standing still and walking, respectively. On the same wearer, the iSEPs captured by the two Floras are highly synchronous, with an average $|\epsilon|$ of 0.9 μs. On different wearers, the $|\epsilon|$ increases to about 1 ms. When the two wearers move, the average $|\epsilon|$ is 0.35 ms smaller than that when they stand still. Note that as shown shortly in Figure 6.36(b), the body orientation affects the time displacement. In Figure 6.33(a), the two wearers stand still facing certain orientations. The time displacement under such body orientation setting is consistently large. Thus, the average time displacement when they move is smaller. However, consistent with intuition, the human body movements increase the variance of $|\epsilon|$, since they create more signal dynamics, as seen in Figure 6.31(b).

Independent grounds

Then, we remove the connection between the two Floras' ground pins, such that they have independent grounds. This setting is consistent with real scenarios, where the wearables are generally not wired. We conduct three tests.

Figure 6.30(b) shows the two Floras' signals when they have no physical contact with any human body. The signals have small oscillations with a frequency of 50 Hz.

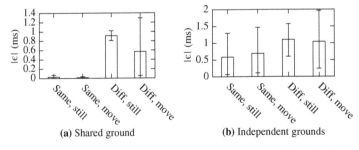

Figure 6.33 Absolute time displacement $|\epsilon|$ between the EMR signals captured by the two Floras in various scenarios. The error bars represent $(5\%, 95\%)$ confidence interval in one minute of data.

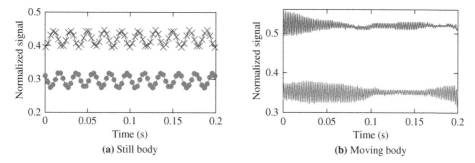

Figure 6.34 iSEPs on the same wearer (independent grounds).

Figures 6.34(a) and 6.34(b) show the signals of the two Floras worn on two wrists of a researcher when he stands still and walks, respectively. Compared with the results in Figure 6.31(a) based on a shared ground, the two signals in Figure 6.34(a) have an offset in their values. This offset is the difference between the electric potentials at the two Floras' grounds. Figure 6.34(b) shows the signals over two seconds that contain about 100 iSEP cycles to better illustrate the changing signal envelopes over time due to the human body movements. Compared with Figure 6.31(b), the two signals in Figure 6.34(b) have different signal envelopes. This is because the electric potentials at the two Floras' grounds, which are induced by the powerline radiation, are not fully correlated in the presence of human body movements.

Figures 6.35(a) and 6.35(b) show the signals of the two Floras worn by two researchers when they stand still and walk, respectively. Salient EMR signals can be observed. Moreover, the human movements cause significant fluctuations of DC lines and the signal amplitudes, as seen in Figure 6.35(b).

We also evaluate the synchrony between the two Floras' signals. Figure 6.33(b) shows the time displacement's error bars that correspond to the scenarios in Figures 6.34(a), 6.34(b), 6.35(a), and 6.35(b), respectively. The average $|\epsilon|$ is below 2 ms. Compared with the results in Figure 6.33(a) that are based on a shared ground, the time displacements increase. This is because of the additional uncertainty introduced by the independent floating grounds of the two Floras. Nevertheless, on the same wearer, the average $|\epsilon|$ is about 0.5 ms only. The body movements increase the 95th percentile of $|\epsilon|$ to 1.5 ms. In Section 6.4.3.2, we use a phase-locked loop to reduce the variations of ϵ.

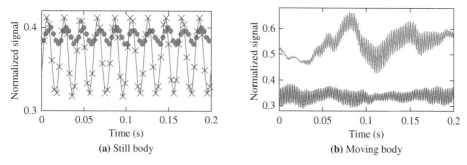

Figure 6.35 iSEPs on different wearers (independent grounds).

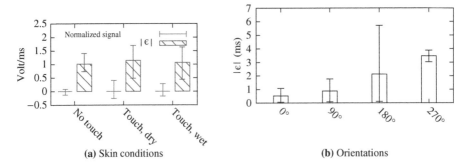

(a) Skin conditions (b) Orientations

Figure 6.36 Normalized range and time displacement $|\epsilon|$ of iSEPs under different skin conditions and body orientations.

We further evaluate the impact of skin condition and body orientation on the iSEPs when the two Floras are worn by two hands of the same wearer. Figure 6.36(a) shows the normalized iSEP signal range and their time displacement on the same still person, when the skin is dry or wet. The iSEP signal is amplified when the sensor has direct contact with the human body. The skin moisture condition does not affect the amplitude and synchrony ($|\epsilon|$) of iSEP much. We have also tested iSEP sensing using two conductive threads for a Flora. Results show that the signal reception is not affected. However, using multiple conductive threads reduces the chance that the wearable loses direct skin contact. Figure 6.36(b) shows the time displacements when the wearer is in different orientations. We can see that the body orientation affects the time displacement. However, the average time displacement $|\epsilon|$ in all tested orientations is below 4 ms.

Summary
From the above measurements, we obtain the following three key observations. First, the human body can act as an antenna that effectively improves the powerline radiation reception. Second, during the human body movements, the iSEP amplitude changes. However, the synchrony between the two iSEP signals captured by the two nodes on the same wearer or different wearers is still acceptably preserved. In Section 6.4.3.2, we condition the iSEP signals to improve the synchrony. Third, the floating ground of a node introduces additional uncertainty, because the powerline radiation can generate a varying electric potential at the ground pin. However, the floating ground does not substantially degrade the synchrony between the two nodes' iSEP signals. All experiments in the rest of this chapter are conducted under the floating ground setting. The above three observations suggest that the iSEP induced by powerline radiation is a good periodic signal that can be exploited for synchronizing wearables.

6.4.3 TouchSync System Design

This section presents the design of TouchSync. Section 6.4.3.1 overviews the workflow. Section 6.4.3.2 presents the signal processing algorithms to generate stable, periodic, and synchronous impulses trains (i.e. Dirac combs) from the iSEP signals. Section 6.4.3.3 presents a synchronization protocol assisted with the Dirac combs. Section 6.4.3.4 solves the integer ambiguity problem to complete synchronization.

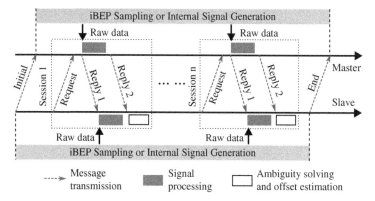

Figure 6.37 A synchronization process of TouchSync.

6.4.3.1 TouchSync Workflow

TouchSync synchronizes the clock of a slave to that of a master. TouchSync focuses on the synchronization between a slave-master pair, which is the basis for synchronizing a network of nodes. A *synchronization process*, as illustrated in Figure 6.37, is performed periodically or in an on-demand fashion. For instance, the wearer(s) may push some buttons on two wearables to start a synchronization process. The period of the synchronization can be determined by the needed clock accuracy and the clock drift rate. During the synchronization process, both the slave and the master continuously sample the iSEP signals and store the timestamped samples into their local buffers. Note that the iSEP signal strength may become too weak to support accurate clock synchronization in certain scenarios. The TouchSync node will switch to a mode that generates an internal periodic signal to drive the clock synchronization, which will be presented in Section 6.4.4. At the beginning of the synchronization process, the slave node sends a message to the master to signal the start of the sensor sampling. As shown in Figure 6.37, a synchronization process has multiple *synchronization sessions*. In each session, the slave and the master exchange three messages: `request`, `reply1`, and `reply2`. The `request` and `reply1` are used to measure the communication delays. After transmitting the `reply1`, the master retrieves a segment of iSEP signal from its buffer to process and transmits the processing results using the `reply2` to the slave. Upon receiving the `reply1`, the slave retrieves a segment of iSEP signal from its buffer to process. Upon receiving the `reply2`, the slave tries to solve the integer ambiguity problem to estimate the offset between the slave's and the master's clocks. If the ambiguity cannot be solved, another synchronization session is initiated; otherwise, the two nodes stop sampling iSEPs and the slave uses the estimated offset to adjust its clock and complete the synchronization process.

6.4.3.2 iSEP Signal Processing

This section presents TouchSync's signal processing pipeline, as illustrated in Figure 6.37, that aims to generate a highly stable, periodic, and synchronous Dirac comb from an iSEP signal with fluctuating dc component and jitters as shown in Figure 6.34 and Figure 6.35. We apply a signal processing pipeline illustrated in Figure 6.38. It consists of three steps that are described in what follows.

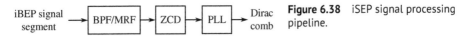

Figure 6.38 iSEP signal processing pipeline.

Band-pass filter (BPF) or Mean Removal Filter (MRF)

We apply a fifth-order/six-tap Infinite Impulse Response (IIR) BPF with steep boundaries of a (45 Hz, 55 Hz) passband to remove the fluctuating DC component and high-frequency noises of the iSEP signal. For too resource-limited wearables, a MRF that subtracts the running average from the original signal can be used instead of the BPF for much lower compute and storage complexities. Its effect is similar to high-pass filtering.

Zero Crossing Detector (ZCD)

This detects the ZCs, i.e. the time instants when the filtered iSEP signal changes from negative to positive. It computes a linear interpolation point between the negative and the consequent positive iSEP samples as the ZC to mitigate the impact of low time resolution due to a low iSEP sampling rate.

Phase-locked Loop (PLL)

We apply a software PLL to deal with the ZC jitters and misdetection caused by significant dynamics of the iSEP signal. The PLL generates an impulse train using a loop and uses an active proportional integral (PI) controller to tune the interval between two consecutive impulses according to the time differences between the past impulses and the input ZCs. The controller skips the time differences larger than 25 ms to deal with ZC miss detection.

6.4.3.3 NTP Assisted with Dirac Combs

TouchSync uses the synchronous Dirac combs at the slave and the master to achieve clock synchronization through multiple synchronization sessions. this subsection presents the protocol for a single synchronization session.

Protocol for a synchronization session

A synchronization session of TouchSync is illustrated in Figure 6.39. We explain it from the following two aspects.

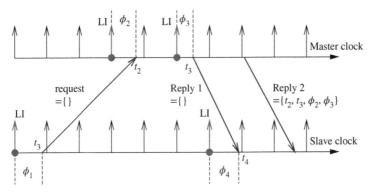

Figure 6.39 A synchronization session of TouchSync. The vertical arrows represent the Dirac comb impulses generated from the iSEP.

Message exchange and timestamping: The session consists of the transmissions of three messages: `request`, `reply1`, and `reply2`. The `request` and `reply1` messages are similar to the two UDP packets used by NTP. Their transmission and reception timestamps, i.e. t_1, t_2, t_3, and t_4, are obtained upon the corresponding messages are passed/received to/from the OS. The master transmits the auxiliary `reply2` message to convey the results of its signal processing, which is detailed below.

Signal processing and clock offset estimation: After the master has transmitted the `reply1` message, the master (i) retrieves from its signal buffer an iSEP signal segment that covers the time period from t_2 to t_3 with some safeguard ranges before t_2 and after t_3, (ii) feeds the signal processing pipeline in Section 6.4.3.2 with the retrieved iSEP signal segment to produce a Dirac comb as illustrated in Figure 6.39, and (iii) identifies the last impulses (LIs) in its Dirac comb that are right before the time instants t_2 and t_3, respectively. The LIs are illustrated by thick red arrows in Figure 6.39. Then, the master computes the elapsed times from the t_2 LI to t_2 and the t_3 LI to t_3, which are denoted by ϕ_2 and ϕ_3, respectively. The ϕ_2 and ϕ_3 are the *phases* of t_2 and t_3 with respect to the Dirac comb. After that, the master transmits the `reply2` message that contains t_2, t_3, ϕ_2, and ϕ_3 to the slave. After receiving the `reply1` message, the slave retrieves an iSEP signal segment that covers the time period from t_1 to t_4 with some safeguard ranges, executes the signal processing pipeline, identifies the LIs right before t_1 and t_4, and computes the phases ϕ_1 and ϕ_4, as illustrated in Figure 6.39. After receiving the `reply2` message, based on $\{t_1, t_2, t_3, t_4\}$ and $\{\phi_1, \phi_2, \phi_3, \phi_4\}$, the slave uses the approach in Section 6.4.3.3 to analyze the offset between the slave's and master's clocks.

Now, we discuss several design considerations for the protocol described above. Touch-Sync uses the `request` and `reply1` messages to measure the clock offset, while the `reply2` is an auxiliary message to convey the timestamps t_2, t_3 and the measurements ϕ_2, ϕ_3. With this auxiliary message, we can decouple the task of timestamping the reception of `request` and the transmission of `reply1` from the signal processing task of generating the Dirac comb and computing ϕ_2 and ϕ_3. On many platforms (e.g., Wear OS and watchOS), continuously sampled sensor data is passed to the application block by block. With the decoupling, the master can compute ϕ_2 and ϕ_3 after the `reply1` is transmitted and the needed iSEP data blocks become available.

Clock offset analysis

We now analyze the offset between the slave's and the master's clocks based on $\{t_1, t_2, t_3, t_4\}$ and $\{\phi_1, \phi_2, \phi_3, \phi_4\}$. Denote by T the period of the Dirac comb. In our region served by a 50 Hz grid, the nominal value for T is 20 ms. To capture the small deviation from the nominal value, T can be easily computed as the average interval between consecutive impulses of the Dirac comb. We define the *rounded phase differences* θ_q and θ_p (which correspond to the `request` and `reply1` messages, respectively) as

$$\theta_q = \begin{cases} \phi_2 - \phi_1, & \text{if } \phi_2 - \phi_1 \geq 0; \\ \phi_2 - \phi_1 + T, & \text{otherwise.} \end{cases} \tag{6.2}$$

$$\theta_p = \begin{cases} \phi_4 - \phi_3, & \text{if } \phi_4 - \phi_3 \geq 0; \\ \phi_4 - \phi_3 + T, & \text{otherwise.} \end{cases} \tag{6.3}$$

As ϕ_k is the elapsed time from the t_k LI to t_k, we have $0 \le \phi_k < T$, for $k \in [1, 4]$. From Equations (6.2) and (6.3), we can verify that $0 \le \theta_q < T$ and $0 \le \theta_p < T$. From our measurements in Section 6.4.1, the times for transmitting the `request` and `reply1` messages can be longer than T. Thus, we use i to denote the non-negative integer number of the Dirac comb's periods elapsed from the time of sending `request` to the time of receiving it at the master, and j to denote the non-negative integer number of the Dirac comb's periods elapsed from the time of sending `reply1` to the time of receiving it at the slave.

We denote τ_q and τ_p the actual times for transmitting the `request` and the `reply` messages, respectively. Thus,

$$\tau_q = \theta_q + i \cdot T - \epsilon, \qquad \tau_p = \theta_p + j \cdot T + \epsilon, \tag{6.4}$$

where ϵ is the time displacement between the slave's and master's Dirac combs. Here, we assume a constant ϵ to simplify the discussion. Therefore, the RTT computed by RTT = $(t_4 - t_1) - (t_3 - t_2)$ must satisfy

$$\text{RTT} = \tau_q + \tau_p = \theta_q + \theta_p + (i + j) \cdot T. \tag{6.5}$$

In Equation (6.5), RTT, θ_q, and θ_p are measured in the synchronization session illustrated in Figure 6.39; i and j are unknown non-negative integers. If the i and j can be determined, the estimated offset between the slave's and the master's clocks, denoted by δ, can be computed by either one of the following formulas:

$$\delta = t_1 - (t_2 - \tau_q) = t_1 - t_2 + \theta_q + i \cdot T - \epsilon, \tag{6.6}$$

$$\delta = t_4 - (t_3 + \tau_p) = t_4 - t_3 - \theta_p - j \cdot T - \epsilon. \tag{6.7}$$

It can be easily verified that the above two formulas give the same result. The analysis in the rest of this chapter chooses to use Equation (6.7). The ϵ is generally unknown. If we ignore it in Equation (6.7) to compute δ, it becomes part of the clock offset estimation error.

Equation (6.5) is an *integer-domain* underdetermined problem. Clearly, from Equation (6.5), both i and j belong to the range $\left[0, \frac{\text{RTT} - \theta_q - \theta_p}{T}\right]$. Thus, Equation (6.5) has a finite number of solutions for i and j. Note that, under the original NTP principle, we have a *real-domain* underdetermined problem of RTT = $\tau_q + \tau_p$ that has infinitely many solutions. NTP chooses a solution by assuming $\tau_q = \tau_p$, which does not hold in general. Thus, by introducing the Dirac combs, the *ambiguity* in determining τ_q, τ_p, and δ is substantially reduced from infinitely many possibilities to finite possibilities. Although we still have ambiguity in the integer domain, our analysis and extensive numeric results in Section 6.4.3.4 show that the ambiguity can be solved.

Note that, in Rabadi et al. (2017), the periodic and synchronous power grid voltage signals collected directly from power outlets are used to synchronize two nodes that have high-speed wired network connections. The approach in Rabadi et al. (2017) uses the elapsed times from LIs (i.e. $\phi_1, \phi_2, \phi_3, \phi_4$) to deal with asymmetric communication delays and improve synchronization accuracy. However, due to the high-speed connectivity, it only considers the case where both i and j are zero. In contrast, with wireless connectivity, i and j are random and often non-zero due to the access time. Estimating i and j is challenging and it is the subject of Section 6.4.3.4.

Integer Ambiguity Solver (IAS)

Before we present the approach to solving integer ambiguity, we make the following two assumptions for simplicity of exposition. First, we assume that the ground-truth clock offset δ_{GT} is a constant during a synchronization process. From our performance evaluation in Section 6.4.6, TouchSync generally takes less than one second to achieve synchronization. Typical crystal oscillators found in MCUs and personal computers have drift rates of 30 to 50 ppm Hao et al. (2011). Thus, the maximum drift of the offset between two clocks during one second is 50 ppm $\times$ 1 s $\times$ 2 = 0.1 ms. This drift is smaller than the millisecond-level time displacement ϵ between two iSEP signals, which dominates the synchronization error of TouchSync. Second, we assume $\epsilon = 0$. In Section 6.4.3.4, we will discuss how to deal with non-zero and time-varying ϵ.

We let i_{min} and i_{max} denote the minimum and maximum possible values for i, respectively; j_{min} and j_{max} denote the minimum and maximum possible values for j, respectively. For instance, from our one-way message transmission time measurements (Section 6.4.1), the BLE's slave-to-master transmission times are always greater than 30 ms. Thus, we may set $i_{min} = 1$, since in our region T is 20 ms. When we have no prior knowledge about the ranges for i and j, we may simply set $i_{min} = j_{min} = 0$ and $i_{max} = j_{max} = \frac{\text{RTT} - \theta_q - \theta_p}{T}$. Section 6.4.3.4 will discuss how the use of the prior knowledge impacts on the integer ambiguity solving.

TouchSync performs multiple synchronization sessions to solve the integer ambiguity problem. In this section, we use $x[k]$ to denote a quantity x in the kth synchronization session. For instance, RTT$[k]$ denotes the measured RTT in the kth session. From Equations (6.5) and (6.7), for the kth synchronization session, we have

$$\begin{cases} \text{RTT}[k] = \theta_q[k] + \theta_p[k] + (i[k] + j[k]) \cdot T; \\ \delta = t_4[k] - t_3[k] - \theta_p[k] - j[k] \cdot T; \\ i_{min} \le i[k] \le i_{max}, \quad j_{min} \le j[k] \le j_{max}. \end{cases} \tag{6.8}$$

If TouchSync performs K synchronization sessions, we have an underdetermined system of $2K$ equations with $(2K + 1)$ unknown variables (i.e. δ and $\{i[k], j[k] | k \in [1, K]\}$). In the integer domain, such an underdetermined system can have a unique solution.

An example of a unique solution

We use the example in Figure 6.40 to illustrate. The unit for time is ms, which is omitted in the following discussion for conciseness. In this example, $T = 20$, $i_{min} = j_{min} = 1$, $i_{max} = j_{max} = 4$, and the ground-truth clock offset $\delta_{GT} = 105$. Two synchronization sessions

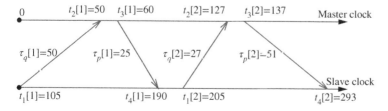

Figure 6.40 An example of solving the integer ambiguity. The transmissions of the auxiliary `reply2` messages are omitted in the illustration.

are performed. The timestamps and the actual message transmission delays are shown in Figure 6.40. The ground-truth values for i and j in the two synchronization sessions are: $i[1] = 2$, $j[1] = 1$, $i[2] = 1$, and $j[2] = 2$. The RTTs can be computed as RTT[1] = 75 and RTT[2] = 78. With any synchronous Dirac combs, from Equations (6.2) and (6.3), the rounded phase differences computed by the two nodes must be $\theta_q[1] = 10$, $\theta_p[1] = 5$, $\theta_q[2] = 7$, and $\theta_p[2] = 11$. For the first synchronization session, Equation (6.8) has two possible solutions only:

$$\{i[1] = 1, j[1] = 2, \delta = 85\}, \{i[1] = 2, j[1] = 1, \delta = 105\}. \tag{6.9}$$

For the second synchronization session, Equation (6.8) has two possible solutions only as well:

$$\{i[2] = 1, j[2] = 2, \delta = 105\}, \{i[2] = 2, j[2] = 1, \delta = 125\}. \tag{6.10}$$

From Equations (6.9) and (6.10), $\delta = 105$ is the only common solution. Thus, we conclude that δ must be 105.

Program for solving integer ambiguity

From the above example, due to the diversity of the ground-truth values of i and j in multiple synchronization sessions, the intersection of the δ solution spaces of these synchronization sessions can be a single value. Thus, the integer ambiguity problem is solved. By contrast, if the ground-truth i and j do not change over multiple synchronization sessions, the ambiguity remains. With application-layer timestamping, the message transmission times are highly dynamic due to the uncertain OS overhead and MAC. Such uncertainties and dynamics, which are undesirable in the original theme of NTP, interestingly, become desirable for solving the integer ambiguity in TouchSync.

From the above key observation, TouchSync performs the synchronization session illustrated in Figure 6.39 *repeatedly* until the intersection among the δ solution spaces of all the synchronization sessions converges to a single value. The pseudocode of the algorithms running at the slave and the master can be found in Algorithms 7 and 8.

Convergence speed

We run a set of numeric experiments to understand the convergence speed of the IAS. We use the number of synchronization sessions until convergence to characterize the convergence speed, which is denoted by K in the rest of the section. We fix $i_{\min}$ and $j_{\min}$ to be zero. For a certain setting $\langle i_{\max}, j_{\max} \rangle$, we conduct 100,000 synchronization processes to assess the distribution of K. For each synchronization session of a synchronization process, we randomly and uniformly generate the ground-truth i and j, as well as θ_q and θ_p within their respective ranges, i.e. $i \in [0, i_{\max}]$, $j \in [0, j_{\max}]$, and $\theta_q, \theta_p \in [0, T)$. Then, we simulate the integer ambiguity solving program presented in Section 6.4.3.4 to measure the K for each synchronization process. In practice, i, j, θ_q and θ_p may not follow the uniform distributions. However, the numeric results here help us understand the convergence speed. In Section 6.4.6.3, we will evaluate the convergence speed in real-world settings.

In Figure 6.41(a), each grid point is the average of all K values in the 100,000 synchronization processes under a certain $\langle i_{\max}, j_{\max} \rangle$ setting. Figure 6.41(b) shows the box plot for K under each setting where $i_{\max} = j_{\max}$. We note that all simulated synchronization

---

Algorithm 7 Slave's pseudocode for a synchronization process

1: Global variables: t_1, t_4, δ's solution space $\Delta = \varnothing$, session index $k = 0$

2:

3: **command** start_sync_session() **do**

4:     $k = k + 1$

5:     $t_1 = $ `read_system_clock()`

6:     send message `request` $= \{\,\}$ to master

7: **end command**

8:

9: **event** `reply1` received from master **do**

10:     $t_4 = $ `read_system_clock()`

11:     **if** iSEP signal strength is good:

12:         wait till iSEP data covering t_1 and t_4 are available

13:         run iSEP signal processing pipeline in Section 6.4.3.2

14:     **else:**

15:         generate internal periodic signal[3]

16:     **endif**

17:     compute ϕ_1 and ϕ_4

18: **end event**

19:

20: **event** `reply2` received from master **do**

21:     compute θ_q and θ_p using Eqs. (6.2) and (6.3).

22:     RTT $= (t_4 - t_1) - (\text{reply2}.t_3 - \text{reply2}.t_2)$

23:     solve Equation (6.8), Δ' denotes the set of all possible solutions for δ

24:     **if** $k == 1$:

25:         $\Delta = \Delta'$

26:     **else:**

27:         $\Delta = \Delta \cap \Delta'$

28:     **endif**

29:     **if** Δ has only one element δ:

30:         use δ to adjust clock

31:     **else:**

32:         start_sync_session() // start a new synchronization session

33:     **endif**

34: **end event**

---

processes converge. From the two figures, even if $i_{\max} = j_{\max} = 10$ (which means that the one-way communication delays are up to 200 ms for $T = 20$ ms), the average K is nine only. Although the K distribution has a long tail, as shown in Figure 6.41(b), 75% of the K values are below 11. This result is consistent with our real experiment results in Table 6.8 of Section 6.4.6.3, where most K values are two only and the largest K is 12.

---

3 Line 15 deals with a situation where the iSEP signal is too weak. The detailed explanation can be found in Section 6.4.6.

Algorithm 8 Master's pseudocode for a synchronization process

---

1: **event** `request` received from slave **do**
2:     t_2 = `read_system_clock()`
3:     ... // execute other compute tasks
4:     t_3 = `read_system_clock()`
5:     send message `reply1 = {}` to slave
6:     **if** iSEP signal strength is good:
7:         wait till iSEP data covering t_1 and t_4 are available
8:         run iSEP signal processing pipeline in Section 6.4.3.2
9:     **else:**
10:       generate internal periodic signal Section 3
11:     **endif**
12:    compute ϕ_2 and ϕ_3
13:    send message `reply2` $= \{t_2, t_3, \phi_2, \phi_3\}$ to slave
14: **end event**

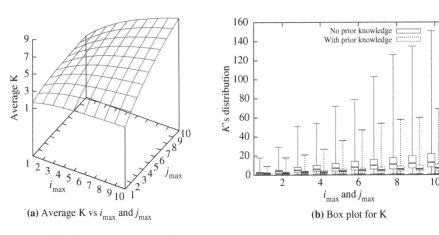

(a) Average K vs i_{max} and j_{max} (b) Box plot for K

Figure 6.41 Convergence speed of IAS.

Discussions

First, we discuss how to address non-zero and time-varying ϵ. From the analysis in Equation (6.8), which is based on $\epsilon = 0$, the difference between two δ solutions is a multiple of T. This can be seen from Equations (6.9) and (6.10). In practice, ϵ can be non-zero and time-varying. It will be a major part of the δ estimation error. From Figure 6.33(b), the $|\epsilon|$ is at most 6 ms. Thus, the resulted variation to the δ solutions will be less than a half and one third of T, in the regions served by 60 Hz and 50 Hz power grids, respectively. Therefore, we can still correctly identify the correspondence among the δ elements in the set intersection operation. Specifically, if two δ elements have a difference smaller than $T/2$, they should be considered the same element in the set intersection operation; otherwise, they are different elements. For this correspondence identification to be correct, the ϵ needs to be smaller than $T/2$. After convergence, the final δ can be computed as the average of the δ elements that are considered the same. We have incorporated this in our implementation of TouchSync.

Second, we discuss how the use of the prior knowledge (i.e. $i_{\min}$, $i_{\max}$, $j_{\min}$, and $j_{\max}$) impacts on the integer ambiguity solving. With the prior knowledge, we may shrink the search range for i and j to speed up the convergence of the IAS. The prior knowledge can be based on the statistical information obtained in offline experiments. For instance, a group of the box plots are the results for the IAS with the prior knowledge of $i_{\max}$ and $j_{\max}$. The IAS can search the i and j within the ranges of $[0, i_{\max}]$ and $[0, j_{\max}]$, respectively. The other group of results are for the IAS without the prior knowledge. Thus, the IAS has to search within the range of $\left[0, \frac{\text{RTT} - \theta_q - \theta_p}{T}\right]$ for both i and j. We can see that, if no prior knowledge is used, the K increases. But the IAS still always converges. Once the IAS converges, the synchronization error of TouchSync mainly depends on the time displacement ϵ.

6.4.4 TouchSync with Internal Periodic Signal

Our design of TouchSync in Section 6.4.3 is based on the periodic and synchronous iSEP signals available to the slave and master nodes. However, in certain scenarios where the wearables are extremely far away from the powerlines or in a Faraday cage (e.g., an elevator cabin), the wearables can hardly sense EMR or iSEP. This section extends the design of TouchSync to deal with such situations by letting the wearables generate internal periodic signals (IPSs) with the same period by themselves. Such IPSs are used to drive the clock synchronization. Section 6.4.4.1 presents our extended design. As the IPSes generated by any two wearables may have a random and bounded time displacement, it is interesting to investigate how this time displacement affects the performance of TouchSync, which is the subject of Section 6.4.4.2.

6.4.4.1 Extended Design

The key extensions made to the design presented in Section 6.4.3 include the coordination between the slave and the master on using the same IPS time period and the generation of their IPSs. Based on the generated IPS, each node follows exactly the procedures described in Section 6.4.3 to estimate the clock offset. In what follows, Sections 6.4.4.1 and 6.4.4.1 present the extensions to the TouchSync protocol and an adaptive period mechanism that assists the convergence of TouchSync, respectively.

Protocol for IPS-based TouchSync
We extend the TouchSync protocol in Section 6.4.3.3 to use IPSs. When the slave is to estimate the offset between its clock and master's, it transmits an initial packet to the master, indicating the beginning of the synchronization process. The packet includes the time period T for generating the IPS. The two devices will record their time instants on transmitting and receiving the initial packet as t_0^{slave} and t_0^{master}. In our approach, these two time instants are used as the first ZCs of the two devices' IPSs. Then, same as the design in Section 6.4.3.3, one request and two reply packets will be exchanged for each synchronization session. Meanwhile, the time instants, i.e. t_1, t_2, t_3, and t_4, as illustrated in Figure 6.39, are recorded by the two nodes. For each recorded time instant, the elapsed time from the corresponding LI is computed. For instance, the slave node computes ϕ_1 as $\phi_1 = (t_1 - t_0^{\text{slave}}) \bmod T$; the master computes ϕ_2 as $\phi_2 = (t_2 - t_0^{\text{master}}) \bmod T$. Then, the slave finds all the possible estimated clock offsets based on the analysis in Section 6.4.3.3.

TouchSync repeatedly runs synchronization sessions until the ambiguity is solved using the IAS presented in Section 6.4.3.4.

In this extended design, the absolute time displacement between the two nodes is $|t_0^{\text{slave}} - t_0^{\text{master}}| \bmod T$. The above protocol transmits an initial packet to establish the initial ZCs at t_0^{slave} and t_0^{master}. As the transmission time of the initial packet is random, the resulting time displacement is also random. An alternative approach is to use NTP to establish the initial ZCs. However, as NTP is susceptible to link asymmetry that is generally true under stochastic wireless link quality, this alternative approach may not give smaller time displacements.

Adaptive period mechanism (APM)

When the IPS time period is larger, the number of IPS time periods elapsed during a synchronization session (i.e. $i + j$) is likely less. Hence, a large IPS period setting can generally reduce the TouchSync session's ambiguity and speed up the convergence. As an extreme example, when the IPS time period is very large such that $i + j = 0$ (i.e. no ambiguity) and TouchSync converges after one session. However, a larger IPS time period setting will lead to larger time displacements and therefore larger clock offset estimation errors. Thus, there exist a trade-off between the convergence speed and clock offset estimation accuracy. Based on this observation, we propose the APM to dynamically increase the IPS period to make sure that the system can always converge. Specifically, the IPS-based TouchSync starts with a small IPS time period setting. Whenever the system cannot converge within a predefined number of synchronization sessions, it increases the IPS time period and restarts IAS. In other words, APM keeps increasing the IPS period until the ambiguity is solved. APM speeds up the convergence at the cost of larger clock offset estimation error bounds. The preset maximum number of synchronization sessions allowed for each IPS time period setting is the knob provided to the system designer for choosing a satisfactory trade-off between the convergence speed and error bound. We evaluate this trade-off via numeric experiments in Section 6.4.4.2.

6.4.4.2 Numeric Experiments

We conduct numeric experiments to evaluate the performance of the IPS-based TouchSync, in terms of clock offset estimation error and convergence speed, when the time displacement between the two self-generated IPSs varies. Specifically, we develop a simulator that can simulate the packet transmissions between the slave and the master. The ground-truth clock offset between the slave and the master is an arbitrary value. The first ZCs of the IPSs at the slave and the master are randomly selected. The transmission delays of the `request` and `reply` packets are uniformly and independently drawn from the range of [0, 100 ms]. TouchSync repeatedly runs synchronization sessions and terminates when either the integer ambiguity is solved or the maximum session number limit of 200 is reached. The numeric experiment results are presented below.

Distribution of clock offset estimation errors

We run 10,000 synchronization processes with the IPS period fixed to 20 ms. Figure 6.42(a) shows the distribution of the clock offset estimation errors. We can see that the errors are bounded by 20 ms, i.e. the IPS period. Now, we use an example in Figure 6.43 to explain this error bound. For this example, we assume that the integer ambiguity has been solved and only focus on the packet transmission from the slave to the master that is used to estimate

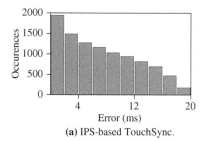

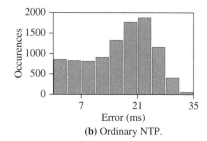

(a) IPS-based TouchSync. **(b)** Ordinary NTP.

Figure 6.42 Distribution of absolute clock offset estimation errors.

Figure 6.43 An example of IPS-based TouchSync (unit: ms).

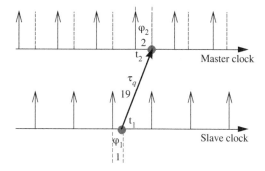

the clock offset between the two nodes. Suppose the one-way transmission delay τ_q is 19 ms and the solved value for i is 0. In Figure 6.43, the two trains of arrows represent the Dirac combs based on the two nodes' IPSs that are not synchronous. To facilitate illustration, we add dashed lines on the master's timeline that are synchronous with the slave's Dirac combs. As illustrated in the figure, $\phi_1 = 1$ ms and $\phi_2 = 2$ ms. From Equation (6.2), we have $\theta_q = 1$ ms. As the two nodes independently generate IPSs, they do not know the value of the time displacement between their IPSs. Thus, from Equation (6.6), the slave node will compute the clock offset as $\delta = t_1 - t_2 + \theta_q$. If the clocks of the two nodes are actually synchronized such that $t_1 - t_2 = -19$, the computed $|\delta|$ of 18 ms is the absolute clock offset estimation error. From this example, if $\phi_1 \to 0$, $\phi_2 \to 0$, and $\tau_q \to 20$ ms, the absolute clock offset estimation will approach to 20 ms. This explains the error bound observed in Figure 6.42(a).

IPS-based TouchSync does not rely on any external signal, which is the same as the NTP. For comparison, we evaluate side-by-side the clock offset estimation errors of the NTP. Figure 6.42(b) shows the results. The absolute clock offset estimation errors are up to 35 ms. Note that, NTP's synchronization error is $(\tau_2 - \tau_1)/2$, where τ_1 and τ_2 are the one-way transmission delays of the `request` and `reply` packets (Section 6.4.1). Hence, unlike IPS-based TouchSync that has an error bound of the IPS time period, NTP's error bound depends on the two one-way packet transmission delays. Over a highly asymmetric link, NTP's error can be much higher than the IPS period.

Convergence speed

As discussed in Section 6.4.4.1, if the system uses a larger IPS period setting, it will need fewer sessions to converge. To demonstrate this, we simulate the IPS-based TouchSync

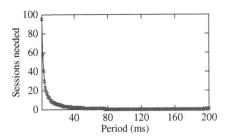

Figure 6.44 Convergence speed under various settings of IPS period.

system under different period settings and record the number of sessions needed for solving the integer ambiguity. Figure 6.44 shows the number of sessions needed for convergence versus the IPS period. The result clearly shows the trade-off discussed in Section 6.4.4.1. From the result, when the IPS period is 10 ms to ensure a 10 ms error bound, a total of 13 synchronization sessions are performed before convergence. If we loosen the error bound to 60 ms, the system needs two sessions only to converge. We note that the convergence speed is affected by the two one-way packet transmission delay, since a long transmission delay will lead to larger values of $i + j$ and thus increase the ambiguity. In Section 6.4.6, we will evaluate TouchSync's convergence speed under real-world settings.

6.4.4.3 Impact of Clock Skews

To simplify discussions, our analysis and the numeric experiments in Section 6.4.4.1 and Section 6.4.4.2 assume that the master and the slave have no clock skews. In practice, small clock skews will result in additional clock synchronization errors for IPS-based Touch-Sync. According to our experiments in Section 6.4.4.2, when the IPS' period is 40 ms, the IPS-based TouchSync takes less than one second before the system converges. The clock drift over a second time duration is often little. For instance, the average clock drift rate of MSP430's clocks is 44.2 ppm Instruments (2018), which will introduce a drift of 44 μs per second. Thus, the time displacement $|t_0^{slave} - t_0^{master}|$ that ranges from 0 to the IPS period dominates the synchronization error of IPS-based TouchSync.

The val.idness of the domination of the time displacement over clock skew depends on the convergence time. However, the convergence time can be monitored and managed. For instance, if we desire to maintain the synchronization error caused by clock skew below 100 μs, we can enforce a convergence time upper bound of 2 s given a drift rate of 50 ppm. If IPS-based TouchSync does not converge within 2 s, the APM presented in Section 6.4.4.1 should be used to increase the IPS period.

In the remainder sections, our analysis ignores clock skews and focuses on estimating the clock offset between the master and slave. Existing clock skew compensation approaches (e.g., Rowe et al. (2009), Li et al. (2011), Hao et al. (2011), Li et al. (2012)) can be integrated with TouchSync to address clock skew and further improve the clock synchronization performance.

6.4.5 Implementation

Designed as an application-layer clock synchronization approach, TouchSync can be implemented as an app or part of an app, purely based on the standard wearable OS calls to sample the iSEP signal, exchange network messages, and timestamp them in the application layer.

To simplify the adoption of TouchSync by application developers, we have implemented TouchSync's platform-independent tasks in ANSI C and provide them in a `touchsync.h` header file Yan et al. (2017b). These tasks include buffer management, iSEP signal processing, and IAS. Other tasks of TouchSync, i.e. sensor sampling, synchronization message exchange and timestamping, are platform dependent. We leave them for the application developer to implement. As these tasks are basics for embedded programming, by using the platform-independent algorithms provided by `touchsync.h`, application developers without much knowledge in signal processing can readily implement TouchSync on different platforms. Our own Arduino and TinyOS programs that implement TouchSync's workflow have about 50 and 150 lines of code only, respectively.

To understand the overhead of TouchSync, we deploy our TinyOS and Arduino implementations to Zolertia's Z1 motes and Floras, respectively. The Z1 mote is equipped with an MSP430 MCU (1 MHz, 8 KB RAM) and a CC2420 802.15.4 radio. Both implementations sample iSEP at 333 Hz. On Z1, we configure the length of the circular buffer defined in `touchsync.h` to be 512. Thus, this circular buffer can store 1.5 s of iSEP data. This is sufficient, because the time periods $[t_2, t_3]$ and $[t_1, t_4]$ that should be covered by the iSEP signal segments to be retrieved from the circular buffer and processed by the master and slave are generally a few milliseconds and below 100 ms, respectively. On Flora, we configure the circular buffer length to be 400 and redefine its data type such that TouchSync can fit into Flora's limited RAM space of 2.5 KB. Table 6.7 tabulates the memory usage of TouchSync and the computation time of different processing tasks. On Z1, a total of 421 ms processing time is needed for a synchronization session. The BPF uses a major portion of the processing time. Flora cannot adopt BPF because of RAM shortage. It uses MRF instead, which consumes much less RAM and processing time. Z1 and Flora are two representative resource-constrained platforms. The successful implementations of TouchSync on them suggest that TouchSync can be readily implemented on other more resourceful platforms.

We use a Monsoon power monitor to measure the energy consumption of TouchSync on a Flora node. With TouchSync running, the node consumes 57.16 mW on average. Without TouchSync, the node's average idle power consumption is 54.99 mW when the node is not in the sleep mode. Thus, TouchSync consumes 2.17 mW or 0.658 mA (with a 3.3 V battery) on average. Assume we would like to run TouchSync on two wearable devices, each with a 150 mAh battery and a 50 ppm crystal, which can continuously run without sleep for a whole day after being fully charged. In order to keep the clock offset within 7 ms, TouchSync should be activated every 140 seconds. Under such settings, the battery time of the wearable is reduced by 39 s per day due to the running of TouchSync. In other words, the execution

Table 6.7 Storage and compute overhead of TouchSync.

Platform	Memory use (KB)		Processing time (ms)			
	ROM	RAM	BPF/MRF	ZCD	PLL	IAS
Z1	10	5	364	9	48	1
Flora	17	1.9[a]	1.3	3	15	0.8

a) Estimated based on buffer lengths.

of TouchSync reduces the battery time by 0.045% only. Note that when the wearable has a sleeping schedule to prolong the battery time (e.g., around 10 days by a duty cycle of 10%), TouchSync can be executed when the wearable is not in the sleep mode.

6.4.6 Evaluation

We conduct extensive experiments to evaluate the performance of TouchSync in various real environments. Each experiment uses two Flora nodes, which act as the TouchSync slave and master, respectively. As Flora does not support BLE master mode, the two Floras cannot communicate directly. Thus, we use a RPi that operates as a BLE master to relay the data packets between the two Floras. This setting is consistent with most body-area networks with a smartphone as the hub. If the hub can sample powerline radiation or iSEP, each wearable can also synchronize with the hub directly using TouchSync. We use the approach discussed in Section 6.4.2.1 to obtain the ground truth clock of each Flora. The details and the results of our experiments are presented below.

6.4.6.1 Signal Strength and Wearing Position

As the intensity of powerline radiation attenuates with distance, iSEPs have varying signal strength. Thus, we evaluate the impact of the iSEP signal strength on the performance of the signal processing pipeline in Section 6.4.3.1. We measure the signal strength as follows. For a full-scale sinusoid signal with a peak-to-peak amplitude of one (normalized using ADC's reference voltage), its standard deviation is $0.5/\sqrt{2} = 0.354$. The signal strength of a normalized sinusoid with a standard deviation of σ is defined as $\sigma/0.354$. Thus, a 100% signal strength suggests a full-scale signal for the ADC. For this experiment, we use a Flora to record an iSEP signal. The strength of this signal is 34%. We feed the signal processing pipeline with this signal to generate a series of *baseline* ZCs. Then, we scale down the amplitude of this signal, re-quantize it, and process it using the pipeline to generate another series of ZCs. We use the mean absolute error (MAE) of these ZCs with respect to the baseline ZCs as the error metric. Figure 6.45 shows the MAE versus the strength of the scaled down signal. When we scale down the signal by 60 times, yielding a signal strength of 0.6%, the ZCs' MAE is 0.14 ms only. This suggests that TouchSync can still detect the ZCs accurately even when the iSEP signal is rather weak.

We evaluate the impact of the wearing position on the synchrony of iSEP signals. A researcher wears a Flora on his left wrist. Then, he conducts four tests by fixing the second Flora to his right wrist, right ankle, forehead, and waist, respectively. Each test lasts for two minutes. Figure 6.46 shows the error bars (5–95% confidence interval) for the

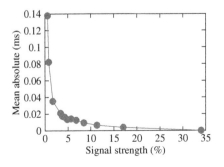

Figure 6.45 Impact of signal strength.

Figure 6.46 Wearing position.

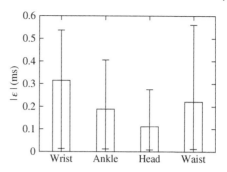

absolute time displacement $|\epsilon|$ between the two Floras in these four tests. The average $|\epsilon|$ values in the four tests are close. This suggests that the wearing positions have little impact on the synchrony of iSEP signals and the synchronization accuracy of TouchSync.

6.4.6.2 Impact of High-Power Appliances on TouchSync

Since the periodic iSEP signal is induced by the EMR field, EMR interference may distort the iSEP signals and result in large and dynamic time displacement between two iSEP signals. This may consequently cause clock synchronization errors. High-power electrical appliances may generate time-varying EMR interference. In this section, we investigate the impact of these appliances on the EMR waveforms and the time displacement between the EMR signals collected from two nearby Floras. We conduct experiments with an electric oven and a microwave oven as the electrical appliances, two representative high-power appliances found in home and office environments. Their rated powers are 800 W and 1050 W, respectively.

In the experiment for each appliance, we place the two Floras close to the tested appliance. The electric oven is placed on a table in a kitchen with no other appliances nearby, whereas the microwave is in an office pantry with other appliances running nearby including a fridge and a water heater. In each experiment, we switch on the tested appliance and switch it off by unplugging it from the power outlet. By investigating the data traces collected by the two Floras when the appliance is on and off, we can understand the impact of the appliance on the EMR and iSEP signals.

Figure 6.47 shows the raw EMR signals, collected from the electric oven and the microwave oven. From Figure 6.47(a), the EMR amplitude decreases significantly when

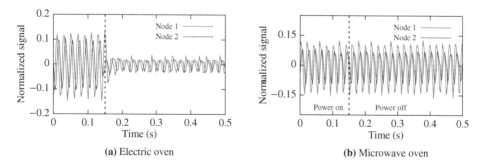

Figure 6.47 EMR signals near an electric oven and a microwave oven.

the electric oven is switched off. Note that the electric oven is a resistive load. Its change of operating status leads to a significant change of electric current through the appliance and the related powerlines. Thus, the operating status of the electric oven significantly affects the nearby EMR. In contrast, from Figure 6.47(b), the EMR amplitude does not change when the microwave is switched off. A potential cause is that the EMR sensed by the Flora is dominated by several other nearby appliances including a fridge and an automatic water heater. Microwave ovens generate EMR at 2.45 GHz. However, such high-frequency EMR cannot be effectively received by the Floras. From Figure 6.47(b), the EMR signal received by the Flora is primarily at the 50 Hz. In other words, the 2.45 GHz EMR emitted by the microwave, if present, does not disrupt the 50 Hz EMR.

We evaluate the impact of the electric and microwave ovens on the synchrony of the EMR signals collected from the two Floras. Figure 6.48 shows the time displacement over the courses in Figure 6.47. From Figure 6.48(a), after the electric oven is turned off, the intensity of time displacement fluctuation becomes larger. This is because, when the oven is switched off, the EMR strength and its signal-to-noise ratio decreases, resulting in more fluctuating time displacement. This suggests that, TouchSync's clock synchronization accuracy is better when a nearby high-power resistive load is operating. However, the mean value of the time displacement is still around zero. From Figure 6.48(b), the time displacement keeps stable. From the results in Figure 6.48, we can see that the nearby high-power appliances introduce little impact on the time displacement between two EMR signals and thus the performance of TouchSync.

6.4.6.3 Evaluation in Various Environments
We evaluate the iSEP signal strength and the accuracy of TouchSync in various indoor environments.

Laboratory
We conduct experiments in a computer science laboratory with about 100 seats and various office facilities (such as lights, computers, printers, projectors, or meeting rooms). Figure 6.49 shows the floor plans of laboratory. We arbitrarily select nine test points, marked by "Lx" in the figure. A researcher carries the Floras to each test point and conducts two

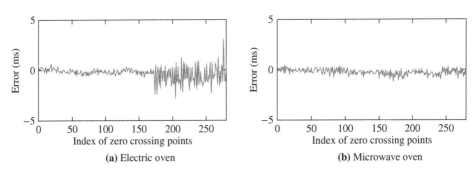

(a) Electric oven (b) Microwave oven

Figure 6.48 Errors introduced into time displacements of the EMR signals.

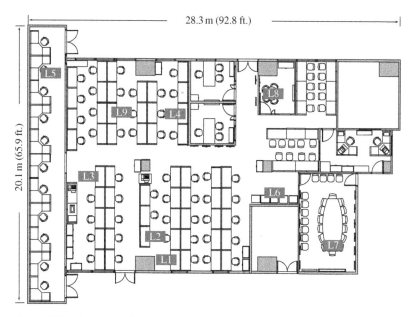

Figure 6.49 Laboratory floor plan with test points marked.

experiments. In the first experiment, the Floras have no physical contact with human body; in the second experiment, the researcher wears the two Floras. Thus, the experiment evaluates the same-wearer scenario. The example applications such as motion analysis, and muscle activation monitoring, belong to this scenario. Each synchronization session takes about 150 ms. A synchronization process completes once the IAS converges.

The first part of Table 6.8 shows the iSEP's signal strength, the number of synchronization sessions until convergence (K), and the clock offset estimation error at each test point. Without skin contact, the signal strength is a few percent only. However, TouchSync can still achieve a 3 ms accuracy. At L7, TouchSync cannot converge because of large and varying ϵ. The skin contact significantly increases the signal strength. Moreover, TouchSync converges after two synchronization sessions in most cases. For $K = 2$, a synchronization process takes less than one second. The absolute clock offset estimation errors are below 1 ms, lower than those without skin contact. However, without skin contact, the accuracy does not substantially degrade. This suggests that, TouchSync is resilient to the loss of skin contact due to say loose wearing.

Home

We conduct experiments in a 104 m^2 three-bedroom home with typical home furniture and appliances. Figure 6.50 show the floor plans of the home, marked by "Hx" in the figure. The second part of Table 6.8 shows the results. Without skin contact, the signal strength results are similar to those obtained in the laboratory. With skin contact, the signal strength increases and the absolute clock offset estimation errors are below 3 ms.

Office
The third part of Table 6.8 shows the results obtained at five test points in a 15 m² office. At test points O1 and O2, the signal strength with skin contact is slightly lower than that without skin contact. This is possible as the two tests were conducted during different time periods and the powerline radiation may vary over time due to changed electric currents. With skin contact, TouchSync gives sub-millisecond accuracy except at O2.

Table 6.8 Signal strength and TouchSync accuracy.

	Test point	Without skin contact			With skin contact		
		Signal strength	*K*	Error (ms)	Signal strength	*K*	Error (ms)
Laboratory	L1	2.6%	3	−0.2	84.7%	2	−0.7
	L2	3.2%	2	0.3	31.5%	3	−0.7
	L3	2.3%	2	−2.5	26.1%	2	0.5
	L4	4.0%	1	−0.6	33.7%	2	0.0
	L5	0.8%	15	1.1	3.3%	2	−0.2
	L6	5.7%	10	−0.4	39.5%	10	−0.0
	L7	2.3%	n.a.	n.a.	3.0%	2	−0.9
	L8	4.6%	2	−1.5	8.3%	2	0.6
	L9	2.6%	1	−1.2	67.4%	2	−0.9
Home	H1	4.2%	2	−1.1	8.9%	2	−0.8
	H2	3.4%	1	−0.9	14.5%	2	−1.0
	H3	4.6%	1	−1.3	44.9%	2	0.2
	H4	7.8%	n.a.	n.a.	39.2%	2	0.3
	H5	3.8%	1	−1.6	3.9%	1	−2.8
	H6	3.9%	4	−4.4	9.9%	2	−2.3
	H7	5.0%	2	−1.9	6.8%	1	−2.9
	H8	8.2%	1	−11.5	54.7%	4	−1.3
	H9	2.9%	1	−2.4	9.1%	1	−1.3
Office	O1	4.0%	n.a.	n.a.	3.3%	4	0.4
	O2	5.6%	1	−7.9	2.9%	2	−1.6
	O3	1.7%	1	−0.4	3.9%	2	−0.3
	O4	5.4%	3	−2.5	5.8%	2	−0.8
	O5	4.8%	6	−6.2	5.6%	12	−0.2
Corridor	C1	3.6%	12	0.1	4.4%	11	0.7
	C2	6.2%	2	0.6	44.2%	2	−1.0
	C3	5.8%	1	−7.6	4.4%	1	−1.1
	C4	1.9%	1	−6.0	2.2%	1	−2.8
	C5	1.9%	2	−3.7	5.8%	1	−1.0

a)   n.a. means that TouchSync cannot converge due to large ϵ.

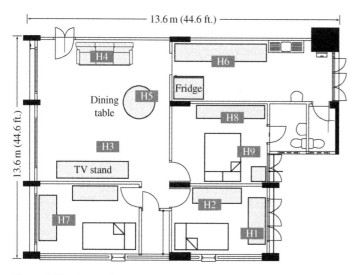

Figure 6.50 Home floor plan with test points marked.

Corridor

We select five test points with equal spacing in a 200 m corridor of a campus building. The fourth part of Table 6.8 gives the results. With skin contact, TouchSync yields absolute clock offset estimation errors of about 1 ms except at C4. The errors with skin contact are lower than those without skin contact.

In summary, with skin contact, TouchSync gives sub-millisecond clock offset estimation errors at 20 test points out of, in total, 28 test points in Table 6.8. All errors are below 3 ms.

From Table 6.8, at 3 out of the 28 test points, the test without skin contact gives higher signal strength than that with skin contact. This is because the two tests are conducted sequentially and the EMR may change over time. Moreover, the signal strength exhibits significant variation at different locations. This is because the EMR decays with the distance from the powerline. Nevertheless, the above results show the pervasive availability of iSEP in indoor environments.

6.4.6.4 TouchSync-over-internet

Tightly synchronizing wearables over long physical distances is often desirable. For instance, in distributed virtual reality applications, tight clock synchronization among participating sensing and rendering devices that may be geographically distributed is essential. Although the synchronization can be performed in a hop-by-hop manner (e.g., wearables ↔ smartphone ↔ cloud), errors accumulate over hops. In particular, tightly synchronizing a smartphone to global time has been a real and challenging problem – tests showed that the synchronization through LTE and wi-fi experiences hundreds of millisecond jitters Lazik et al. (2015). In contrast, TouchSync can perform end-to-end synchronizations for wearables distributed in a geographic region served by the same power grid. The basis is that the 50/60 Hz power grid voltage, which generates the powerline radiation, is highly synchronous across the whole power grid Viswanathan et al. (2016). TouchSync can synchronize wearables directly with a cloud server in the same region. The smartphone merely relays the messages exchanged among the wearables and the cloud server if the wearables cannot directly access Internet. The cloud server can use a sensor directly plugged in to a power outlet to capture the power grid voltage. Owing to the internet connectivity, the end-to-end synchronization scheme greatly simplifies the system design and implementation.

We conduct a proof-of-concept experiment of TouchSync-over-internet as follows. Two researchers carry a Flora-RPi setup each to two buildings that are about 10 km apart. The RPi is attached with an Adafruit GPS receiver to obtain ground-truth coordinated universal time (UTC) with microsecond accuracy. The two nodes, one as TouchSync master and the other as TouchSync slave, communicate through a tunnel established by ngrok 1.7, an open-source reverse proxy often adopted for creating IoT networks. Figure 6.51 shows the distributions of the two one-way delays over the ngrok tunnel. We can see that the ngrok exhibits significant dynamics. We evaluate TouchSync-over-internet eight times during a day. Figure 6.52 shows the box plots of the time displacements (i.e. ϵ) between the iSEPs captured by the two nodes. We can see that ϵ varies from −2 ms to 9 ms during the day. From the building managements, the two rooms where the master and slave nodes are located draw electricity from the R and Y phases of the power grid, respectively. There is a phase difference of $20/3 = 6.7$ ms between these two phases. Moreover, from power engineering, the difference between the power grid voltage phases at different geographic locations is non-zero and time-varying. These factors lead to the non-zero and time-varying ϵ in Figure 6.52. The dotted line in Figure 6.52 shows the synchronization errors of TouchSync-over-internet (i.e. δ) in various experiment runs. They are within the range of ϵ, since ϵ is the major source of TouchSync's synchronization error. The largest δ is 7 ms. The integer ambiguity solver converges within 4 to 13 synchronization sessions.

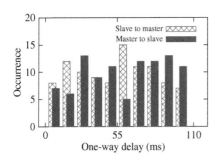

Figure 6.51 One-way delays over a ngrok tunnel.

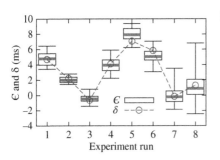

Figure 6.52 Accuracy of TouchSync-over-Internet.

6.5 Conclusion

This chapter reviews two important time services in IoT systems, i.e. data timestamping and clock synchronization. Two approaches are introduced to solve the timestamping in civil environments and clock synchronization on wearables, respectively. The first approach answers in the affirmative the question of whether powerline EMR can produce continuous ENF fluctuations that give an accurate and robust fingerprint indicative of time. A natural timestamp is proposed based on this fingerprint. A signal processing pipeline is designed for extracting reliably an ENF trace from often weak and noisy EMR wireless signals. An implementation of the proposed system is conducted on two common sensor platforms that contrast in cost and available nodal resources. By extensive measurements at six deployment sites in a city, the EMR provides natural timestamps with median time errors down to 50 ms. Based on the results, three practical applications are discussed of EMR natural timestamping, namely, time recovery, run-time clock verification, and clock synchronization that is secure against the packet delay attack. The second approach proposed in this chapter is an application-layer clock synchronization method for wearables. An explanation of the electrostatics of iSEP with supporting measurement results is introduced. Based on the understanding, the design of TouchSync is shown to synchronize the clocks of wearables by exploiting the wearers' iSEP. Different from existing WSN clock synchronization approaches that find difficulties in being applied on diverse IoT platforms due to their need for hardware-level packet timestamping or non-trivial extra hardware, TouchSync can be readily implemented as an app based on standard wearable OS calls. Extensive evaluation shows TouchSync's synchronization errors of below 3 ms and 7 ms on the same wearer and between two wearers 10 km apart, respectively. With the capability to run on the application layer, more wearable applications are possible, such as virtual reality gaming, remote surgery, and synchronous audio streaming.

References

Adafruit. Adafruit FLORA, 2018. https://www.adafruit.com/category/92.

Apple Inc. Device synchronization over bluetooth, 2015. https://www.google.com/patents/US20150092642.

BusinessWire. Symmetricom annouces general availability of industry's first commercially-available chip scale atomtic clock, 2017. http://bit.ly/2ctnwjB.

Roberto Casas, Héctor Gracia, Álvaro Marco, and Jorge Falcó. Synchronization in wireless sensor networks using bluetooth. In *Proceedings of the Third Workshop on Intelligent Solutions in Embedded Systems*, 2005.

Marie Chan, Daniel EstèVe, Jean-Yves Fourniols, Christophe Escriba, and Eric Campo. Smart wearable systems: Current status and future challenges. *Artificial intelligence in medicine*, 56(3):137–156, 2012.

Yin Chen, Qiang Wang, M Chang, and A Terzis. Ultra-low power time synchronization using passive radio receivers. In *The 10th International Conference on Information Processing in Sensor Networks (IPSN)*, pages 235–245, Chicago, IL, USA, 2011. IEEE. ISBN 9781450305129.

Mihail C Dinescu, Joseph Mazza, Adam Kujanski, Brian Gaza, and Michael Sagan. Synchronizing wireless earphones, April 7 2015. https://www.google.com/patents/US9002044.

Jeremy Elson, Lewis Girod, and Deborah Estrin. Fine-grained network time synchronization using reference broadcasts. In *Proceedings of the 5th USENIX Symposium on Operating Systems Design and Implementation*, 2002.

Tom Fawcett. An introduction to roc analysis. *Pattern recognition letters*, 27 (8):861–874, 2006.

Niklas Fechner and Matthias Kirchner. The humming hum: Background noise as a carrier of enf artifacts in mobile device audio recordings. In *The 8th International Conference on IT Security Incident Management and IT Forensics*, pages 3–13, Münster, Germany, 2014. IEEE.

Network Time Foundation. NTP: The network time protocol, 2017a. http://www.ntp.org.

Raspberry Pi Foundation. Raspberry pi, 2017b. https://www.raspberrypi.org.

Raspberry Pi Foundation. Raspberry Pi 3 Model B, 2017c. http://bit.ly/1WTq1N4.

Saurabh Ganeriwal, Ram Kumar, and Mani B. Srivastava. Timing-sync protocol for sensor networks. In *Proceedings of the First ACM Conference on Embedded Networked Sensor Systems*, 2003.

Saurabh Ganeriwal, Christina Pöpper, Srdjan Čapkun, and Mani B Srivastava. Secure time synchronization in sensor networks. *ACM Transactions on Information and System Security*, 11(4), 2008.

Ravi Garg, Avinash L Varna, and Min Wu. Seeing ENF: natural time stamp for digital video via optical sensing and signal processing. In *ACM Multimedia (MM)*, pages 23–32, Scottsdale, AZ, USA, 2011. ACM.

Gartner. Gartner says worldwide wearable device sales to grow 17 percent in 2017, 2017. http://gtnr.it/2AAHfgG.

Catalin Grigoras. Digital audio recording analysis–the electric network frequency criterion. *International Journal of Speed Language and the Law*, 12(1):63–76, 2005.

Tian Hao, Ruogu Zhou, Guoliang Xing, Matt W Mutka, and Jiming Chen. WizSync: Exploiting Wi-Fi infrastructure for clock synchronization in wireless sensor networks. In *Proceedings of the 32nd IEEE Real-Time Systems Symposium*, 2011.

IEEE. IEEE standard for a precision clock synchronization protocol for networked measurement and control systems. *IEEE Std 1588-2008 (Revision of IEEE Std 1588-2002)*, pages 1–300, 2008.

Texas Instruments. Efficient Multiplication and Division Using MSP430, 2017. http://www.ti.com/lit/an/slaa329/slaa329.pdf.

Texas Instruments. MSP430 LFXT1 oscillator accuracy. http://www.ti.com/lit/an/slaa225a/slaa225a.pdf, 2018.

Patrick Lazik, Niranjini Rajagopal, Bruno Sinopoli, and Anthony Rowe. Ultrasonic time synchronization and ranging on smartphones. In *Proceedings of the 21st IEEE Real-Time and Embedded Technology And Applications Symposium*, 2015.

Liqun Li, Guoliang Xing, Limin Sun, Wei Huangfu, Ruogu Zhou, and Hongsong Zhu. Exploiting FM radio data system for adaptive clock calibration in sensor networks. In *The 9th International Conference on Mobile Systems, Applications, and Services (MobiSys)*, pages 169–182, Washington, DC, USA, 2011. ACM.

Yang Li, Rui Tan, and David KY Yau. Natural timestamping using powerline electromagnetic radiation. In *Proceedings of the 16th International Conference on Information Processing in Sensor Networks*, 2017.

Yang Li, Rui Tan, and David KY Yau. Natural timestamps in powerline electromagnetic radiation. *ACM Transactions on Sensor Networks (TOSN)*, 14(2):1–30, 2018.

Zhenjiang Li, Wenwei Chen, Cheng Li, Mo Li, Xiang-Yang Li, and Yunhao Liu. Flight: Clock calibration using fluorescent lighting. In *The 18th Annual International Conference on Mobile Computing and Networking (MobiCom)*, pages 329–340, Istanbul, Turkey, 2012. ACM.

Konrad Lorincz, Bor-rong Chen, Geoffrey Werner Challen, Atanu Roy Chowdhury, Shyamal Patel, Paolo Bonato, and Matt Welsh. Mercury: a wearable sensor network platform for high-fidelity motion analysis. In *Proceedings of the 7th ACM Conference on Embedded Networked Sensor Systems*, 2009.

Martin Lukac, Paul Davis, Robert Clayton, and Deborah Estrin. Recovering temporal integrity with data driven time synchronization. In *The 8th ACM/IEEE International Conference on Information Processing in Sensor Networks (IPSN)*, pages 61–72, San Francisco, CA, USA, 2009. IEEE.

Miklós Maróti, Branislav Kusy, Gyula Simon, and Ákos Lédeczi. The flooding time synchronization protocol. In *The 2nd ACM Conference on Embedded Networked Sensor Systems (SenSys)*, pages 39–49, Baltimore, MD, USA, 2004. ACM.

T. Mizrahi. Security requirements of time protocols in packet switched networks, 2014. https://tools.ietf.org/html/rfc7384.

Frank Mokaya, Roland Lucas, Hae Young Noh, and Pei Zhang. Myovibe: Vibration based wearable muscle activation detection in high mobility exercises. In *Proceedings of the 2015 ACM International Joint Conference on Pervasive and Ubiquitous Computing*, 2015.

Frank Mokaya, Roland Lucas, Hae Young Noh, and Pei Zhang. Burnout: a wearable system for unobtrusive skeletal muscle fatigue estimation. In *Proceedings of the 15th International Conference on Information Processing in Sensor Networks*. IEEE, 2016.

Dima Rabadi, Rui Tan, David KY Yau, and Sreejaya Viswanathan. Taming asymmetric network delays for clock synchronization using power grid voltage. In *Proceedings of the 2017 ACM on Asia Conference on Computer and Communications Security*, 2017.

Anthony Rowe, Vikram Gupta, and Ragunathan Raj Rajkumar. Low-power clock synchronization using electromagnetic energy radiating from ac power lines. In *The 7th ACM Conference on Embedded Networked Sensor Systems (SenSys)*, pages 211–224, Berkeley, CA, USA, 2009. ACM.

Hui Su, Adi Hajj-Ahmad, Min Wu, and Douglas W Oard. Exploring the use of enf for multimedia synchronization. In *International Conference on Acoustics, Speed and Signal Processing (ICASSP)*, pages 4613–4617, Florence, Italy, 2014. IEEE.

Markus Ullmann and M Vogeler. Delay attacks – implication on ntp and ptp time synchronization. In *IEEE International Symposium on Precision Clock Synchronization for Measurement, Control, and Communication*, 2009.

U.S. Navy. Extremely low frequency transmitter site Clam Lake, Wisconsin, 2003. http://www.fas.org/nuke/guide/usa/c3i/fs_clam_lake_elf2003.pdf.

Sreejaya Viswanathan, Rui Tan, and David KY Yau. Exploiting power grid for accurate and secure clock synchronization in industrial IoT. In *The 37th IEEE Real-Time Systems Symposium (RTSS)*, pages 146–156, Porto, Portugal, 2016. IEEE.

Allen J Wood and Bruce F Wollenberg. *Power generation, operation, and control*. John Wiley & Sons, Hoboken, NJ, USA, 2012.

Zhenyu Yan, Yang Li, Rui Tan, and Jun Huang. Application-layer clock synchronization for wearables using skin electric potentials induced by powerline radiation. In *Proceedings of the*

15th ACM Conference on Embedded Network Sensor Systems, SenSys '17. Association for Computing Machinery, 2017a. ISBN 9781450354592. doi: 10.1145/3131672.3131681.

Zhenyu Yan, Yang Li, Rui Tan, and Jun Huang. Touchsync implementation, 2017b. https://github.com/yanmarvin/touchsync.

Zhenyu Yan, Rui Tan, Yang Li, and Jun Huang. Wearables clock synchronization using skin electric potentials. *IEEE Transactions on Mobile Computing*, 18(12):2984–2998, 2018.

Zolertia. Z1, 2017. http://zolertia.io/z1.

7

Conclusion

The digital world is driven by big data and powered by intelligent algorithms. For people-centric applications, data is collected from everyone's own devices and ubiquitous public facilities, and then utilized for better satisfying a customer's personal preferences and requirements. For industrial applications, data is generated from a variety of IoT devices in different business sectors and service scenarios, and then processed for supporting more efficient, productive, and autonomous operations. In order to fully exploit massive IoT data, understand customers' real needs and create commercial values, domain-specific knowledge, operation procedures and disruptive business models should be digitalized and integrated with a series of intelligent algorithms for different service scenarios and working situations. As a result, more and more data-driven, cross-domain innovation ecosystems will be established and quickly become the cornerstones for a smarter society. They are more collaborative and inclusive, and can simultaneously consider different requirements from multiple perspectives, creatively identify feasible approaches with various objectives, and effectively produce a lot of social and economic benefits.

More complex data and intelligent algorithms demand more network connections, communication bandwidth, computing power, and storage space. These trends in user requirements and technology advancements motivate us to propose the data-driven multi-tier computing network architecture as the most important ICT civil infrastructure for supporting the development and evolution of the digital world. Based on it, the CATS framework and different task offloading/scheduling algorithms can effectively coordinate distributed network resources not only in the cloud, but also on the devices, at the edge, and inside the network (fog), for collaboratively analyzing heterogeneous IoT data and satisfying diverse user requirements on service delay, energy consumption, data privacy, and monetary cost. Further, the FA2ST framework and a series of service orchestration and provisioning algorithms can flexibly combine dispersive micro-services with a variety of objectives, resources and capabilities at global, regional, local, and device levels, for jointly composing much more comprehensive, sophisticated, and intelligent IoT applications and services. In addition to these challenging problems, this book investigates and applies a light-weight privacy protection algorithm and a wide-area clock synchronization algorithm in practical IoT systems, which are crucially important for managing and guaranteeing mission-critical IoT applications with reliable QoS satisfaction. Eventually, a truly adaptive and intelligent IoT system is made of everything, managed by everything, and serving everything, as all kinds of distributed but democratized things/devices/nodes

Intelligent IoT for the Digital World: Incorporating 5G Communications and Fog/Edge Computing Technologies, First Edition. Yang Yang, Xu Chen, Rui Tan, and Yong Xiao.
© 2021 John Wiley & Sons Ltd. Published 2021 by John Wiley & Sons Ltd.

will contribute to and benefit from this collaborative multi-tier computing network architecture.

In the digital world, everything and everyone have always-updated records as their digital twins, which may not be desirable from our current perspective, but in the long term, could enable more productive businesses, more efficient governance, more sustainable society, and safer and greener environments. The well-known philosophy of "user-centric" for product design and service management is gradually upgraded to "everything/everyone-centric", thanks to fine-granularity data, feasible algorithms, and ubiquitous 3C resources at per-device/per-user level. As the former European Consumer Commissioner, Meglena Kuneva, said in 2009: "Personal data is the new oil of the internet and the new currency of the digital world." Ideally, private personal data and valuable business data should be protected and analyzed by trusted computing resources and AI algorithms at the locations close to where it is generated, thus effectively fulfilling everyone's customized service requirements at anywhere and anytime. In other words, most requests from the demand side are best satisfied by feasible local/regional resources from the supply side in real time. This simple and agile approach not only minimizes data transmissions, communication bandwidth, energy consumption, and service delay, but more importantly, improves data privacy, network security, service reliability, user experience, and satisfaction.

As the newly promoted factor of production, data's full potential for scientific innovation, wealth creation, and social progress are yet to be discovered, studied, and exploited for building a better digital world. To realize this dream, we still have many challenging problems to solve, such as:

1. How to securely and efficiently store, retrieve, protect, and delete massive IoT data in distributed, dynamic networks.
2. How to precisely integrate diverse IoT data at various temporal-spatial scales and from different application sources.
3. How to accurately comprehend massive IoT data, extract useful information, and discover new knowledge under multiple rules.
4. How to effectively identify and combine obscure opinions and perspectives from IoT data, and seek the common ground for actions.
5. How to proactively prevent, control and track the duplication, dissemination, and trading of IoT data under certain regulations and laws.
6. For people-centric applications, how to balance a user's privacy with his/her own convenience when personal data need to be utilized.
7. For industrial applications, how much and how widely data should be shared among business partners for cross-domain innovations and services.
8. How to seamlessly incorporate machine intelligence with human intelligence for timely analyzing IoT data and making critical decisions in various applications.
9. How to properly consider data accuracy, integrity, rareness, neutrality, traceability, verifiability, and auditability into the measure of its value?
10. How to promptly build a secured and trusted environment over anonymous computing nodes in the neighborhood for mission-critical IoT applications.
11. If differential privacy and confidential computing are adopted, what are the trade-offs between user privacy protection and system performance and service experience?

For infinite raw material of data in the digital world, we firmly believe multi-tier computing networks provide the powerful platform and practical tools, while algorithms consist of the right methods and specific procedures to efficiently process and refine raw data for various intelligent IoT applications. They collaboratively contribute and lead to the ultimate dominance of data-driven, everything/everyone-centric, cross-domain innovation ecosystems in human society. This book is dedicated to the study of computing network and service architecture, and a series of corresponding algorithms for improving resource utilization, system performance, service orchestration, user experience, and application management. We hope you find these topics and discussions interesting and informative. It would be more exciting if some of our work could assist you in identifying new problems, creating new ideas, and developing new solutions for the digital future.

Index

Intelligent IoT for the Digital World: Incorporating 5G Communications and Fog/Edge Computing Technologies,
First Edition. Yang Yang, Xu Chen, Rui Tan, and Yong Xiao.
© 2021 John Wiley & Sons Ltd. Published 2021 by John Wiley & Sons Ltd.